T29/SK

Dr Ante Bilić

liječnik - stomatolog

25.03.81.

Ante Bilić

Current Topics in Microbiology and Immunology

81

Edited by

W. Arber, Basle · W. Henle, Philadelphia · P.H. Hofschneider,
Martinsried · J.H. Humphrey, London · J. Klein, Dallas · P. Koldovský,
Düsseldorf · H. Koprowski, Philadelphia · O. Maaløe, Copenhagen ·
F. Melchers, Basle · R. Rott, Gießen · H.G. Schweiger, Wilhelmshaven ·
L. Syruček, Prague · P. K. Vogt, Los Angeles

Springer-Verlag
Berlin Heidelberg New York 1978

Lymphocyte Hybridomas

Second Workshop on "Functional Properties of Tumors of T and B Lymphocytes"

Sponsored by the National Cancer Institute (NIH)
April 3—5, 1978 Bethesda, Maryland, U.S.A.

Edited by
F. Melchers, Basel M. Potter, Bethesda N. Warner, Albuquerque

With 85 Figures

Springer-Verlag
Berlin Heidelberg New York 1978

Editors

F. Melchers, Basel Institute for Immunology, 487 Grenzacherstraße,
CH-4005 Basle, Switzerland

M. Potter, Division of Cancer Biology and Diagnosis, Laboratory of Cell Biology,
National Cancer Institute, Bethesda, MD 20014/USA

N. L. Warner, Dept. of Pathology, University of New Mexico, School of Medicine,
Albuquerque, N.M. 87131/USA

ISBN 3-540-08810-5 Springer-Verlag Berlin Heidelberg New York

ISBN 0-387-08810-5 Springer-Verlag New York Heidelberg Berlin

Library of Congress Cataloging in Publication Data. Workshop on Functional Properties of Tumors of T and B
Lymphocytes, 2d, Bethesda, Md., 1978. Lymphocyte hybridomas. (Current topics in microbiology and immunology;
v. 81) Bibliography: p. 268. Includes index. 1. Immunoglobulins-Congresses. 2. T cells-Tumors-Congresses.
3. B cells-Tumors-Congresses. 4. Cell hybridization-Congresses. I. Melchers, Fritz, 1936–   II. Potter, Michael.
III. Warner, Noel Lawrence, 1934–   IV. Title. V. Series. [DNLM: 1. B-Lymphocytes-Congresses. 2. T-Lympho-
cytes-Congresses. 3. Neoplasms, Experimental-Immunology-Congresses. 4. Hybrid cells-Congresses. 5. Antigens,
Neoplasm-Congresses. 6. Antibodies, Neoplasm-Congresses. W1 CU82K v. 81/QZ206 W925 1978L]. QR1.E6
vol. 81 [QR186.7] 576′.08s [599′.02′9] ISBN 0-387-08810-5 78-16343

Table of Contents

Indexed in ISR

Preface

F. Melchers, M. Potter, N.L. Warner

HISTORICAL

Prior to 1964 many workers had experienced considerable difficulty in establishing continuous lines of murine plasmacytoma in culture. Pettingill and Sorenson in 1966 grew X5563 after a painstaking period of adaptation. During a visit to the Salk Instiute in 1965 Leo Sachs began experiments with K. Horibata,E.S. Lennox and M. Cohn that ultimateley convinced most workers that murine plasmacytomas could be grown in vitro with ease. They adapted MOPC-21* (called P3 at the Salk Institute) to in vitro culture and showed the cells produced a large amount of the IgG1 myeloma protein. Horibata, Lennox and Sachs did not report the important finding though a manuscript was written and accepted. The reason for the failure to publish the paper was that the original line was not recovered from the freezer and knowing there would be many requests for it the paper was withdrawn. As it later turned out, the failure to recover the line may have been trivial, for Horibata and Harris (1970) soon re-established a continuous line of P3 now named P3K. This line was distributed to other workers including Cesar Milstein in Cambridge. Soon after the success with P3, Scharff and colleagues at Albert Einstein established MPC-11* in culture.

Up to now continuous growth of human myeloma cells in vitro has remained a formidable problem. Success in this area of lymphocyte cultures would certainly open the way for obtaining human lymphocyte hybridomas producing human monoclonal antibodies - which may well revolutionize their clinical applications.

Initially both P3 and MPC-11 murine cell lines were used extensively to study Ig synthesis in vitro. The cloning efficiency of MPC-11 was very high (higher than 50 %)and this permitted Scharff to study the stability of MPC-11 lines in regards to Ig synthesis in culture. It also opened the way to using the plasmacytoma cells in mutagenesis experiments. The high rate of loss of H chain synthesis and secretion ranging from 1×10^3 per cell per generation was a striking finding. Non-secreting mutants were also found to occur in high frequency. All of this work formed the background for the dramatic discoveries of Köhler and Milstein that are the basis for this workshop.

Hybridization studies in other systems had not offered great promise for immunologists. Most of these though involved fusion of lymphocytes with cell types which were dissimilar in respect to their status of differentiation. Such hybrid cells of dissimilar parental lines, in particular plasmacytoma-fibroblast hybrids, showed that Ig-synthesis was suppressed. Milstein nonetheless derived an 8-azaguanine resistant subline of P3, now termed P3/X63-Ag8. This line was used first to study the fusion of two plasmacytoma cells. When this revealed that the hybrid cells produced both myeloma proteins, Köhler

---

*both induced at the National Cancer Institute.

and Milstein then went directly to the important and daring experiment of fusing a normal plasma cell with X63. They and all of us were rewarded with a spectacular result: the hybrid of an anti-SRBC normal plasma cell and X63 produced both the MOPC-21 myeloma protein and SRBC antibody. Thus, the specific antibody production of a normal B-cell became immortal through hybridization with a neoplastic cell.

FUSIONS OF PLASMACYTOMAS WITH B-LYMPHOCYTES OR PLASMA CELLS

Cell Lines Used for Fusion

Table 1 summarizes the principal tumor lines that have been used to date. In only a few instances have direct comparisons been made in a single experiment, and in these cases it appears that similar fusion frequencies and successes of establishing hybrids were obtained. In the overall experience of many investigators it was again apparent that X63, NS-1, and certain MPC-11 lines have been used successfully with some variations in results. In view of the production of heavy chains from myeloma parent lines, it might be desirable to fuse to a non-secreting myeloma line such as NS-1, particularly in cases where the myeloma protein and mixed molecules of myeloma and antibody proteins interfere with the specificity of the monoclonal antibody. Other Ig-secreting or non-Ig-synthesizing myeloma lines should be screened for this purpose.

Thus far, only mouse plasmacytoma lines are available. The development of human, primate, rat, and hamster Ig-secreting tumors would greatly expand the numbers and types of useful hybridomas.

Table 1: Characteristics of tumor cell lines* used in fusion experiments

Complete Name	Short Name	Strain	Characteristics
P3 X63 Ag8	X63	BALB/c	Plasmacytoma (γ1, k)
P3-NSI-1-Ag4-1	NS-1	BALB/c	Plasmacytoma (k, non-secreting)
MPC11-X45-6TG	X45	BALB/c	Plasmacytoma (γ_2b, k)
BW5147	BW	AKR	Thymic Lymphosarcoma (Ly^-, 2^-, 3^-, 6^+)
EL-4	EL-4	C57BL	Thymic Lymphosarcoma

*The cell lines of these tumors are available from the Salk Institute Cell Distribution Center, La Jolla, California.

Technology of Fusion

In general, little discussion or divergence of opinions was evident in this conference on the actual procedures for fusion. Most investigators used polyethylene glycol (PEG) as fusing agent. PEG lots vary considerably in degree of purity and suitability for in vitro use. Each should be tested for toxicity with cell lines before use. The optimal molecular weight of PEG for fusion was not clearly established. Preparations of widely different molecular weights (between 1000 and 4000) were effective.

X

Several variations in the actual procedure of dispensing and locating the cells in PEG were described, some of which certainly improved the hybridoma success rate, and these are described in detail in the articles.

Cell numbers used for fusion similarily were not a major issue, although variations in the frequencies of successful fusion events yielding hybridomas appeared more dependent on cell numbers at lower cell input. Most investigators dealing with this problem agreed that under the present conditions for fusion — and independent of the fusing agent or the technique of locating myeloma cells and normal cells next to each other — the best frequencies are one hybridoma in 2 to 5×10^4 normal cells. Single clones of normal B-cells, activated and grown "in vitro" to approximately 10^3 cells are, therefore, too small to be immortalized through fusion. Klinman et al., however, using the spleen focus growth of single B-cell clones, reported successful hybridoma production from such in "in vivo" grown single clones.

Three specific points might be noted as the result of the discussions at the workshop:

(i) It is not yet clear whether fusion occurs to a specific subpopulation of cells within the myeloma or lymphoma clone, although it is clear that heterogeneity in cell cycle and cell differentiation states exist within tumor lines, indicating a potential for such restriction.

(ii) Although the subject of the relative differentiation states between normal and neoplastic cells is considered frequently throughout this volume, no information is available on whether it is an absolute requirement for obtaining functional hybrids expressing the desired characteristic normal function of lymphocytes that the neoplastic tumor cell must belong to the same differentiation subpopulation as the normal fusing cell. It is certainly clear that this similarity or identity is not necessary for fusion itself to occur and for hybrids to be clones although these hybrid lines may or may not exhibit biological functions.

(iii) The designation of a tumor line by a specific name does not necessarily infer that all such tumor lines of identical name need be equally suitable for fusion. Sublines, variants, mutants, and revertants of such lines will frequently occur, many of which may be quite unsuitable. In an effort to minimize this problem, the available sublines that have been shown suitable for fusion are to be centered in the Salk Institute Tumor Distribution Center for further distribution to investigators. The moral in this area for newcomers to the field is to "use a proven fusable tumor line" and to learn the tissue culture rules for how to keep the line.

The Normal Target Cell of Fusion

It is perhaps on this issue, that the most debate and uncertainty still exists. Cell fusions were successfully described where unfractionated total spleen cell preparations were used.

Mitogen-induced polyclonal stimulation of the total B-cell repertoire followed by cell fusion produced large numbers of Ig producing and specific antibody forming clones. (See Andersson and Melchers, and Goldsby et al., this vol.) This method also offers the potential to immortalize the entire repertoire of antibody producing cells, which could provide the means to analyze precisely the range of specificities that are present in such a repertoire as derived from various different genetic backgrounds. The exact differentiated type of the normal target cell for fusion in the spleen has not been identified in detail, although it seems likely that an activated B-lymphocyte in its blast stage is a preferred partner of the myeloma cell lines in the fusion reaction. Since the success of obtaining functional hybridomas would clearly be enhanced if the normal fusing population were to be specifically pre-selected for specificity, it is to be cautioned that this pre-selection may need to be not only for the required antigen reactive specificity, but perhaps also for a specific differentiation stage.

In attempting to analyze this problem, it was stressed in the discussion that interpretations of the differentiation type of the normal cell involved in the fusion, cannot be necessarily derived from assessment of the functional state of the hybridoma product. This latter status will clearly be complicated by unknown factors determining gene expression (repression and de-repression) in the hybrid cell, and its particular chromosomal complement. Direct answers to this problem will only be obtained when isolated purified normal cell populations of different types are used for fusion.

The feasibility of hybridizing mouse myeloma cells with normal Ig producing cells of other species was discussed in several reports (see Howard et al.; Levy et al.). Ig-secreting hybridomas were obtained from fusions of mouse myelomas and rat Ig producing cells. Hybridomas of human CLL cells and mouse myeloma cells were reported but these did not secrete large quantities of the human Ig. The use of human lymphoblastoid lines may be a fruitful source of human Ig producing cells. Recent studies by Zurawski et al. and Steinitz et al. (this vol.) indicate that human Ig producing cell lines can be established. However fusion experiments have not yet been reported.

Use of Feeder Layers

Frequently throughout the conference discussion was raised on the effect of the addition of other cell types either at the time of fusion and immediately thereafter, or during the period of stabilization and isolation of the specific clones. This problem clearly relates to other tissue culture studies indicating a "feeder" effect from the addition of certain cell types to those under culture. In general, feeder cells can be used as monolayers, incorporated in a basal agar layer, mixed in with the whole suspension, or used to prepare conditioned media. It was evident that the effect of such feeder cells has not been well characterized in this fusion system, and many diverse opinions were expressed.

It is probably reasonable to conclude at this stage that the presence of feeder cells will most likely be advantageous, and that where the deliberate addition of such cells has not resulted in growth-supporting effects may be because such cells are already present within the normal fusing cell population.

Many examples were brought out in this conference, of approaches to pre-selection which have improved the success of obtaining functional hybridomas. These include:

(i) Biophysical separation procedures that select for either large cells from polyclonal stimulation or for specifically enriched immune cells.

(ii) Rosetting methods with appropriate antigens attached to sheep erythrocytes.

(iii) PLL bound monolayer preparations of erythrocytes bearing appropriately bound antigen.

(iv) Clonal isolation of specifically reactive cells by the Klinman spleen fragment assay.

(v) Specifically stimulated continuous cell cultures.

It should be noted that the actual frequency of cell fusions that occur will thus be dependent at least in part on the frequency of the appropriate normal cell type within the population that is capable of fusion with the particular neoplastic cell. Until this cell type is exactly characterized, some caution might thus be appropriate in refining further pre-selection methods.

Selection of Fused Cells

Virtually all studies reported to date have used the appropriate 8-azaguanine resistant cell lines and subsequent selection of the hybrids in HAT medium, with azaserine and ouabain also being used in some studies. Several aspects of this were occasionally noted in that particularly some of the T cell tumors are most sensitive to growth inhibition by thymidine which may be a potential problem. Post selection by the use of the fluorescence activated cell sorter potentially may be of considerable advantage when normal cell parental surface markers can be used for selection, as in this instance, drug resistant lines of tumor cells need not be used.

Stability of Hybrids

Probably the major technical problem in this field is the difficulty to maintain a functional stable hybridoma. Although numerous examples were presented of functionally active and specific hybridomas that have been maintained for many months, the predominant pattern seems to be of loss of activity from the hybrid after a few months of culture. In general it is most likely that this loss of activity is associated with specific chromosome losses, although the possibility of true mutants of variants emerging cannot be eliminated. Whether any of these events could be associated with an actual change in the functional specificity of the hybridoma product has not been determined.

One possible mechanism for subtle changes in specificity could involve Ig chain associations. In situations where four chain types are expressed and antibody specificity is primarily determined by the H chains derived from normal cells slightly different specificities could result from different H-L combinations. Random associations of Ig chains can result in the formation of mixed molecules, at least within IgG subclasses, although not between IgM and IgG. Ig chain loss is also a relatively frequent event and usually if not invariably follows the pattern of H chain loss prior to L chain loss.

It has been suggested that recloning early in the propagation of the specific clone is advantageous to maintaining a stable clone, although even with this procedure, clones have been found to be stable for many months, and then rapidly lose H chain and thereafter L chain production. Stability of malignant growth and of expression of immune functions such as Ig-secretion may well depend on the preservation of special sets of chromosomes from the malignant and from the normal cell. If chromosome losses occur at random, recloning has to be followed up for many parallel clones expressing function so that the chances for the loss and the preservation of the "right" genetic information are increased.

Clearly the resolution of the problem of stability has not yet been found, and all efforts should be made to preserve lines at an early stage, during their functionally active state by freezing.

Criteria for Hybrid Characterization

In general the same criteria as are usually applied in the field of somatic cell hybridization are relevant here, namely identification of parental gene products in the hybrid clone, either cell surface markers or enzyme activities, and not simply reliance on a polyploid state. The immunologist has an advantage in this field with the availability of specific immunological properties — both the specific immune product of the normal cell, and many other cell surface markers.

Analysis of Hybrids and their Products

The latter aspect above, namely screening for cell surface marker products on hybridomas, is potentially complicated by many problems previously encountered in the typing of murine tumors, notably the presence of anti-viral antibodies in murine alloantisera. This problem has been particularly discussed in this volume in relation to I-J typing of T cell hybridomas.

The availability of monoclonal antibodies of defined specificity has also re-instituted considerable interest in the functional activities of antibody molecules in terms of precipitating, agglutinating and hemolytic activities. In particular the hemolytic abilities of monoclonal antibodies were extensively discussed as this method is frequently used for screening of specific antibody producing clones, and it has become evident that the ability to detect such activity will be dependent upon various factors, including antigen density on the target cell surface, and the presence or absence of additional "natural" sublytic or partially lytic antibodies in the complement source.

Assessment of the homogeneity of monoclonal antibodies was frequently raised as a discussion point in this conference. In general, it could be concluded that the optimal approaches of immuno-chemistry should be applied, ranging from polyacrylamide gel electrophoresis, and isoelectric focusing, to 2 dimensional peptide map analysis or Ig chain sequencing. At a minimum, IEF patterns or PAGE gel analysis should be performed. In view of the potential for random chain associations, it is pertinent to note that in a hybrid synthesizing both chains of tumor and normal origin, 16 different recombinant molecules could potentially be secreted.

Production of Monoclonal Antibodies

In view of the potential instability of functional hybridomas, particular efforts should be made early on to produce maximal amounts of potentially valuable and useful monoclonal antibodies. This raises the question as to the efficiency of Ig secretion by the cultured hybridoma versus its Ig production in vivo when transplanted as a tumor. The in vitro method has the advantage of not being potentially contaminated by other antibodies in normal mouse serum, and many of the cultured hybridoma lines have been described to produce reasonable Ig levels, up to 100 µg/ml of culture fluid. In general most cultured hybridomas have been transplantable to syngeneic mice, and others of different species origins can frequently be transplanted in athymic nude mice.

Current Applications and Future Potential of Functional Hybridomas

Basically the fusion of normal and malignant lymphocytes opens the possibility of obtaining monoclonal antibodies and other specific products of lymphocytes to any known or unknown antigenic specificity including many that are of practical relevance to both experimental and clinical immunology, molecular biology and cell biology.

Table 2 represents a summary of the monoclonal antibodies and the T-cell hybrid lines that have been produced to date and are described in some detail in this book.

Monoclonal antibodies to viral antigens. Many classification schemes of microorganisms are based on antigenic analysis, and in this instance the use of monoclonal reagents will provide a far greater resource of exquisitely specific reagents that could be used to clearly define, for example, antigenic variations in a given virus type as derived from various geographic locations. These reagents could thus potentially serve as the ideal reference typing reagents, providing sufficient amounts of antibody.

Of even greater potential is the therapeutic use of monoclonal antibodies in protective studies, and preliminary successful results of this nature were reported at this conference. In similar fashion, several laboratories are currently considering the use of such reagents prepared against parasitic antigens, where considerable potential exists for the application of this technique.

Monoclonal antibodies to defined soluble antigens. Many successful monoclonal antibodies have been developed against defined antigens such as haptenic determinants, and as such provide an ideal means to study the diversity of antibodies against simple restricted antigens. In several isntances a broader range of idiotypes in a primary response have been detected by the hybridoma method, several of which have in fact been shown to be identical to particular myeloma proteins.

The reagents developed in this field may well be of value in analyzing the genetic basis of the immunoglobulin molecule, including both aspects of V_H region sites and allotypic C_H region determinants. Potentially however this approach can be used to better define many diverse molecular structures, and in turn interpret their genetic derivation.

Monoclonal antibodies to cell surface antigens. In view of the now well established heterogeneity of functionally distinct lymphocyte subpopulations, considerable interest has developed in attempts to prepare antisera to lymphocyte differentiation antigens that identify these distinct subpopulations, and in turn, can then be used to further characterize and isolate these subpopulations and their marker molecules. Several laboratories have turned to the hybridoma approach, and as reported in this volume, have demonstrated that monoclonal antibodies can be produced to various differentiation antigens that are specific for particular immune or human hematopoietic cell populations. It is clearly evident that this approach may well result in the detection of many previously unrecognized cell surface specific antigens and define much better those which we currently know.

This approach has also been extended into the production of monoclonal antibodies to cell surface antigens that appear to define allelic forms of such cell surface molecules. This has involved both allogeneic and xenogeneic immunizations, and has initially concentrated on alloantigens determined by loci in the major histocompatibility complex. It is already apparent that new specificities will be defined by this approach, and that such monoclonal reagents will have enormous value as standard typing reagents both for experimental studies in animals, and for clinical HLA typing of humans.

Again this approach can be potentially applied to many other cell surface antigens of many cell types, and with the current expanding list of Ly T and Ly B differentiation loci, will hopefully provide the specific reagents, free of potentially contaminating anti-viral antibodies, that can be used to further elucidate these and other similar lymphocyte cell membrane components.

Monoclonal antibodies may eventually prove to be of considerable value in relation to therapeutic studies of graft rejection, and a preliminary report in this volume holds out considerable promise in this field.

Clinical application of hybridoma reagents also involves the field of tumor associated antigens, where considerable confusion has frequently resulted from the use of relatively poorly defined anti-tumor antisera. Such monoclonal reagents will be invaluable both for immunodiagnosis, and also for better experimental definition of the nature of tumor associated antigens.

Detection of new or rare Ig gene products. The fusion of neoplastic plasma cells with normal antibody producing cells may well involve normal cells producing rare or even as yet undescribed Ig chain products. Clearly if pre-selection for such rare types can first be made, the chances of producing such fusions and thereby obtaining large amounts of such Ig molecules, will be considerably enhanced. Several examples of this aspect were discussed during the conference and included IgM chains of different molecular weight (? IgM subclass), murine IgE, IgG3, and λ_1 and λ_2 containing immunoglobulins.

HYBRIDS OF T-LYMPHOCYTES

Although very promising approaches to this problem were presented, this area is not as developed as with the plasmacell-plasmacytoma fusions. Two major advances were described. First, T-lymphoma-T cell hybrids have been made and these hybrid cells have interesting biological properties. Second, the ability of several workers to grow and clone peripheral T cells in continuous culture has opened many new possibilities.

Functional T-Cell Hybridomas

In the case of T cell lymphomas, the AKR thymic lymphocytic tumor BW has proved to be the most useful in forming hybrids. Tumors EL-4 and

S49 have also been used, but with less success, although some functional hybridomas have resulted. Again it is clear that a wider range of parental T-cell tumors need be screened.

The ability to fuse the tumors with functionally relevant normal cells and to select functional hybrids, is less certain. Several examples of antigen specific functional suppressor T-cell hybridomas were described (see Simpson) clearly indicating the potential of this field. The general issue that is still not certain is how specific is the like-like requirement in terms of obtaining a functional hybrid. Since only a few T cell tumors have so far been used in this field, it is possible that a wider range of T cell tumors for fusion may result in further functional hybridomas of different T cell functions.

In contrast to the relatively straightforward approaches to specificity analysis with monoclonal antibodies, the functional assessment of specificity of T cell hybridomas is further complicated by a lack of knowledge of the potential range of specificities of the T cell repertoire. Opposing current views exist concerning the relationship of such receptors to the B cell Ig repertoire. Several of the studies on specificity described in this conference have in fact used solubilized products from the T cell hybridomas, which potentially offer means of finally resolving the nature of the T cell receptor site. Should functional hybridomas of other T cell populations become available, it will clearly be of considerable interest to attempt to compare the antigen specific receptor sites in these different populations.

Continuous Lines of Activated T Cells

In general, antigenic stimulation in vitro of T or B cells has led to immune responses that are usually quite short lived, and the cultures rapidly die out. New approaches to this aspect are described in this volume (Nabholz, Fathman, Hengartner) in which either continuous stimulation with antigen or continuous culturing of activated T cells in condition media have led to long term and possibly permanent cell lines. The cell lines described in these studies apparently do not obey the usual rule for normal cells in culture to terminate after a relatively modest number of cell divisions.

The studies described indicate these approaches with lymphocytes will be most fruitful with cells of the T cell lineage, and probably those of several subpopulations including cytotoxic T cells.

Continuous cell lines of specifically activated T cells may provide extremely useful targets for cell fusion, to ensure an immortalized cell line.

CONCLUDING REMARKS

Monospecific antibodies have been the dream of many immunologists for long. The dream was contrasted by the reality that a given animal would respond to several determinants on one antigen with the production of several antibody species against each determinant. Lymphocyte hybridoma have made the dream become reality. It is, therefore, predictable — and it has already been sensed at the workshop — that lymphocyte hybridoma cultures are likely to replace many rabbits in many research laboratories.

However, the impression was gained from this conference, that many of these research laboratories are likely to enter this field with little experience in the fields of somatic cell genetics, tissue culture technology or even immunology. It is our hope that the many exciting reports in this book will stimulate the application of lymphocyte hybridomas to many biological and biomedical research problems. We also hope that the practical information contained in the many contributions to this book will make it easier for those who enter and follow to avoid time and fund consuming deviations in their own efforts to enlarge Table 2 by their specific lymphocyte hybridomas. We feel that the book will enable the interested reader to judge where this fast evolving field of research stands in the Spring of 1978, where it shows its difficulties and unresolved problems and where it can be lead to by all those which are already working in it and all those which are eager to enter it.

ACKNOWLEDGEMENTS

Our sincere thanks are due to the National Cancer Institute for its support of the workshop series entitled "Functional properties of tumors of T- and B-lymphocytes" of which the present one on "Lymphocyte hybridoma" was the second.

We are grateful to all contributors to this book to keep to the deadline for submission of their manuscripts and, thereby, to facilitate a rapid publication of the workshop proceedings. The articles appear in the form in which the authors have submitted them; no editing for style or content has been done.

We thank Springer-Verlag for ensuring a rapid publication of the proceedings.

Finally our thanks go to Ms. Karen Meinzen, Kappa Systems, Inc., Arlington, Va. for her invaluable help and never-ending efforts to organize this workshop and its proceedings in the most efficient and charming way.

Table 2: List of hybridomas produced by participants in the Workshop

Investigator Address	Immunization Normal Cell	Myeloma$^\tau$	Specificity of Hybridoma (L and H chain class)
G. Köhler Basel Institute	Mouse	X63	SRBC TNP Arsonyl
H. Koprowski C. Croce W. Gerhard Wistar Institute		X63	Influenza A Influenza B Para Influenza-1 Herpes Simplex Herpes Zoster Rabies SV-40 Polyoma SV-40 transformed cells Polyoma transformed cells Human melanoma Human colorectal carcinoma
U. Hämmerling SKI, New York	(A/Thy.1.1xAKR/H-2b)Fl α ASLI	X63	Thy-1.2
G.J. Hämmerling University of Cologne	AKR α C57BL/Ly1.1	X63	Thy-1.2
	(B6 x C3H/H-2I)Fl α I29	X63	Ly-b2(?)
	(B6 x A/TL)FL α A thymus	X63	TL
	A.TH α A.TL	X63	Ia (?Ia7)
M. Reth G.J. Hämmerling K. Rajewsky	C57BL/6 α NP-CG alum pertussis 1x i.p.	X63	NP (4-hydroxy-3-nitro-phenacetyl (γ_1,λ)
			NP (γ_1,λ) NP (γ_1,λ) NP μ,λ NP μ,λ
	α NP-Strep, 9x i.v.		NP γ_3,λ NP μ,λ NP γ_{2a},λ NP γ_{2a},κ NP γ_{2a},κ NP γ_1,κ NP. γ_{2a},κ NP γ_{2b},κ NP γ_3,κ
	NIP-CG al.pertussis boost NP-CG i.p.		NIP γ_1,λ
	NP-CG 2x i.p.		chick gamma glob.
	NP-CG 1x i.p.		chick gamma glob.
	NP-CG 1x i.p.		chick gamma glob.

Investigator Address	Immunization Normal Cell	Myeloma$^\tau$	Specificity of Hybridoma (L and H chain class)
K. Rajewsky G. Hesberg H. Lemke G.J. Hämmerling	A/J Strep.A, 2x	X63	A-carbohydrate (γ_{2a}, κ) (A-CHO) (α, κ) " (μ, κ) " (μ, κ) " (μ, κ)
	A/J anti-A5A-idiotype 6x Strep.A	X63	" (γ_{2a}, κ) " (γ_{2a}, κ) " (γ_{2a}, κ) " (μ, κ)
	A/J anti-A5A-idiotype 2x Strep.A	X63	" (γ_{2a}, κ) " (μ, λ) " (γ, κ) " (μ, κ)
G.J. Hämmerling	B10.A Hen egg Lysozyme 3x		Lysozyme
	A/J Hen egg Lysozyme 3x		Lysozyme "
	C57BL/6 (T,G)-A--L 3x		(T,G)-A--L
	BALB/c SRBC 3x i.p	X63	SRBC γ_1 " γ_1 " γ_1
I. Boettcher G.J. Hämmerling Schering AG Berlin	BALB/c Ovalbumin Al(OH)$_3$, 5x		Ovalbumin IgE
H. Lemke G.J. Hämmerling C. Hohmann K. Rajewsky	BALB/c CBA spleen 4x	X63	H-2K^k, prob. H-2.11 (γ_{2a}, κ) H-2K^k, " H-2.25 (γ_{2a}, κ) H-2K^k, " H-2.5 (γ_{2b}, κ)
	BALB/c CBA spleen 4x		H-2K^k Ia-A^k
G.J. Hämmerling U. Hämmerling	(C3H/AnxB6.Ly2.1)Fl ERLD 5x	X63	? (Autoantibody reacts with X63
R. Wallich G.J. Hämmerling	Wistar rat Eb DBA/2 T Lymphoma 3x	X63	Murine T cells Murine T and B cells
U. Hämmerling G.J. Hämmerling H. Lemke	BALB/c AKR spleen + thymus 2x	NS-1	Ia-A^k
V. Oi P. Jones J. Goding	CWB α C3H spleen	NS-1	I-A $(\gamma 2b)$* I-A $(\gamma 2a)$
S. Black L. Herzenberg Stanford University	BALB/c α CKB spleen	NS-1	H-2K^k $(\gamma 2a)$* I-A^k $(\gamma 3)$ I-A^k $(\gamma 2b)$ I-A^k $(\gamma 2b)$ I-A^k $(\gamma 2b)$

XX

Investigator Address	Immunization Normal Cell	Myeloma[t]	Specificity of Hybridoma (L and H chain class)
	BALB/c α BW pertussis + complexes	NS-1	Ig 1^b* Ig 1^b Ig 1^b Ig 1^b Ig.5a (γ2a) Ig.5b (γ1)
D. Zagury P. Cazenave Univ. Pierre and Marie Curie Pasteur Inst.	Mouse α horse-radish peroxidase	X63	Horseradish peroxidase
G. Buttin Univ. Paris	BALB/c α MOPC460 myeloma protein	X63	MOPC460 idiotype
M. Gefter M.I.T.	DBA/z α TNP	X63	TNP (μ) TNP (γ)
	DBA/z α TNP	X45	TNP (μ) TNP (γ)
T. Springer G. Galfre D. Secher C. Milstein Harvard and MCR Cambridge	Nylon wool purified red blood cell depleted spleen cells B10 mice	NS-1	mouse macrophage specific Ag (IgG) mouse lymphocyte 210K mol. wt. (IgG) mouse heat stable species specific Ag (IgG) Forssman Ag (IgM)
R. Goldsby	LPS stimulated cells	NS-1	DNP
	BALB/c α hemoglobin S		Hemoglobin S, A
	BALB/c α human RBC		Human RBC
	Human PBL (mixed donors)		Human PBL
K. Jin Kim K. Schroer NIAID, NIH	BALB/c α Type III pneumococcal poly-saccharide	X45	Type III Pneumococcal polysaccharide
D. Gottlieb J. Grove Wash. Univ., St. Louis	BALB/c α embryonic chick brain BALB/c α embryonic chick muscle		Embryonic chick brain Embryonic chick muscle
B.L. Clevinger D. Hansburg J. Davie Wash. Univ., St. Louis	Mouse α 1 3 dextran (see MSS)		α 1 3 dextran
L. Claflin Univ. of Michigan	BALB/c α R36A	X45	Phosphocholine (IgM, M511 type)
P. Estess J.D. Capra	A/J α KLH-azophenyl arsonate	NS-1	Phenylarsonate

Investigator Address	Immunization Normal Cell	Myeloma[T]	Specificity of Hybridoma (L and H chain class)
B.B. Knowles D. Solter Wistar Inst.	BALB/c α F9 teratocarcinoma		Ag on teratocarcinoma and pre-implantation stage mouse embryos
R. Kennett N. Klinman Univ. of Penn.	Immunization with DNP hemocyanin	X63	DNP NP
K. Denis R. Kennett Univ. of Penn.	LPS stimulated neonatal cells	X45	NP (μ)
K. Denis N. Klinman Univ. of Penn.	Neonatally derived DNP stimulated spleen fragments	X45	DNP (μ)
R. Kennett Univ. of Penn.	Mice immunized with concentrated super-nates from human B lymphoblastoid cell lines	X63	Human B cells
R. Kennett Univ. of Penn.	Human neuroblastoma coated with anti human B-cell antiserum	X63	Human neuroblastoma
R. Kennett Univ. of Penn.	Human CLL cells coat-ed with human B cell antiserum	X63	Human CLL
R. Polin K. Kennett	Mice immunized with Ricin	X63	Ricin
G. Slaughter M. Cancro R. Kennett	Mice immunized with human alkaline phosphatase	X63	Human alkaline phosphatase
K. Bechtol R. Kennett		X63	mouse testis
L.A. Manson R. Kennett Wistar Inst.	C57BL/10 anti L5178Y (DBA/2 MCA induced lymphoma) Primary immunization several months previous Boosted IV 2 days before	X63	12 hybridomas with specificities for: EL-4 + L5178 (μ) EL-4 + L5178 ($\mu+\gamma$) L5178
F. Melchers Basel Institute	Mouse LPS stimulated blasts	X63	SRBC(μ) = 2 hybrids TNP NIP = 10-30 hybrids HRC = 5 hybrids unkn. (μ) c 1000 hybrids unkn. (γ) c 30 hybrids
J. Paslay K. Roozen J.C. Bennett J. Kearney Univ, of Alabama	LPS stimulated spleen cells		RBC unknown (IgM)

Investigator Address	Immunization Normal Cell	Myeloma[τ]	Specificity of Hybridoma (L and H chain class)
R. Ceppellini M. Trucco Basel Institute	Mouse anti human EB virus transformed lymphoblastoid lines	NS-1	Human β2 microglobulin (several lines) non-specific HLA- (probably HLA-β2M complex) subclass of human B cells 5-10% Ig+cells
Inga Melchers P. Goodfellow J. Levinson Stanford University	BALB.B anti (T,G)-A-L	X63	(T,G)-A-L (IgM and IgG) (3 stable lines)
	C3H anti PCC4	NS-1	reacts with cell lines carrying C-type RNA virus
I. Melchers P. Goodfellow J. Levinson Stanford University	Rat α mouse pre-implantation embryo	X63	mouse embryo specific
D. Gasser R. Kennett Univ. of Penn.	Lewis α Lewis/KGH Lewis α KGH		AgB
T.J. McKearn Univ. of Chicago	Rabbit α SRBC	X63	SRBC Pig RBC viral glycoprotein (Hep 12)
	Rat α Rat	X63	AgB-3 (BN Rats)
J.C. Howard G. Galfre C. Milstein MRC Institute of Animal Physiology Cambridge	AO rat α DA rat (MHC H-1 α H-1)	X63	H-1A (K/D-like) determinants At least 3 non-competitive determinants involved
	(AOxHO.B2)F1 rat α OVG rat (MHC H-1 α H-1^c)	NS-1	Class and subclass not determined. Reactive against H-1A (K/D-like) determinant.
E. Simpson S. Kontiainen Univ. College London	In vitro primed T cells	BW5147	None with cytotoxic function 8 with KLH suppressor function. At least 2 are specific. 12 with TGAL help specificity unknown.

Footnotes and abbreviations:

τ = see Table 1 for complete name

α = anti

*Author submitted identifying clone or hybridoma accession numbers. These were omitted from the table.

Fusion of Mouse Myeloma and Spleen Cells

D.E. Yelton, B.A. Diamond, S.-P. Kwan, M.D. Scharff

A. Introduction

Kohler and Milstein (1) launched a revolution in serology when they
showed that it was possible to obtain continuous cell lines making
homogeneous antibodies by fusing mouse myeloma cells to spleen cells
from immunized animals. This approach not only promises to provide a
stable source of monospecific antibodies, but, since the antibody forming cell line is malignant, it can be injected into the peritoneal
cavity of mice and large amounts of ascites containing high concentrations of antibody can be obtained (1). With improvements in the techniques of fusing mouse myeloma cells (1,2,3), the basic approaches described by Kohler and Milstein (1,4,5) have gained wide acceptance as
evidenced by the many contributions to this monograph. This is merely
the tip of the iceberg, since our laboratory alone has provided drug
marked mouse myeloma cells for fusion to spleen cells to over 100 different investigators.

In spite of the successes reported in this monograph and elsewhere,
many technological problems remain and a number of fundamental questions
are unanswered:

1. Which drug marked malignant cell lines should be used to fuse to
spleen cells? This includes such problems as the relative fusion frequency of different cell lines, the importance of the state of differentiation of the malignant partner and spleen cell one hopes to "cap-
ture" in the fusion, the potential of some malignant partners to immediately extinguish antibody production or to introduce long term instability of expression, and the dominance of the malignant partner so
that the cell line can be injected into animals to produce large amounts
of antibody.

2. When is the optimum time after immunization to fuse? Included in
this are questions about the optimum state of differentiation of the
spleen cell partner and the potential role of accessory cells (helper
and suppressor T cells, macrophages) in facilitating or inhibiting the
growth of antibody forming hybrids.

3. How can the number and frequency of antibody forming spleen cells
be increased? The solution to this problem involves finding methods
for enriching antibody forming cells at the optimum stage of differentiation, determining the role of continued exposure to antigen and the
role, if any, of accessory cells. Enriched populations of antibody
forming cells might make it possible to obtain homogeneous antibodies
against poor immunogens such as alloantigens or minor antigenic determinants and would decrease the need for excessive tissue culture
work and brute force screening of hybrids. Enrichment after fusion
may also be useful.

4. Is it possible to increase the frequency of fusion? This will be
essential if enrichment of antibody forming cells is to be useful.
For example, if a spleen (10^8 cells) contains 10^5 antibody forming

1

cells and the fusion frequency is between 1×10^{-5} and 1×10^{-6}, then one would require all of the antibody forming cells from 1 to 10 spleens in order to obtain a single antibody producing hybrid.

5. When is the best time to screen and what is the best method of screening?

6. What is the basis for the instability of antibody forming hybrids and is it possible to increase stability?

These and other technical and conceptual problems will be addressed by a number of contributors to this monograph. In the subsequent sections we will discuss some of our own attempts to investigate a few of these questions.

I. Comparison of Drug Marked Myeloma Cells to be Used in Fusions with Spleen Cells

Most investigators attempting to generate antibody forming hybrids from mouse spleen cells have used drug marked mouse myeloma cells. The two major cell lines used have been the IgG_1 kappa producing $P_3X63Ag8$ and one of its non-secreting variants derived by Kohler and Milstein (1), and the IgG_{2b} kappa producing MPC-11 (45.6.TG1.7) derived in our laboratory (2). Both cell lines have defective hypoxanthine guanine phosphoribosyl transferase (HGPRT), are killed in HAT selective medium and can be complemented by fusion with $HGPRT^+$ cell lines (1,2). When fused with other myeloma cell lines making a variety of immunoglobulin classes, both $P_3X63Ag8$ and 45.6.TG1.7 continue to express their own heavy and light chains (6,7). Furthermore, when variants of both cell lines which have lost the ability to produce their heavy and/or light chains are fused to other producing myelomas, the resulting hybrids reflect the phenotype of the parents, i.e., chains which were not being synthesized by the P_3 or MPC-11 parent are not turned on nor are the chains expressed by the producing parent extinguished (6,7). In P_3, free light chains which are not associated with heavy chains are degraded intracellularly and not secreted (8). If a light chain producing variant of P_3 is fused to a myeloma producing another heavy chain, hybrid molecules containing P_3 light chains may now be secreted (4).

We have compared the ability of $P_3X63Ag8$ and 45.6.TG1.7 to generate spleen x myeloma hybrids and to produce antibody forming cells. The fusion technique is exactly as described previously (3) except that the ratio of spleen to myeloma cells is approximately 5:1. Polyethylene glycol is the "fusagen" and the particular batch used gives a maximum number of hybrids at a concentration of 30% for 8 minutes (3). Spleen cells were divided into two identical batches and fused simultaneously with the two cell lines. Following fusion 5×10^4 myeloma cells were distributed into each well of a small microtiter dish (Linbro Scientific Co., Hamden, Conn. #IS-FB-96 TC) plate in approximately .1ml of medium (2). The number of wells containing hybrids which could be grown up to mass culture is shown in Table 1. While these experiments were done in slightly different ways, they illustrate a large number of experiments which have led us to two conclusions: 1) there is significant variation from fusion to fusion even though we try to control the conditions, and 2) within this variation, we do not feel that $P_3X63Ag8$ or 45.6.TG1.7 give significantly different yields of hybrids. However, the P_3 hybrids do grow more quickly and can therefore be screened and processed in a shorter period of time.

2

Table 1. Frequency of hybrids with two different myeloma cell lines

Cell Line	Number of wells positive for hybrids per 10^8 spleen cells	
	Exp. A	Exp. B
45.6.TG1.7	336	500
P3X63Ag8	450	120

We have also determined the relative frequency of antibody forming hybrids generated by each myeloma cell line. Spleen cells from an animal immunized with sheep red blood cells (SRBC) were fused as described in the previous experiment, and the frequency of wells containing hybrids and the number of wells in which 10% or more of the cells rosetted with SRBC was determined (Table 2). As will be described below, we have found that this frequency of rosetting is closely correlated with the production of anti-SRBC antibody. In this fusion P3X63Ag8 yielded approximately 30% more hybrids and 11% of the wells contained antibody producing hybrids while 13% of the 45.6.TG1.7 wells were positive for antibody forming cells. This is consistent with our experience with many independent fusions and we have concluded that the two cells are equally effective in generating antibody forming hybrids.

Table 2. Relative frequency of antibody producing hybrids obtained with light chain producing variants of 45.6.TG1.7 and P363Ag8

	Number of wells		
	containing hybrids	screened	containing antibody forming hybrids
Exp. A			
45.6.TG1.7	121	93	12 (13%)
4T00.1L1	188	177	0
P3X63Ag8	165	160	17 (11%)
Exp. B			
P3X63Ag8	70	49	28 (57%)
P3X63Ag8U1	115	89	43 (48%)

Many people have found it inconvenient to use a myeloma producing large amounts of heavy and light chains and would prefer to use non-producing, or at least non-secreting myelomas. Since variants arise at a high frequency (9,10), such cell lines are easy to obtain and Milstein and his colleagues have used them successfully (11). However, we have observed two problems with such variants. First, some variant subclones of the 45.6.TG1.7 cell line do not fuse as efficiently as the parent and the yield of hybrids is low (Yelton, unpublished). Secondly, 4T00.1L1, the first light chain producing (H^-L^+) variant of 45.6TG1.7 which we tested, was effective in producing hybrids but the number which produced antibody was greatly reduced (Exp. A, Table 2). The spleen cells were from an animal immunized with SRBC and the fusions were carried out under identical conditions and at the same time. A much larger number of 4T00.1L1 x spleen hybrids from a number of independent fusions have been examined. They not only fail to make antibody, but very few synthesize any immunoglobulin heavy chains. This was especially surprising since 20 out of 20 hybrids between 4T00.1L1 and P3 continued to synthesize and secrete the P3 IgG1 heavy chain (Yelton, unpublished). Similarly, Siebert, Harris and Gefter (12) have shown that hybrids between 4T00.1L1 and MOPC 315 also continue to synthesize the 315 IgA heavy chain. These authors have, however, found that 4T00.1L1 is not really H⁻L⁺, but rather continues to synthesize very small amounts of the MPC-11 heavy chain (12). They do not find any heavy chain in other MPC-11 heavy chain minus variants (13). 4T00.1L1 may, therefore, be unusual in that it has suppressed the synthesis of its own heavy chain, will effectively extinguish heavy chain synthesis by spleen cells, but does not extinguish immunoglobulin synthesis by other myeloma cells. Whatever the defect, the major conclusion is that variants must be tested in myeloma x spleen fusions before being used routinely. Useful variants can be found, since Milstein and his colleagues have used such a variant (11). We have isolated another non-secreting variant from P3X63Ag8 which is as effective as the parent in generating antibody producing hybrids (Exp. B, Table 2). The unusually high frequency of antibody forming clones in Exp. B may be related to a very high frequency of plaque forming spleen cells.

II. Enrichment

In order to decrease the amount of screening and the number of hybrids which must be carried, we have attempted to enrich for antibody forming cells. The publication of a technique for increasing the frequency of fusion called the "pancake" method (14) suggested the possibility that we could enrich and fuse simultaneously. The technique studied by Greeley, Berke and Scott (20) has been adapted. SRBC were attached to the end (the face opposite the cap) of polylysine treated T-25 Falcon tissue culture flasks, spleen cells from SRBC immunized mice were gently rocked in the flasks, and the nonadherent cells were removed. Between 10 and 15% of the spleen cells including 90% of the plaque forming cells attached to the SRBC monolayer. Myeloma cells were centrifuged directly onto the spleen cells (which were attached to the SRBC) and treated with PEG (14). The bottles were gently washed and then allowed to incubate in normal growth medium for 16 hrs. at 37⁰ during which time most of spleen and myeloma cells including heterokaryons spontaneously detached from the monolayer. These were placed in HAT selective medium as in other fusions. The fusion frequency was approximately 1 in 10^6 spleen cells which is similar to that which we achieve in suspension (3) for spleen x myeloma fusions (see above) and there appears to be a few fold enrichment for antibody forming cells. While this approach reduces the number of clones which must be screened, it is clear that in its current form this technique will not achieve the enrichment or increase in fusion frequency which are needed if poor immunogens are to be used.

III. Loss Due to Segregation and Overgrowth

Once antibody forming hybrids are identified, the production of antibody is often lost as the hybrids are being grown to mass culture. This is probably due to a number of factors. First, if the antibody forming hybrid is one of a number of hybrids in a given well or culture, it may be overgrown. We examined two solutions to this problem. If the antigen is on a SRBC or can be attached to it, it is relatively easy to determine the approximate percentage of antibody forming cells in the clone by rosetting. We have found that the medium from cultures with more than 10% rosette positive cells almost always contains sufficient secreted antibody to be easily detected by the radioimmunoassay

4

of Jensenius and Williams (17). In stable clones of antibody forming hybrids where all or most of the cells are producing antibody, only 50% of the cells may rosette at a given time. Rosetting therefore underestimates the percentage of positive cells. Nevertheless, the isolation of rosette positive cells through Ficoll-Isopaque (15) has allowed us to enrich for the antibody forming cells in mixed cultures. A second approach is to clone in soft agar, overlay with either sub-class specific antibody or antigen-coated SRBC and complement (1), recover and further characterize the appropriate clones.

Even if one has a cloned hybrid line, there may be sudden or gradual loss of antibody production (17). This is illustrated for one of our hybrids in Figure 1. Clone 147 is an IgG$_1$ producing hybrid making

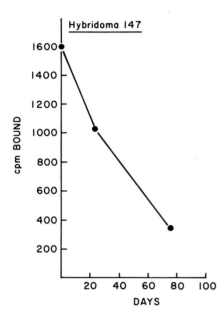

Fig. 1. Loss of antibody production by clone 147. The amount of mouse immunoglobulin which bound to the EL-4 target cell was determined at various times after the isolation of clone 147 by the radioimmune assay (17).

antibody against a surface antigen of the EL-4 lymphoma cell line. Following isolation, the hybrid clone produced progressively less anti-body. The hybrid was recloned in soft agar. Approximately 400 sub-clones picked in groups of 2-4, grown to mass culture and screened by the radioimmunoassay (17). This sort of "sib-selection" allowed us to screen many clones. Only 2% of these cultures were producing large amounts of antibody. These were again recloned, individual clones picked, and 20% of these were positive. Using this sort of cloning procedure, it is possible to obtain stable, antibody-producing hybrids.

It is reasonable to assume that most of the loss of antibody production is due to chromosome loss and segregation of the genes for the antibody heavy and/or light chains (2,6). In addition we have found that sub-clones of apparently stable populations of antibody forming hybrids

5

are heterogenous with respect to the amount of immunoglobulin produced. In one instance a hybrid was recloned, overlayed with subclass
specific antibody and clones surrounded by large and small amounts of
antigen-antibody precipitate were identified. The "maxiproducers"
were devoting 30% of the protein synthetic activity to antibody while
15% of the newly synthesized protein was antibody in the "mini-pro-
ducers". Some hybrids which are making the largest amounts of antibody begin to die after a number of generations in culture. We suspect that in at least some cases these have terminally differentiated.

The hybrid-positive wells initially contain a variety of cell types.
In studies to be reported elsewhere (18), we have isolated macrophage
x myeloma hybrids from wells also containing antibody secreting hybrids.
In addition to having Fc receptors, complement receptors and secreting
lysozyme and plasminogen activator, the medium from these cells partially suppresses antibody production by the antibody forming hybrid
with which they coexisted (18). It is also possible that normal accessory cells or other hybrids facilitate the growth of antibody producing hybrids.

Finally, myelomas may frequently generate variants whose immunoglobulin is changed in its affinity for antigen (19). Similar antigen binding mutants may arise and overgrow antibody forming hybrids.

B. Concluding Remarks

Although the fusion of mouse myeloma cells to antibody forming spleen
cells has already been effective in producing homogeneous antibodies,
many of the technical problems have not been solved and some fundamental questions have not been answered. It seems certain that within
a few years these uncertainties will be resolved, and cell fusion will
become the preferred methodology for obtaining highly specific and
reliable serological reagents.

Acknowledgments. D.E. Yelton is a medical scientist trainee supported
by Grant 5T5 GM1674 from the National Institute of General Medical
Sciences. B.A. Diamond is a trainee of the National Cancer Institute
(CA 09173). S.-P. Kwan is a fellow of the Cancer Research Institute,
Inc. This research was supported by grants from the National Institutes of Health (AI 10702 and AI 5231) and the National Science Foundation (PCM75-13609).

We would like to thank Ms. Theresa Kelly and Regina Flynn for their
expert technical assistance. We would also like to thank Dr. C. Milstein
for providing us with the P$_3$X63Ag8 cell line.

References

1. Kohler, G., and Milstein, C.: Continuous cultures of fused cells
   secreting antibody of predefined specificity. Nature 256, 495-497
   (1975)
2. Margulies, D.H., Kuehl, W.M., Scharff, M.D.: Somatic cell hybridization of mouse myeloma cells. Cell 8, 405-415 (1976)

3. Gefter, M.L., Margulies, D.H., Scharff, M.D.: A simple method
   for polyethylene glycol-promoted hybridization of mouse myeloma
   cells. Somatic Cell Genetics 3, 231-236 (1977)
4. Kohler, G., Howe, S.C., Milstein, C.: Fusion between immunoglobulin-secreting and non-secreting myeloma cells. Eur. J. Immunol.
   6, 292-295 (1976)
5. Kohler, G., Milstein, C.: Derivation of specific antibody-producing
   tissue culture and tumor lines by cell fusion. Eur. J. Immunol.
   6, 511-519 (1976)
6. Milstein, C., Adetugbo, K., Cowan, N.J., Kohler, G., Secher, D.S.,
   Wilde, C.D.: Somatic cell genetics of antibody-secreting cells:
   Studies of clonal diversification and analysis by cell fusion.
   Cold Spring Harbor Symp. Quant. Biol. 41, 793-803 (1976)
7. Margulies, D.H., Cieplinski, D., Dharmgrongartama, B., Gefter,
   M.L., Morrison, S.L., Kelly, T., Scharff, M.D.: Regulation of
   immunoglobulin expression in mouse myeloma cells. Cold Spring
   Harbor Symp. Quant. Biol. 41, 781-791 (1976)
8. Baumal, R., Scharff, M.D.: Synthesis, assembly and secretion of
   γ-globulin by mouse myeloma cells. V. Balanced and unbalanced
   synthesis of heavy and light chains by IgG-producing tumors and
   cell lines. J. Immunol. 111, 448-456 (1973)
9. Coffino, P., Scharff, M.D.: Rate of somatic mutation in immunoglobulin production by mouse myeloma cells. Proc. Natl. Acad.
   Sci. USA 68, 219-223 (1971)
10. Adetugbo, K., Milstein, C., Secher, D.S.: Molecular analysis of
    spontaneous somatic mutants. Nature 265, 299-304 (1977)
11. Williams, A.F., Galfre, G., Milstein, C.: Analysis of cell sur-
    faces by xenogenic myeloma-hybrid antibodies. Differentiation
    antigens of rat lymphocytes. Cell 13, 663-673 (1977)
12. Siebert, G., Harris, J., Gefter, M.: Regulation of immunoglobulin
    biosynthesis in the murine plasmacytoma MPOC 315. In press
13. Siebert, G. and Gefter, M.: Personal communication
14. O'Malley, K.A., Davidson, R.L.: A new dimension in suspension
    fusion techniques with polyethylene glycol. Somatic Cell Genetics
    3, 441-448 (1977)
15. Parish, C.R., Kirov, S.M., Bowern, N., Blanden, R.V.: A one-step
    procedure for separating mouse T and B lymphocytes. Eur. J. Immunol.
    4, 808-815 (1974)
16. Coffino, P., Baumal, R., Laskov, R., Scharff, M.D.: Cloning of
    mouse myeloma cells and detection of rare variants. J. Cell.
    Physiol. 79, 429-440 (1972)
17. Jensenius, J., Williams, A.F.: The binding of anti-immunoglobulin
    antibodies to rat thymocytes and thoracic duct lymphocytes. Eur.
    J. Immunol. 4, 91-97 (1974)
18. Diamond, B.A., Newman, W., Bloom, B.R., Scharff, M.D.: Manuscript
    in preparation
19. Cook, W.D., Scharff, M.D.: Antigen-binding mutants of mouse mye-
    loma cells. Proc. Natl. Acad. Sci. USA 74, 5687-5691 (1977)
20. Greeley, E., Berke, G., Scott, D.W.: Receptors for antigens on
    lymphoid cells. I. Immunoadsorption of plaque-forming cells to
    poly-L-lysine-fixed antigen monolayers. J. Immunol. 113, 1883-
    1890 (1974)

Anti-Viral and Anti-Tumor Antibodies Produced by Somatic Cell Hybrids

H. Koprowski, W. Gerhard, T. Wiktor, J. Martinis, M. Shander, C.M. Croce

A.   Introduction

Monoclonal, monospecific antibodies produced by hybrid cultures
between mouse myeloma cells (P3 x 63 Ag8) and lymphocytes of
various origins (Table 1) have aided us in studies of: a) the
antigenic determinants of a number of viruses, b) transforming proteins in cells transformed by oncogenic viruses, c) antigens expressed by human tumor cells and d) the genetics of antibodies.

Table 1. Species and source of lymphocytes hybridized with P3 x 63
Ag8 myeloma cells

Species	Source of Cells
Mouse	Spleen
Rat	Spleen
Human	Buffy Coat Cerebrospinal Fluid Lymphocytes EBV-transformed B cell lines Myeloma

The aim of producing monoclonal anti-viral antibodies was to characterize the antigenic relationship among viruses at the level of
individual determinants, to identify variant viruses that were previously unrecognizable in tests using heterogenous populations of
antibodies produced in vivo (1, 2) and to facilitate the isolation
of viral components for their biochemical characterization. The use
of monoclonal antibodies produced by hybridomas were found to be a
particularly suitable technique for the identification and characterization of transforming proteins of cells transformed by oncogenic
viruses (3). Although attempts to identify antigenic determinants
of human tumor cells have so far been unsuccessful, the production
of hybrid cultures secreting antibodies against either human melanoma or colorectal carcinoma has permitted us to undertake a
preliminary study of cross-reactive specificities between human
tumor cells and normal cells. Finally, the production of human
lymphocyte-mouse myeloma hybrids that secrete human immunoglobulins
in culture makes it possible to investigate the genetics of human
antibodies.

Table 2 shows the specificities of monoclonal antibodies produced
by hybridomas between P3 x 63 Ag8 myeloma cells and either mouse or
human lymphocytes. The techniques for fusion and selection of
hybrids that are described in detail elsewhere (1, 2, 3) have been
used throughout the study with only minor modifications.

8

Table 2. Antibodies produced by hybridomas between either mouse or human cells and myeloma

Donor	Viral Antigens									Other Antigens
	Infl. A	Infl. B	Para 1	Rabies	SV40	SV40 TM	Py TM	HSV	HZ	Melanoma
Mouse	+	+	+	+	+	+	+	+		+
Human				+					+	

Infl. = Influenza; Para = Parainfluenza 1; SV40 TM = Simian Virus 40-Transformed Mouse Cells; Py = Polyoma; HSV = Herpes Simplex 1; HZ = Herpes Zoster

B. Hybridomas Producing Anti-Influenza Antibodies

I. Frequency of Hybrids Secreting Anti-Influenza Antibody

Polyethylene glycol (PEG) 1000-induced fusion of influenza virus-primed BALB/c spleen cells with P3 x 63 Ag8 (4) myeloma cells generated somatic cell hybrids with high frequency and reproducibility. For instance, hybrids secreting anti-influenza antibody arose in 100% of the cultures that initially contained 15 - 17 x 10^5 spleen cells (Table 3). Obviously, it must be assumed that under the above conditions most hybrid cultures were derived from multiple fusion events. Thus, in order to obtain an estimate of the frequency

Table 3. Frequency of hybrids secreting anti-influenza antibody

Virus Strain Used for Immunization	Number of Spleen Cells per Culture x 10^{-5}	Hybrid Cultures Per Total Number of Cultures	Percentage of Hybrid Cultures Secreting Anti-viral Antibody
PR/8/34 (HON1)	17	90/90	100
WSN/33 (HON1)	15	114/120	100
PR/8/34 (HON1)	3	56/192	43
USSR/77 (H1N1)	3	42/116	57
HK/68 (H3N2)	2.3	40/60	53

Spleen cells were obtained from BALB/c mice 3 days after secondary intravenous injection with the indicated influenza A virus. Somatic cell hybridization with P3 x 63 Ag8 myeloma cells (spleen cell/myeloma cell ratio: 5 - 10/1) was performed with PEG-1000. Cultures of more than 10^6 and less than 10^6 spleen cells were seeded separately in individual wells of FB16-24TC and IS-FB-96TC Linbro plates, respectively.

of antiviral antibody-producing hybrids, cultures initially containing 3 x 10^5 or less spleen cells were seeded in individual wells of Linbro IS-FB-96TC plates. Under these conditions, 30% to 65% of the cultures developed hybrids and, on the average, 50% of these hybrid cultures secreted antiviral antibodies (Table 3). The latter observation agrees well with the previous observation that 62% of randomly selected hybrid clones derived from several initially polyclonal mass cultures produced antiviral antibodies (1, 2). Considering the fact that secondary virus-committed precursor B cells occur at a frequency of roughly 1/4000 splenic BALB/c cells (5) and assuming that cell division and/or recruitment are very unlikely to generate a 4000-fold increase of virus-committed splenic B cells in the 3 days after antigenic boost in vivo, these results indicate that antigenically stimulated (i.e., virus-committed) B cells are more likely than non-stimulated B cells to produce a successful hybrid. The incidence of antiviral hybrids was independent of the virus strain used for immunization (Table 3).

II. Comparison of the Specificity of Hybridoma and Splenic Antibodies

Somatic cell hybridization has been an effective method of generating large quantities of monospecific antibodies that can be applied to accurate antigenic analysis of complex antigens. However, the general usefulness of this approach depends on the feasibility of producing a spectrum of hybrids broad enough to express an unbiased cross-section of the total antibody specificity repertoire. To determine whether such a broad spectrum of hybrids can be obtained, the specificities of anti-PR8 influenza virus antibodies secreted by hybridomas were compared to those of monoclonal antibodies produced by splenic precursor B cells of BALB/c mice in the splenic fragment culture system (2, 5).

Table 4 shows that both splenic and hybridoma anti-PR8 antibodies expressed the specificities for the three viral surface antigens, hemagglutinin (H), neuraminidase (N) and chicken host (CH) derived component, at similar frequencies. The anti-H antibodies were

Table 4. Specificity of secondary monoclonal BALB/c anti-PR8 antibodies obtained in the splenic fragment system and by somatic cell hybridization.

Method of Antibody Production[1]	Number and Specificity[2] of anti-PR8 Antibodies			
	H	N	ChHC	Unknown[3]
Splenic Fragments	98 (72%)	11 (8%)	16 (12%)	12 (8%)
Hybridization	19 (66%)	1 (3%)	4 (14%)	5 (17%)

[1]Spleen cells from BALB/c mice that had been primed with PR8 3 - 6 months previously were adoptively transferred to lethally irradiated syngeneic recipients and stimulated in vitro with PR8 in the splenic fragment culture system (5) or were used for hybridization 3 days after stimulation in vivo with PR8. The splenic fragment clones were derived from experiments with spleen cells from 5 individual mice; the hybrids represent a random selection of cloned hybrids from fusions performed with spleen cells from 6 individual mice.

[2]The specificity of the antibodies (H, anti-hemagglutinin; N, anti-neuraminidase; ChHC, anti-chicken host component) was determined in the RIA (6).

[3]Antiviral antibodies to which no obvious specificity could be

10

assigned. In the splenic fragment system, such antibodies have always been assumed (probably incorrectly) to be polyclonal in origin.

further tested for the fine specificity of the antibody combining site by analysis of their crossreactivity with a panel of antigenically related influenza virus H molecules (Table 5). This analysis shows

Table 5. Reactivity of splenic and hybridoma anti-H antibodies

| | HSW | HO | | | | | | | | | | | | H1 | | Number of Antibodies Expressing Indicated Reactivity | |
| | | PR8 | | | | | WSN | WWE | BH | MEL | HIC | BEL | WEI | CAM | FM1 | | |
RT	SW	W	M	V2	V3	V6										Splenic	Hybridoma
1	-	+	+	+	+	+	-	-	-	-	-	-	-	-	-	7	4
2	-	+	-	+	+	+	-	-	-	-	-	-	-	-	-	3	1
3	-	+	+	-	+	+	-	-	-	-	-	-	-	-	-	3	0
4	-	+	+	+	-	+	-	-	-	-	-	-	-	-	-	4	5
5	-	+	+	-	-	+	-	-	-	-	-	-	-	-	-	1	1
6	-	+	+	-	-	-	-	-	-	-	-	-	-	-	-	0	1
7	+	+	+	-	+	+	-	-	-	+	-	-	-	-	-	0	1
8	-	+	+	+	+	+	-	-	-	-	-	-	+	-	-	1	1
9	-	+	+	+	+	+	+	+	-	+	-	-	-	-	-	0	1
10	-	+	+	+	+	+	+	+	+	+	-	-	-	-	-	3	2
11	-	+	+	+	+	+	-	-	-	-	+	-	-	+	+	0	1
12	-	+	+	+	+	+	+	+	+	-	+	-	-	+	+	0	1
Other reactivity types																74	0

Splenic: rows 1–6 = 18; rows 7–12 + Other = 78. Hybridoma: rows 1–6 = 12; rows 7–12 = 7.

The monoclonal antibodies were tested in the RIA (1, 2) for their reaction with the influenza A viruses SW/31, WSN/33 and WSE/33 and 5 antigenically distinct PR/8/34 viruses (V2, V3 and V6 represent antigenic variants of W, the PR8 strain used in this laboratory, M represents a PR8 strain obtained from a different laboratory) BH/35, MEL/35, Hickcox/40, BEL/42, Weiss/43, CAM/46, and FM/1/47. Negative (-) reactivity indicates that equal to or less than 10% and positive (+) reactivity that more than 10% of the antibody in the test sample bound to the indicated virus. The amount of antibody in a given test sample that bound to the homologous virus PR8 (W) was defined as 100%.

that the 19 anti-H (PR8) hybridomas expressed 11 distinct specificities (RT), six of which coincided with splenic RT. In particular, the hybridomas expressed all but one (RT3) of the splenic antibody

11

specificities within the repertoire of H (PR8)-specific BALB/c antibodies (RT1 through RT6). Furthermore, both splenic and hybridoma
antibodies expressed RT1 and RT4, most frequently. On the other
hand, there is an obvious discrepancy between the incidence of H
(PR8) specific and cross-reactive anti-H antibodies: of 96 splenic
antibodies, 18 (19%) were specific for determinants expressed only by
the homologous H (PR8) and all or some of its variants, whereas
81% reacted with H determinants present (in identical or cross-
reactive form) on PR8 and one or several heterologous viruses (Table
5). In contrast, 63% of the hybridoma antibodies reacted with H
(PR8)- specific determinants and 37% with heterologous virus H determinants. Since the incidence of crossreactive hybridomas can be
increased by using a heterologous virus for secondary stimulation, it
seems likely that this discrepancy between splenic and hybridoma
antibody specificities after homologous boost merely reflects a
difference in the kinetics of differentiation of strain-specific
and crossreactive B cells. Thus, using appropriate stimulatory
modifications, it seems quite feasible to generate a spectrum of
hybrids that express the complete BALB/c anti-H (PR8) specificity
repertoire.

C. Hybridomas Producing Antirabies Antibodies

Rabies virus-committed B cells were isolated from the spleens of
mice that had been primed with vaccine containing inactivated rabies
(ERA strain) 1 month before receiving an intravenous booster inoculation of the same antigen (7). When spleens were processed 3 days
after the booster, an overwhelming number of hybrid cultures (with
P3 x 63 Ag8 myeloma) produced rabies antibodies. However, when
spleens were obtained 10 days after the booster, the number of hybrid
cultures producing rabies antibody was very low. In general, 75%
of hybrid cultures produced rabies antibodies binding to rabies
virus in RIA; 58% of these also produced virus-neutralizing antibody
and 15% of antibodies reacted only with rabies virus nucleocapsids
detected in the cytoplasm of rabies-infected fixed cells. Virus-
neutralizing antibodies were also identified by staining membranes
of rabies-infected live cells in an immunofluorescence assay.

Up to now, only minor antigenic differences have been described
between the numerous rabies virus strains isolated throughout the
world. However, hybridoma antibodies that react with one strain of
rabies may not interact at all with another strain. As shown in
Table 6, of 52 hybridomas secreting rabies antibodies (detected
in RIA) nine reacted in virus-neutralization, cytotoxicity and membrane immunofluorescence assays with all three strains of rabies,
ERA, CVS and HEP-Flury, fifteen reacted with the ERA and HEP strains,
two with ERA and CVS, three with ERA only and one with HEP virus only.
The latter results were rather interesting in light of the fact that
ERA strain vaccine was used for priming and booster inoculation of the
mice whose spleens were used for production of hybrid cultures.
Twenty-eight hybrid cultures produced antibodies reacting with nucleocapsids of all 3 virus strains; and of those 22 reacted only with
nucleocapsids (Table 6) and with no other viral components.

Hybrid cultures secreting rabies virus-neutralizing antibodies
protect mice against the lethal effect of rabies virus. As shown
in Table 7, such hybrid cultures implanted into BALB/c mice protected
all mice against death by intracerebrally injected rabies virus.

Table 6. Antibodies produced by antirabies hybridomas

| Number of Hybridomas | Reactivity Against Virus Strains in Different Assays | | | | | | | | | | | |
| | ERA | | | | CVS | | | | HEP | | | |
	VN	CT	M	NC	VN	CT	M	NC	VN	CT	M	NC
7	+	+	+	-	+	+	+	-	+	+	+	-
2	+	+	+	-	±	±	+	-	+	+	+	-
2	+	+	+	-	+	+	+	-	-	-	-	-
11	+	+	+	-	-	-	-	-	+	+	+	-
2	+	+	+	+	-	-	-	+	+	+	+	+
2	+	+	+	+	-	-	-	+	-	-	-	+
1	+	+	+	-	-	-	-	-	-	-	-	-
2	-	±	+	+	-	-	-	+	+	+	+	+
1	-	-	-	-	-	-	-	-	+	+	+	-
22	-	-	-	+	-	-	-	+	-	-	-	+

VN = Virus Neutralization; CT = Cytotoxic Test; M = Membrane Fluorescence; NC = Nucleocapsid Fluorescence.
Spleen cells were obtained from mice 3 days after secondary i.v. injection with ERA rabies vaccine. Somatic cell hybridization performed as described in Table 3. Antibody determination performed on media of hybrid cultures grown in wells of FB16-24TC Linbro plates. ERA, CVS and HEP (Flury) refer to fixed strains of rabies virus of various origins.

Table 7. Protective effect of hybridoma antibody in mice exposed to rabies virus by intracerebral inoculation

Treatment	Ratio of Protected Mice
Hybridoma* (VN+)	6/6
None	0/6
Hybridoma* (VN-)	0/6
None	0/6

VN+ Producing neutralizing antibody
VN- Producing non-neutralizing antibody
*1 x 10^6 hybridoma cells were subcutaneously implanted in adult BALB/c mice; 4 days later implanted and control mice were injected intracerebrally with 100 infectious doses of rabies virus. Sera of mice showed approximately 500-fold increase in antirabies antibody over tissue culture media.

D. Hybridomas Producing Anti-SV40 and Anti-Herpes Simplex Virus Antibodies

We have also hybridized spleen cells derived from mice immunized and boosted with either SV40 or herpes simplex Type I virus with P3 x 63 Ag8 mouse myeloma cells and we have studied the hybrid cells for

13

the production of antiviral antibodies. African Green Monkey Kidney cells were infected with SV40 at the multiciplity of 1 PFU per cell and the cells were fixed in acetone after 48 and 72 hours. Hybridoma culture fluid was added on the coverslips containing the infected cells for 1 hour and a rabbit anti-mouse immunoglobulin antiserum labeled with fluoresceine was then added to detect the presence of antiviral antibodies. Specific SV40 nuclear fluorescence was detected using culture fluids derived from 14 out of 22 independent hybridomas. Studies are in progress to determine against which viral proteins and against which viral antigenic determinants the hybridoma antibodies are directed. Similar results were obtained using culture fluids from somatic cell hybrids with spleen cells derived from mice immunized with HSV-1.

E. Hybridomas Producing Antibodies Specific for the Tumor Antigen of SV40

SV40 transformed cells produce a nuclear tumor (T) antigen (8) that is coded by the A gene of the virus. Recent studies indicate that the T antigen is not a single SV40 gene product and that two early SV40 T antigens are produced in SV40 transformed or infected cells: one that has a molecular weight of 94,000 daltons and is called large T and one that has a molecular weight of 17,000 daltons and is called small T (9). These proteins appear to be coded by overlapping DNA sequences. In order to determine whether the large and small T antigens share crossreacting antigenic determinants, and to study the properties of these molecules, we have produced somatic cell hybrids between P3 x 63 Ag8 mouse myeloma cells and spleen cells derived from mice immunized with SV40 transformed syngeneic or allogeneic mouse cells. Only 25 independent hybrids out of approximately 300 that were analyzed produced antibodies specific for SV40 T antigen. The titers of the culture fluids of some of the hybridomas are given in Table 8.

Table 8. Titer of anti-SV40 T-Antigen antibodies produced by hybridomas in culture fluids, serum and ascites.

Hybridomas	Culture Fluid	Titer	
		Serum	Ascites
A25.1#1B3	1:5	1:6400	1:3200
B16.1#2A2	1:2	ND*	ND
B16.1#2C4	1:10	ND	ND
B16.1#2D5	1:2	ND	ND
B16.1#1B6	1:20	1:6400	1:3200
B16.1#1C5	1:10	ND	1:3200
B16.1#2A3	1:5	1:6400	ND
B16.1#1B2	1:2	ND	1:100
A17.2#1**	negative	1:50	ND

* ND   mice were not injected with hybridoma cells

** This hybrid was originally a producer of anti-SV40 T antigen antibodies, but became negative following subculture.

14

As shown in the table, injection of the hybrids into syngeneic host as solid tumors or ascites resulted in production of ascites or sera with extremely high titers of anti-SV40 T antigen antibodies. These antibodies were also tested on cells transformed by another papovavirus, BK virus, that has been isolated from patients that underwent immunosuppressive therapy. The BK virus T antigen has been shown to crossreact with SV40 T antigen (3). As shown in Table 9, only 4 out of 12 anti-SV40 antigen hybridoma antibodies reacted against BK virus T antigen, indicating that the two T antigens shared only some, but not all, antigenic determinants. We have also initiated experiments of immunoprecipitation

Table 9. Cross reactivity between SV40 and BK virus T antigens

Source of Antibodies[1]	Test Cells	
	LN-SV	HKBK-DNA-4
Anti-T serum[2]	+	+
A17.2#1[3]	+	-
A25.1#1B3	+	+
B16.1#1B2	+	-
B16.1#1B6	+	-
B16.1#1C1	+	-
B16.1#1C5	+	-
B16.1# A2	+	+
B16.1# A3	+	-
B16.1# A5	+	-
B16.1# C4	+	+
B16.1# D2	+	-
B16.1# D5	+	+

[1]Except where noted, undiluted culture fluids from hybrid cultures were used for test.

[2]1:100 dilution of control serum from SV40 immune mouse.

[3]1:20 dilution of serum from animal bearing a tumor induced by A17.2#1 hybrid cells. Culture cells from the A17.2#1 tumor were found to produce anti-SV40 T antigen antibodies that did not cross react with BK virus T antigen.

and characterization of the SV40 T antigens recognized by the different hybridoma antibodies. Preliminary evidence indicate that some of the antibodies are specific for the large T antigen and some react with both large and small T antigens suggesting that the large and small T antigens share antigenic determinants. Studies are in progress to further characterize immunologically these tumor antigens.

F. Hybridomas Producing Antibodies Against Human Tumor Antigens

Cells obtained from human melanomas and human colorectal carcinomas grown in tissue culture (10 - 11) and hybrid cultures between a

15

human melanoma and mouse IT22 cells were used for immunization of mice and as immunoadsorbents in RIA with media from hybrid cultures.

Table 10. Crossreactivity of hybridoma antibodies with human melanomas, colo-rectal carcinomas, and normal human tissue.

Donors Immunized With:	Hybridoma No.	Melanoma						Colo-rectal Carcinoma					Normal Human		
		690	691	489	843	1614	1694	480	948	403	1116	837	WI38	HFF	FS2
691 Human Melanoma	2	+	+	+	+	+	+	+	+	–	+	+	+	+	+
	4	+	+	+	+	+	+	–	+	–	–	–	+	+	+
	5	+	+	+	+	+	–	+	+	–	+	+	+	+	+
	6	+	+	+	+	+	+	–	–	–	+	+	–	–	–
	9*	+	+	–	+	+	–	+	+	–	–	+	+	+	+
	11*	+	+	–	+	+	+	+	+	+	–	+	+	+	+
	12*	+	+	+	+	+	+	+	+	–	+	+	+	+	+
	13	+	+	–	–	–	+	–	–	–	–	–	–	–	–
	19*	+	+	+	+	+	–	–	–	–	–	–	+	–	–
691 x IT22 Hybrid	2*	–	+	+	+	–	+	–	–	–	–	ND	–	–	–
	4*	+	+	–	+	–	+	–	–	–	–	ND	–	–	–
	6*	+	+	+	+	+	+	–	+	–	–	ND	+	+	+
480 Human Colo-rectal Carcinoma	1	+	+	–	+	+	+	+	+	+	+	+	+	+	+
	3	–	–	–	–	+|	–	+	–	–	–	–	–	–	–
	4	+	+	+	+	+	+	+	–	+	+	+	+	+	+

* Also reactive with 691 x IT22 hybrid; ND = Not Done

Spleen cells obtained from mice after primary intraperitoneal injection of 3×10^7 tumor cells and secondary i.v. booster of 1×10^6 cells. RIA with gluteraldehyde fixed cells. Results expressed as mean cpm's of test samples minus mean of triplicate of cpm of control P3 x 63 Ag8 culture.

16

Live tumor cells were used for primary and secondary immunization. Of 29 hybrid cultures obtained after fusion of spleen cells from mice immunized with a human melanoma (B), 9 (Table 10) secreted antibodies that reacted in RIA with human melanoma. Antibodies secreted by hybridomas interacted better with the melanoma (691) used for immunization and with melanoma (690) than with other melanomas but differed from each other in their cross-reactivity with other cells. Roughly, the anti-melanoma antibodies may be classified into three categories: 1) antibodies reacting only with a malanoma but not with any other cells tested (Table 10); 2) antibodies reacting with all melanomas and with some colorectal carcinomas, and 3) antibodies reacting with all melanomas, all normal human cells and all but one of the colorectal carcinomas. Categories 1) and 3) were also obtained with antibodies secreted by hybridomas produced after immunization of mice with human melanoma x IT22 hybrid cells. After mice were immunized with human colorectal carcinomas, three out of eight hybrid cultures produced anti-colorectal carcinoma antibody. Of the three, one reacted only with homologous colorectal carcinoma but not with any other cells, whereas the other two showed broad cross-reactivity with other colorectal carcinomas, human melanomas and normal human cells. Clones of hybrids showed remarkable homogeneity; reacting antibodies expressing similar specificities as antibodies secreted by parental culture.

Growth of melanoma tumors in nude mice was suppressed by prior implantation of hybridomas producing anti-melanoma antibodies in mice and sera obtained from these mice showed a 500-1000 fold increase in binding capacity to melanoma cells over tissue culture media.

G.   Genetics of Human Antibodies

We have also produced somatic cell hybrids between P3 x 63 Ag8 mouse myeloma cells and lymphocytes derived from the cerebrospinal fluid of a patient with a herpes zoster encephalitis. Mouse-human hybridomas were obtained in HAT selective medium and were tested for the production of human antibodies by immunoprecipitation of the radio labeled human antibodies using rabbit anti-human IgM, IgG, μ, γ, and κ antisera and by polyacrilamide gel electrophoresis. The hybrids were found to segregate into clones that produced human antibodies or antibody chains and clones that did not. Karyotype of one of the human IgM producing clones is shown in Figure 1. Karyologic and isozyme analysis of the hybrid clones should make it possible to assign the genes for human antibody heavy and light chains to their specific chromosomes.

H.   Conclusion

The experiments described in this paper indicate that it is possible to select hybridomas producing monoclonal antibodies specific for viral antigens and that such antibodies provide enormously sensitive tools for studying these complex antigens. The hybrids producing antibodies against the tumor antigens of oncogenic DNA tumor virus can be used to purify and characterize these antigens and to study the genetic regulation of their expression. The fact that the small and large SV40 T antigens share antigenic determinants, as determined using the hybridoma monoclonal antibodies strongly suggest that they are coded by the same nucleotide sequences. The results obtained with antibodies produced by hybrids between mouse myeloma

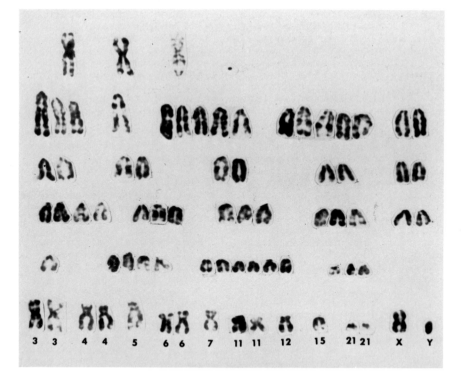

Figure 1. Karyotype of a cell of mouse-human hybrid clone producing
human IgM antibodies. The human chromosomes present in the hybrid
are in the last row. All other chromosomes derived from the
P3 x 63 Ag8 cell parent.

cells and spleen cells derived from mice immunized with human melanoma cells suggest that some of them might bind melanoma specific
surface antigens. Studies to assess this point are presently
in progress. We have also produced hybrids between mouse myeloma
cells and human lymphocytes, some of these were found to produce
either human IgG or IgM. We believe that the karyological and
isozyme analysis of the hybrids should result in the chromosome
assignment of the genes coding for human antibody chains.

I. References

1. Koprowski, H., Gerhard, W., Croce, C. M.: Production of
    antibodies against influenza virus by somatic cell hybrids
    between mouse myeloma and primed spleen cells. Proc. Natl. Acad.
    Sci. USA 74, 2985 - 2988. (1977).
2. Gerhard, W., Croce, C. M., Lopes, D., and Koprowski, H.:
    Repertoire of antiviral antibodies expressed by somatic cell
    hybrids. Proc. Natl. Acad. Sci. USA 75, In Press. (1978).
3. Martinis, J., and Croce, C. M.: Somatic cell hybrids producing
    antibodies specific for the tumor antigen of simian virus 40.
    Proc. Natl. Acad. Sci. USA, In Press.

4. Kohler, G., and Milstein, C.: Derivation of specific antibody producing tissue culture and tumor lines by cell fusion. Eur. J. Immunol. 6, 511 - 519. (1976).
5. Gerhard, W., Braciale, T. J. and Klinman, N. R.: The analysis of the monoclonal immune response to influenza virus. I. Production of monoclonal antiviral antibodies in vitro. Eur. J. Immunol. 5, 720 - 725. (1975).
6. Gerhard, W.: The analysis of the monoclonal immune response to influenza virus. II. The antigenicity of the viral hemagglutinin. J. Exp. Med. 144, 985 - 995. (1976).
7. Wiktor, T. J., Croce, C. M. and Koprowski, H.: Manuscript in preparation.
8. Black, P., Rome, W., Turner, H. and Huebner, R.: A specific complement fixing antigen present in SV40 tumor and transformed cells. Proc. Natl. Acad. Sci. USA 50, 1148 - 1156. (1963).
9. Crawford, L. V., Cole, C. N., Smith, A. E., Pancha, E., Tegtmeyer, P., Rundell, K. and Bery, P.: Organization and expression of early genes of simian virus 40. Proc. Natl. Acad. Sci. USA 75, 117 - 121. (1978).
10. Leibovitz, A., Stinson, J. C., McCombs III, W. B., McCoy, C. E., Mazur, C., and Mabry, N. D.: Classification of human colorectal adenocarcinoma cell lines. Cancer Research 36, 4562 - 4569. (1976).
11. Koprowski, H., Steplewski, Z., Herlyn, D., and Herlyn, M.: Production of monoclonal antibody against human melanoma by somatic cell hybrids. Proc. Natl. Acad. Sci. USA. In Press.

Analysis of V Gene Expression in the Immune Response by Cell Fusion

T. Imanishi-Kari, M. Reth, G.J. Hämmerling, K. Rajewsky

Introduction

In this paper we describe an experimental approach in which the cell hybridization technique is used for the analysis of V gene expression in the immune response.

The experimental system is the immune response of C57BL/6 mice against the hapten 3-nitro-4-hydroxy-phenylacetyl (NP). Several characteristic features of this response have been established previously by Mäkelä and coworkers (1, 2, 5) and ourselves (3, 4). The primary anti-NP response of C57BL/6 mice is restricted to lambda chain-bearing antibodies (3, 5) which exhibit characteristic patterns in isoelectric focusing (2, 6) and in terms of fine specificity of hapten binding (so called heteroclicity, i.e. preferential binding to a cross-reactive rather than the sensitizing hapten). These antibodies can be idiotypically defined as the NP^b idiotype (3, 5). Every single individual invariably expresses this antibody population with its full set of idiotypic determinants, and in genetic crosses heteroclicity as well as NP^b idiotype are inherited as a single genetic unit in close linkage to the heavy chain allotype (1, 3, 5). In the secondary and the hyperimmune response the antibody population becomes more heterogeneous. Kappa-bearing antibodies appear and the frequency of the NP^b idiotype drops from approximately 90% to 10-20% (4 and Imanishi-Kari unpublished results).

In contrast to previous investigations in which only a few attempts were made to study single antibody species (5), the present approach aims at the elucidation of the spectrum of individual antibodies appearing in the primary and hyperimmune anti-NP response.

We obtained these antibodies from cloned hybrid cell lines which were generated by fusion of myeloma cells with NP sensitized C57BL/6 spleen cells.

Methodology and Fine Specificity Characteristics of the Hybrid Cell Anti-NP Antibodies

The details of the isolation and characterization of the various anti-NP antibodies of C57BL/6 origin are described in separate publications (3, 7). The protocol can be summarized as follows:

Primary antibodies to NP were induced by injecting C57BL/6 mice intraperitoneally with alum-precipitated NP_{12}-chicken gamma-globulin (NP_{12}-CG; 100 µg) and inactivated pertussis vaccine as an adjuvant. The mice were bled 18-28 days later. For hyperimmunization the mice were intravenously injected with 10^9 NP-conjugated group A streptococcal particles (NP-Strep.A) on day 1 and given similar injections on days 8, 10, 12, 15, 17 and 19 (7). In order to establish cell lines producing mono-clonal anti-NP antibodies from the primary response, spleen cells from C57BL/6 mice primed 7 days previously were fused with the myeloma cell line P3-X63-Ag8, in the presence of polyethylene glycol (molecular weight 2000 or 4000). Spleen cells from hyperimmunized mice were fused 4 days after the last injection to P3-X63-Ag8 myeloma cells in the presence of Sendai Virus as described by Köhler and Milstein (8, 9) with minor mo-difications. Both procedures are described in detail by Hämmerling et al. in this book and by Reth et al. (7). The P3-X63-Ag8 myeloma cell line was a gift from Dr. C. Milstein. It is a subline of the BALB/c myeloma MOPC 21 and produces an IgG_1 myeloma protein carrying kappa light chains.

Hybrid cells were selected in hypoxanthine-aminopterin-thymidine (HAT) medium and analysed for the production of anti-NP antibody in a NP spe-cific hemolytic plaque assay or in a sensitive plastic plate radio-immuno assay (10) in which microtiter plates were coated with NIP-BSA and incubated with the culture supernatant. Bound immunoglobulin was then detected with an ^{125}-I-labelled rabbit F(ab)$_2$ anti-mouse immuno-globulin. Several cell lines were isolated, some of which produced an anti-NP antibody. Some of the positive lines were cloned in soft agar. Finally, 14 lines were serially transplanted subcutaneously or intra-peritoneally in (BALB/c x C57BL/6)F_1 mice. Large quantitites of anti-NP antibodies appeared in the serum and/or ascitic fluid of the reci-pient mice.

The various anti-NP antibodies were purified by adsorption to NIP or NP-conjugated bovine serum albumin (BSA) coupled to Sepharose 4B and subsequent elution from the immunosorbent with free NIP-caproic acid (NIP-cap).

In Table 1 we summarize some properties of the various hybrid cell an-tibodies. The data include (i) chain composition as determined with class- and type specific antisera in double diffusion and an indirect hemagglutination assay (sensitisation of NP-coupled sheep red cells with subagglutinating amounts of anti-NP antibody and subsequent agglutination with defined anti-Ig antisera), (ii) affinity for the NIP hapten as determined in a modified Farr assay (1, 7) and (iii) the affinity for NIP-cap, NNP-cap and NClP-cap (in the cross reacting hap-tens position 5 of the aromatic ring is substituted with iodine, NO_2 or chlorine) as compared to that for NP-cap as measure of heteroclici-ty. The corresponding data for primary anti-NP antisera are shown for comparison. The antibodies were grouped into families on the basis of their fine specificity of hapten binding. The table shows the order of affinities by which a series of cross reacting haptens are bound to the antibodies. The binding patterns are very characteristic for each family in quantitative terms (7).

It is satisfying to see that all isolated antibodies from the primary response belong either to the IgM or the IgG_1 class and carry lambda light chains, in perfect agreement with the situation encountered in primary anti-NP sera (3,4,5).The most striking result of the analysis is the finding of antibody families characterized by light chain type and fine specificity of hapten binding. The members of the largest fa-mily (which comprises anti-NP antibodies from both the primary and the hyperimmune response) carry lambda light chains and exhibit the characteristic "heteroclitic" fine specificity. The other hybrid cell antibody families are from the hyperimmune response. The members of these families are non heteroclitic, are of different class- and sub-classes and carry kappa light chains. This again is in perfect agree-

Table 1. Families of Hybrid Cell Antibodies of C57BL/6 Origin with Specificity for the Hapten NP

Antibody family	1° and 2° response	Antibodies	H-chain class-subclass	L-chain type λ (κ)	I_{50}NIP-cap(M) (a)	K_{rel} (b)	Fine specificity (c)
C57BL/6 serum anti-NP antibodies (d)	1°	No. of members not known	μ, γ1	λ (κ)	1.6×10^{-7}	4.6	NNP ≥ NIP > NP ≥ NC1P
I	1°	B53-8	γ1	λ	2.4×10^{-6}	5.8	NNP ≥ NIP > NP ≫NC1P
		B53-12	γ1	λ	3.3×10^{-7}	11	NNP ≥ NIP > NP ≥ NC1P
		B1-48	γ1	λ	4.5×10^{-7}	12	"
		B1-52	μ	λ	3.0×10^{-7}	10	"
		B1-8	μ	λ	1.0×10^{-7}	20	"
	2°	S34	μ	λ	7.7×10^{-9}	7	"
		S24	γ3	λ	5.5×10^{-8}	3.5	"
		S43	γ2a	λ	2.4×10^{-8}	12	"
II	2°	S8	γ2a	κ	5.8×10^{-9}	0.9	NP ≥ NIP ≥ NNP ≫ NC1P
		S92	γ2a	κ	2.0×10^{-9}	1.2	"
III	2°	S10	γ1	κ	1.3×10^{-7}	0.2	NP > NIP ≥ NNP≫ NC1P
		S36	γ2a	κ	3.5×10^{-7}	0.1	"
		S74	γ2b	κ	2.2×10^{-6}	0.3	"
IV	2°	S19	γ3	κ	2.8×10^{-7}	0.2	NP > NIP > NNP ≥ NC1P

(a) Molar hapten concentration resulting in 50% inhibition of binding to N^{125}IP-cap at 2×10^{-9} M.

(b) $K_{rel} = \dfrac{I_{50}\text{NP-cap}}{I_{50}\text{NIP-cap}}$

(c) Order of haptens according to their affinity to antibody

(d) Obtained from C57BL/6 mice 3 weeks after priming with 100 μg alum-precipitated $NP_{12}CG$ plus pertussis.

ment with the situation encountered in hyperimmune anti-NP sera (2, 4 and Imanishi-Kari unpublished results).

Idiotypic Analysis of the Hybrid Cell Anti-NP Antibodies

In order to determine possible relationship of the variable portions of the individual anti-NP hybrid cell antibodies idiotypic analysis these antibodies was performed. Two types of anti-idiotypic antisera were used, namely   anti-idiotypic antisera raised in rabbits or guinea pigs against the primary anti-NP serum antibodies (the NP^b idiotype) of C57BL/6 mice and various anti-idiotypic antisera against the individual hybrid cell anti-NP antibodies. The detailed characterization of the anti-idiotypic antisera will be the subject of separate publications. In this communication only the results which were obtained with a rabbit anti-idiotypic antiserum against primary anti-NP antibodies (rabbit anti-NP^b idiotype) will be reported. This antiserum was shown to react specifically with the majority (> 80%) of primary C57BL/6 anti-NP antibodies but it did not bind C57BL/6 normal Ig or antibodies of unrelated specificity (4). Furthermore, using this antiserum it was shown that every individual C57BL/6 mouse expressed a similar set of idiotypic determinants in its primary anti-NP antibody population and that the idiotypic marker defined by the antiserum was linked to the Ig-1^b allotype (4).

The idiotypic analysis of the hybrid cell antibodies is exemplified in Fig. 1 with a few members of the lambda-bearing heteroclitic antibody family. For this study it was attempted to inhibit by hybrid cell anti-NP antibodies the binding of radiolabelled NP^b to the rabbit anti-idiotypic antiserum. It can be seen in Fig. 1 that the reaction of rabbit anti-NP^b with labelled NP^b was completely inhibited by unlabelled NP^b

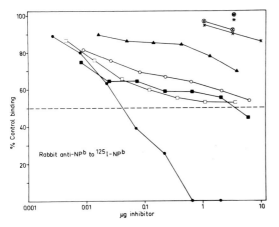

Fig. 1. Analysis of the inhibition of the binding of ^{125}I-NP^b to rabbit anti-NP^b antiserum. Inhibitors: ●—● C57BL/6 primary anti-NP-Ig (NP^b); hybrid cell anti-NP antibodies o—o B53-12 (γ_1, λ); ■—■ S34 (μ,λ); □—□ B1-52 (μ,λ); ▲—▲ B53-8 (γ_1, λ); ✕--✕ C57BL/6 pooled normal Ig; ⊗—⊗ C57BL/6 anti-CG; ✳ MOPC 104E and ⊗ MOPC 21.

23

idiotype indicating that the rabbit anti-NP^b antiserum detects the full set of idiotypic determinants on the primary anti-NP serum antibodies. However, when hybrid cell anti-NP antibodies were used for inhibition only partial inhibition was observed (Fig. 1). It is also apparent from Fig. 1 that various degrees of inhibition can be observed with the individual heteroclitic anti-NP antibodies indicating that they differ from each other. Altogether these findings suggest that the hybrid cell anti-NP antibodies are not identical to the NP^b idiotype but are related to it.

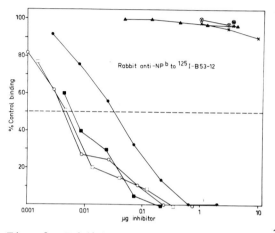

Fig. 2. Inhibition of the binding of ^{125}I-B53-12 to rabbit anti-NP^b antiserum. Inhibitors: ●—● C57BL/6 primary anti-NP-Ig (NP^b); o—o B53-12 (γ_1, λ); ■—■ S34 (μ, λ); □—□ B1-52 (μ, λ); ▲—▲ B53-8 (γ_1,λ); x—x C57BL/6 pooled normal Ig; ⊘--⊘ C57BL/6 anti-CG; ✗ MOPC 104E; ⊛ MOPC 21.

In other experiments (data not shown) it was observed that most of the heteroclitic hybrid cell anti-NP antibodies can be bound by the anti-NP^b antiserum. These reactions could be inhibited by some but not all heteroclitic hybrid cell anti-NP antibodies and also by the NP^b idiotype. In Fig. 2 an example of such an analysis is shown. The binding of rabbit anti-NP^b idiotype to radiolabelled hybrid cell anti-NP antibody B53-12 can be inhibited not only by the homologous unlabelled B53-12 but also by some heteroclitic hybrid cell anti-NP antibodies such as B1-52 and S34 and by the NP^b idiotype, but not by B53-8. These results suggest that some of the hybrid cell heteroclitic anti-NP antibodies share idiotypic determinants which are also present in the NP^b idiotype. In further experiments (data not shown) anti-idiotypic sera were raised against the individual hybrid cell anti-NP antibodies and subsequently assayed in a checker board type experiment against the hybrid cell anti-NP antibodies. Although our studies are not yet complete the available results strongly suggest that within each family the members differ from each other in detail. They seem to share cross reactive determinants but express also individual (private) idiotypic determinants.

Concluding Remarks

Every single feature of the primary anti-NP response in C57BL/6 mice such as its idiotypic and isoelectric focusing analysis, its genetic determination, its heteroclitic quality and its association with lambda light chains had suggested the presence of a very restricted spectrum of antibody species. However, the cell hybridization technique reveals that the NP^b idiotype represents a large family of interrelated but non-identical antibodies. It remains to be shown whether the same will hold true for the various other "major" idiotypes that have been established as genetic markers of antibody V genes.

It is tempting to look at the various members of the lambda-bearing families of anti-NP antibodies as products of a diversification process by which they might have arisen from a single germ line encoded ancestor antibody in ontogeny. Such a hypothesis can be tested by straightforward structural and functional experiments which might also reveal the rule by which V genes are expressed and diversified in the immune system.

Acknowledgements. We are grateful to Dr. C. Milstein for providing the P3-X63-Ag8 cell line, to Drs. H.C Morse and U. Hämmerling for gifts of reagents, to Dr. J.F. Kearney, A. Radbruch, B. Liesegang and Christa Müller, who kindly helped with the typing of the hybrid cell anti-NP antibodies, to Sigrid Irlenbusch and Maria Palmen for skilful technical assistance and to Åsa Böhm for her patience in the preparation of this manuscript.

This work was supported by the Deutsche Forschungsgmeinschaft through Sonderforschungsbereich 74.

References

1. Imanishi, T., Mäkelä, O.: Inheritance of antibody specificity. I. Anti-(4-hydroxy-3-nitrophenyl)acetyl of the mouse primary response. J. Exp. Med. 140, 1498-1510 (1974)
2. Mäkelä, O., Karjalainen, K.: Inheritance of antibody specificity. IV. Control of related molecular species of one V_H gene. Cold Spring Harbor Symp. Quant. Biol. 41, 735-741 (1977)
3. Jack, R.S., Imanishi-Kari, T., Rajewsky, K.: Idiotypic analysis of the response of C57BL/6 mice to the (4-hydroxy-3-nitrophenyl)acetyl group. Eur. J. Immunol. 7, 559-565 (1977)
4. Krawinkel, U., Cramer, M., Melchers, M., Imanishi-Kari, T., Rajewsky, K.: Isolated hapten binding receptors of sensitised lymphocytes. III. Evidence for idiotypic restriction of T cell receptors. J. Exp. Med. in press
5. Karjalainen, K., Mäkelä, O.: A Mendelian idiotype is demonstrable in the heteroclitic anti-NP antibodies of the mouse. Eur. J. Immunol. in press.
6. McMichael, A.J., Phillips, I.M., Williamson, A.R., Imanishi, T., Mäkelä, O.: Inheritance of genes coding for an antibody showing characteristic isoelectric spectrum. Immunogenetics. 2, 161-173 (1975)
7. Reth, M., Hämmerling, G.J., Rajewsky, K.: Analysis of the repertoire of anti-NP antibodies in C57BL/6 mice by cell fusion. Eur. J. Immunol. in press

8. Köhler, G., Milstein, C.: Continuous culture of fused cells secreting antibody of predefined specificity. Nature 256, 495-497 (1975)
9. Köhler, G., Milstein, C.: Derivation of specific antibody producing
   tissue culture and tumor lines by cell fusion. Eur. J. Immunol. 6,
   511-519 (1976)
10.Rajewsky,K., von Hesberg,G., Lemke, H., Hämmerling,G.J.: The isolation of thirteen cloned hybrid cell lines secreting mouse strain
   A derived antibodies with specificity for group A streptococcal
   carbohydrate. Ann. Immunol. (Inst. Pasteur) 129 C,389-400 (1978)

Production of Hybrid Lines Secreting Monoclonal Anti-Idiotypic Antibodies by Cell Fusion on Membrane Filters

G. Buttin, G. LeGuern, L. Phalente, E.C.C. Lin, L. Medrano, P.A. Cazenave

A. Introduction

Following the demonstration by Köhler and Milstein (1) that permanent cell lines secreting antibodies of predefined specificity can be obtained by cell fusion of myeloma cells with spleen cells of immunized mice, much effort has been devoted to improve the efficiency of hybrid formation between cells of the immune system and to enlarge the spectrum of monoclonal antibodies which such hybrid lines("hybridomas") can supply. The high efficiency of polyethylene-glycol (PEG) as a fusing agent for fibroblasts and its ability to yield hybrids in cell combinations recalcitrant to Sendai virus (2,3,4) has led most groups to substitute this simple chemical for the agglutinating virus. But PEG concentration is critical for the yield of viable hybrids ; this variable is difficult to control locally in most reported fusion procedures which rely on "gentle pipetting" for the removal of the culture medium and time-controlled resuspension of the cell pellets in a PEG solution. This prompted us to develop a simpler method which insures the formation of an intimate layer of the two parental cell types essentially free of culture medium, closely mimicking the conditions used for the efficient fusion of cells growing attached to a support. We describe this method, analyze some parameters which control its efficiency for the production of intraspecific and interspecific hybrids and demonstrate that it can be successfully utilized for the isolation of hybridomas secreting anti-idiotypic antibodies.

B. Materials and Methods

Cell Lines and Media   The myeloma line P3/X63-Ag8 (5) - used as one parent in all fusion experiments - is a clone of MOPC 21(6,7) of Balb/c origin. It is resistant to 20µg/ml of 8-azaguanine and dies in hypoxanthine, aminopterin and thymidine containing HAT medium. Unless otherwise indicated, the medium was Eagle medium reinforced by doubling the aminoacids, vitamins and glucose, supplemented with pyruvate (1mM), glutamine (2mM) and 10% heat inactivated (56°C, 30min) horse serum (ERH medium). Cells were grown in Petri dishes incubated at 37°C in a 10% CO_2 atmosphere.

Karyotype Analysis   The karyotype was analysed on metaphase plates of mitotic cells accumulated during a 3hr incubation in the presence of 1µM colchicine. The spreads were fixed in methanol : acetic-acid (3:1) and stained with Giemsa.

Materials and Chemicals   The filter support screen, mesh undersupport and Teflon gasket of the filtering centrifuge tubes were obtained from the Millipore filter Corp. (Bedsford,Ma) as parts of the XX62 025 50 filtration set. The inner holder assembly was made of autoclavable

Delrin to fit with the size of our centrifuge tubes. The polycarbonate filter membranes (3.0μ pore size ; 25mm Ø) were obtained from Nucleopore Corp.(Pleasanton, Calif.). PEG 6000 was from J.T.Baker Chemicals (Deventer, Holland).

Tumors and Myeloma Proteins MOPC 460 and McPC 870 tumors were obtained from Dr.M.Potter (NIH,Bethesda), J558 and MOPC 315 from Dr.B.Blomberg (Basel Institute for Immunology). They were maintained by serial subcutaneous passage using the trochar method (6) in Balb/c J mice. For production of large amounts of ascites the tumors were grown in(Balb/c x DBA/2) F1 hybrids primed with pristane (8). J558 protein was isolated from whole ascite by affinity chromatography using dextran B-1355 conjugated to serumalbumin-Sepharose (9), M460 and M315 proteins on DNP-Lys-Sepharose columns (10). M460 Fab fragment was prepared after papaïn hydrolysis (11) by chromatography on a DEAE-cellulose column equilibrated in 5mM phosphate buffer pH 8.0. The purity of the fragment has been tested by electrophoresis on SDS acrylamide gels (12). To isolate anti-idiotypic antibodies against M460, the protein was covalently attached to AH-Sepharose (13).

Anti-Idiotypic Immunizations Rabbit antisera to M460 protein were passed successively through columns of AH-Sepharose (14) with the covalently attached immunoglobulin (Ig) fraction of an McPc 870 ascite and the Ig fraction of normal Balb/c mouse serum.
Balb/c mice used as donors of cells for fusion were given five weekly injections of 150μg of M460 protein copolymerized with keyhole limpet hemocyanin by means of glutaraldehyde (15,16).Unless otherwise indicated five or six days before excision of popliteal lymph nodes, mice were boostered by footpad injection of the antigen emulsified in complete Freund's adjuvant (FCA).

Radioimmunoassay Fab M460 fragment was labeled by the chloramin T method (17). The specific activity ranged from 20 to 50 Ci/mmole. M460 idiotype-anti-idiotype reactions were studied using an indirect precipitation method (18). To a dilution of rabbit anti-idiotypic serum was added 5 ng of ^{125}I M460 Fab in 300μl PBS-BSA buffer containing normal rabbit serum. After 24 hr at 4°C, antigen-antibody soluble complexes were precipitated with 10μl of goat antiserum against normal rabbit serum. This technique was used to detect anti-idiotypic activity of each cell culture supernatant.

Anti-β-Galactosidase Immunization A rabbit of al/al, b4/b4 genotype was immunized by a single injection in the hind footpads of 0.5mg of highly purified β-galactosidase (a gift of Dr.A.Ullman) emulsified in FCA.

Identification of Rabbit Immunoglobulins Rabbit Ig chains were detected in supernatant cultures of hybridomas by inhibition of al[+] b4[+] ^{125}I-labeled IgG fixation on insoluble polymers of anti-al and anti-b4 sera (19). By this method 50ng/ml of allotype can be detected.
Anti-β-galactosidase activity was detected in supernatant cultures by the following method. To 100μl of supernatant (5X concentrated) were added 10[2]U of enzyme in PBS-BSA buffer containing normal rabbit serum and 10μl of anti-rabbit Fc goat serum. After 18hr at 4°C, test tubes were centrifuged and the enzymatic activity revealed in the precipitate (20). An indication of the sensitivity of this method is given by the possibility to detect anti-β-galactosidase activity in 100μl of a l : 50 000 dilution of anti-β-galactosidase rabbit serum (one ml of this serum precipitated 4.10^5U of enzyme).

Analysis of Secreted Radioactive Labeled Products The ^{14}C-lysine labeling of the proteins secreted by P3/X63-Ag8 and hybridomas, the isoelectrofocusing (IEF) of the reduced and non-reduced products were done as previously described (21). SDS acrylamide gel electrophoresis was also performed to determine molecular weights (12).

C. Production of Intraspecific and Interspecific Hybridomas

I. Fusion Protocol

10^5-10^7 lymphoïd cells from an immunized animal and 2.10^6-4.10^6 myeloma cells-both collected in serum-free medium are mixed(total volume : 2ml), loaded in the upper compartment of a filtering centrifuge tube (fig.1) and sedimented by 5min centrifugation at 1000 r.p.m.in the 29030 rotor of a Janetzki T32 C centrifuge.Fusion is triggered by dipping the filter for 30-45 seconds at room temperature into a 45%(W/V) solution of PEG 6000 in serum-free medium (3).The filter is transfered to a Petri dish containing 5ml of ERH medium which is incubated for 24hr at 37°C, during which time most cells detach from the membrane. The cell suspension in the dish is used to rinse the filter by pipetting, then distributed as 0.1ml fractions into the 48 wells of 2 Linbro plates (FB16-24TC, Linbro Scient.Co., Hamden, Conn.), each containing 0.3ml of ERH medium. After 24hr incubation at 37°C, 0.4ml of selection medium(see below)is added to each well ; two days later, 0.4ml of selection medium is further added and incubation is continued for 2-3 weeks. Growth in the wells is observed under the microscope. Samples taken from the wells in which a culture developped (within 10-20 days) are reinoculated into Petri dishes containing 2ml of selection medium. The cell population from each well which survived such a transfer is recorded below as one "hybridoma".

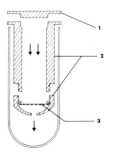

Fig.1. The filtering centrifuge tube. 1.Delrin cap ; 2.Delrin filter holder assembly ; 3.Millipore filter support, undersupport and gasket.

II. Factors Controlling the Yield of Mouse-Mouse Hybridomas

Fusing Agent   Control experiments showed that the centrifugation procedure does not alter the viability of the myeloma cells. PEG 6000 is an efficient fusing agent (table 1) at concentrations between 42-45% for exposure times of 30 to 45 seconds (survival of 3.10^6 myeloma cells treated alone is above 50%). At higher concentrations or upon longer treatments, lysis of most myeloma cells becomes apparent. Lower PEG concentrations yielded less than 1 hybridoma per 4.10^6 lymph node cells. PEG 1500 and PEG 1000 appear to be more toxic than PEG 6000 under these conditions.

Table 1. Recovery of hybridomas in HAT or H-Aza selective media
Lymph node cells from two mice immunized against M460 myeloma protein
were pooled. After 30 seconds PEG (45%) treatment of the indicated number of cells, each suspension was distributed into 48 wells completed
as indicated in "Fusion protocol", 24 with HAT medium, 24 with H-Aza
medium (see below).

| P3/X63-Ag8 | Lymph node cells | Hybridomas | | recovered |
		HAT	H-Aza	Total
2×10^6	1×10^6	3	4	7
2×10^6	4×10^6	6	12	18
4×10^6	1×10^6	4	4	8
4×10^6	1×10^6	4	5	9
4×10^6	5.2×10^6	14	18	32
5×10^6	1×10^5	O	1	1

Cell Number Table 1 illustrates the influence of the number of parental cells on the recovery of hybridomas in a typical experiment. A more
complete study remains desirable to determine the most efficient ratio
of the two parental types at the highest number of cells which does not
cause clogging of the filter.

Selective Medium P3/X63-Ag8 dies in HAT medium because it cannot uti-
lize hypoxanthine (HGPRT assay not shown). The substitution for HAT medium of ERH medium supplemented with hypoxanthine ($50\mu M$) and azaserine
($10\mu M$) as an inhibitor of purine synthesis (H-Aza medium) increased mo-
derately but consistently the number of wells containing viable cell
populations (table 1). The increase does not correspond to the proliferation of azaserine-resistant myeloma cells : cells randomly picked from
20 wells all were 8-azaguanine sensitive, demonstrating the recovery of
active HGPRT expected for authentic hybridomas.

Table 2. Recovery of hybridomas upon fusion of myeloma cells with
spleen or lymph node cells from a mouse immunized against M460 myeloma
protein. (see text). PEG (45%) treatment was for 30 seconds.

P3/X63-Ag8	Lymph node cells	Spleen cells	Hybridomas
2×10^6	2×10^6	−	O
2×10^6	−	2×10^6	8
2×10^6	−	4×10^6	10
2×10^6	−	10×10^6	31

Lymphocyte origin As a rule, we utilized as one parent cells from the
popliteal lymph nodes of mice boosted 5 or 6 days earlier in the footpads with the antigen. Table 2 describes the results of an experiment
in which the last immunization was peritoneal. The popliteal lymph nodes
were not stimulated to respond, as judged from their very small size
(total cell recovery : 2×10^6). The yield of hybridomas from fusions
involving the spleen cells of this animal was quite comparable to what
we commonly observe for fusions with lymph node cells in the routine
immunization procedure. In contrast, the lymph node cells from this
mouse yielded no hybridoma, although no abnormal killing of the myelo-

ma cells by PEG treatment was observed. This negative result must be taken with care because large unexplained fluctuations are occasionnally observed in hybridoma production, but it was reproduced, suggesting that the origin of the lymphocyte population may determine the yield of viable hybrids (see : concluding remarks).

III. Mouse-Rabbit Hybridomas

We attempted to isolate hybridomas by fusion of the P3/X63-Ag8 line with lymph node cells of rabbits immunized 10 days earlier in the hind footpads against β-galactosidase. Hybridomas were recovered at an average frequency of 1 per $2x10^6$ rabbit lymphocytes in fusions between $2x10^6$ myeloma and $2x10^6 - 10^7$ rabbit cells. We repeatedly observed in these experiments that increasing the number of rabbit lymphocytes above $4x10^6$ actually decreased the yield of hybridomas. Karyotype analysis of one hybridoma grown in culture for three weeks is shown on fig.2 : its hybrid nature is confirmed both by the increase in chromosome number (132-88) as compared to the P3/X63-Ag8 karyotype (61-66) and by the presence of recognizable rabbit chromosomes.

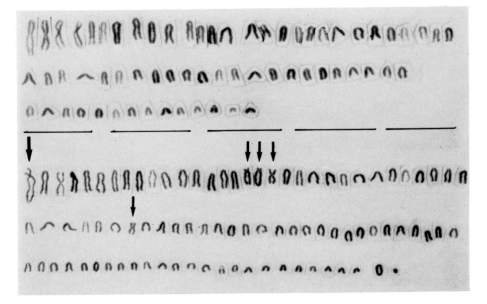

Fig. 2. The karyotypes of P3/X63-Ag8 (top) and of a mouse-rabbit hybridoma (bottom). Arrows designate chromosomes clearly recognized as being of rabbit origin. The rabbit karyotype (not shown) was analysed on metaphase plates of lymph node cells stimulated "in vitro" by the Nocardia mitogen.

No hybridoma secreting complete molecules of anti-β-galactosidase antibodies was identified among 23 tested. 3 of them secreted rabbit κ chains and 1 secreted H chains, as judged from the presence in the culture supernatants of material bearing respectively the b4 and a1 allotypic specificities (see : methods).

31

D. Production of Monoclonal Mouse Anti-Idiotypic Antibodies

Hybridomas between P3/X63-Ag8 and lymph node cells from Balb/c mice immunized against the MOPC 460 myeloma protein(also of Balb/c origin) were further cultured in selective medium ; the supernatants were screened for the presence of antibodies directed against the idiotypic determinants of this antigen : to this end, 110 supernatants were tested for their capacity to inhibit the binding of ^{125}I-M460 Fab to rabbit antiidiotypic antibodies.

Table 3. Anti-idiotypic activity of hybridoma supernatants. Supernatants (100μl) or sera (dilution in 100μl) were incubated with 5ng of ^{125}I-M460 Fab(100μl) for 24hrs at 4°C in PBS-BSA buffer supplemented with normal rabbit serum (1%). Then, 100μl of rabbit anti-idiotypic serum dilution (1:200) and 10μl of goat antiserum were added to each test tube. After an additional incubation of 18 hrs at 4°C, the tubes were centrifuged, the supernatants discarded and the radioactivity in the precipitate was counted. All experiments were performed in quadruplicate.

Inhibitor	% of the total radioactivity in the precipitate
Nil	50
Balb/c normal serum 1μl	48
0.5μl	50
Balb/c anti-M460 serum 0.2 μl	15
0.05μl	25
0.02μl	40
Balb/c hybrid line supernatants	
21	50
22	50
51	5
63	40
69	35

The results(table 3) show that the method is specific (no displacement of M460 by 1μl of normal Balb/c serum) and very sensitive : 0.02μl of Balb/c serum containing approximately 1mg/ml of anti-idiotypic antibodies displaced 20% of ^{125}I-M460 Fab. Three hybridoma culture supernatants (designated here as 51, 63 and 69) were able to inhibit the fixation of ^{125}I-M460 Fab on the rabbit anti-idiotypic antibodies, suggesting that these hybridomas secrete anti-idiotypic antibodies against M460. One of them (n°51) has been more extensively studied.

Karyotype As determined after more than four weeks of continuous culturing, the karyotype of hybridoma 51 was characterized by a modal number of 102 chromosomes.

Analysis of the Products Secreted by Hybridoma 51 IEF analysis of the ^{14}C-lysine labeled products (fig.3) indicated that-besides the γ and κ chains of P3/X63-Ag8 origin-hybridoma 51 secretes new chains(H and L) responsible for the anti M460 activity. Hybridoma 51 was subcloned by the limit dilution technique. All subclones(40/40)manifested the anti M460 activity ; 7/7 subclones also exhibited the same pattern of new bands as hybridoma 51, demonstrating its monoclonal origin. The mobility of the new H chain in SDS electrophoresis(not shown) corresponds to a γ-chain.

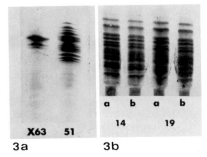

3a   **3b**

Fig.3. Autoradiographs of labeled components secreted by P3/X63-Ag8 (X63), hybridoma 51(51) and two subclones of 51(n°14,19),analysed by IEF. pH range was 3.5-10(left) and 5-8(right).a and b are duplicate samples labeled in the presence of 10% and 0% horse serum in the incubation medium.

Tumors Subcutaneous injection of about 10^7 cells of hybridoma 51 into normal Balb/c mice resulted in the formation of solid tumors.Using the M460 immunoadsorbant it was possible to isolate 6mg of anti M460 product per ml of transplanted mice serum ; the spectrotype of this non reduced and reduced product is shown in Fig.4. Solid tumors are maintained by serial subcutaneous passage in Balb/c mice ; ascitic tumors have also been obtained.

Immunochemical Characterization Immunochemical studies of the isolated product of hybridoma 51(to be described elsewhere) suggest that it is an IgG_1 κ.

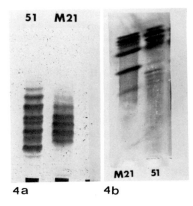

4a   **4b**

Fig.4. IEF spectrogram of the anti-M460 product of hybridoma 51, eluted from the M460 immunoadsorbant. Left: non-reduced product(pH range: 5-8) ; Right : reduced product(pH range : 3,5-10 in 6M urea).The M21 protein (M21) is used as a control.

Anti-Idiotypic Activity of the Product of Hybridoma 51 The serum of hybridoma 51 transplanted Balb/c mice can bind ^{125}I-M460 Fab $(4.10^{-3}μl$ of serum bind 1.5 ng of labeled Fab). The specificity of the binding was studied (table 4) with several reagents, some of which bear the DNP or TNP haptens that M460 binds with a high degree of specificity (6) . The results clearly demonstrate the anti-idiotypic activity of hybridoma 51 against a ligand-modifiable idiotope.

Table 4. Inhibition of the binding of ^{125}I M460 Fab to hybridoma 51
antibody by several reagents.

100μl of inhibitor were mixed with 100μl of hybridoma 51 serum (diluted 1:3000) and incubated 18 hr at 4°C before the addition of 100μl of
^{125}I-M460 Fab(5ng) and 20μl of rabbit antiserum against murine IgG Fc
fragment. All dilutions were made in PBS-BSA buffer containing 2% AKR
ascitic fluid as carrier. After an additional incubation at 4°C (24hrs),
the tubes were centrifuged and the radioactivity counted in the pellets. All assays were performed in quadruplicate.

Inhibitor	% of the total radioactivity in the precipitate
Nil	50
IgAκ 7S M460(binds DNP and TNP)20ng	0
IgGκ M21 4000ng	50
IgAλJ558(binds α(1-3)dextran)2000ng	50
IgAλ M315(binds DNP and TNP)4000ng	50
Normal Balb/c serum (1μl)	50
DNP-glycine 16mM	0
TNP-glycine 14mM	0
TNP-lysine 10mM	5

E. Concluding Remarks

A simple method is described which allows the recovery of viable hybridomas between myeloma cells and lymphoid cells from a preimmunized mouse at a frequency comparable to or better than those previously descri-
bed(22,1).It should be useful for the fusion of any cell types growing
only in suspension. Myeloma x lymph node cell hybridomas were generally
recovered at a frequency of 1/2.10^5 lymphocytes, occasionally 1/3.10^4
lymphocytes. Hybridoma recovery depended on the number of parental
cells within the same experiment (involving lymph node cells from the
same animal or pooled cells from more than one mouse) but large fluctuations were observed from experiment to experiment. We have so far been
unable to obtain frequencies of viable hybrid production comparable to
those observed in the intraspecific fusion between two fibroblastic
lines growing on a support (4 and our observations) or between two myeloma lines (23). This may indicate that many parameters remain to be
varied before the optimal conditions for cell fusion are defined. Alternatively, as previously suggested (24,25)the proliferation and differentiation status of the lymphoid cells in response to the immunization
program may be a determining factor in the yield of viable hybridomas.
Preliminary observations reported above are consistent with this interpretation. The presence in the mixed cell populations of cytotoxic cells
or cell products must also be considered. Experiments are in progress
to clarify these points

Because myelomas cannot be induced in rabbits, mouse x rabbit hybridomas secreting rabbit Ig's would be of special interest. Hybridomas which
secrete rabbit Ig chains have been produced (without sacrificing the
animal) by fusing rabbit lymph node cells with P3/X63-Ag8 cells. The
stability of these lines remains to be determined ; a fast segregation
of rabbit chromosomes is suggested both by the karyotypic analysis of
several hybridomas and by the observation that either type of chain

(H or L) but no complete rabbit Ig molecules were detected in the supernatants.

The heterogeneity of conventional anti-idiotypic sera directed against several idiotopes complicates the analysis of idiotypy. We show that homogeneous anti-idiotypic antibodies produced by hybridomas can be obtained in large quantities. They should prove powerful tools to study various problems, such as the idiotypic-antiidiotypic regulation of the immune system, observed in rabbits(26,27) and in mice (Le Guern and Cazenave, to be published).

Acknowledgments. The authors are grateful to Dr.C. Milstein for the gift of the P3/X63-Ag8 line and to Drs M.Potter and B.Blomberg for the gift of myeloma tumors ; they thank Dr.M.Stanislavski for his help in immunochemical characterizations, Dr.J.Oudin for advice, and Dr.F.Dray for making accessible his laboratory facilities.

This work was supported by grants from the "Fondation pour la Recherche Médicale Française", the "Ligue Nationale Française contre le Cancer", the University Pierre et Marie Curie and the C.N.R.S.(ATP : Réponse immunitaire).

References

1. Köhler, G., Milstein, C. : Continuous cultures of fused cells secreting antibody of predefined specificity. Nature 256,495(1975)
2. Pontecorvo, G. : Production of mammalian somatic cell hybrids by means of polyethylene-glycol treatment.Somat.Cell.Genet. 1, 397 (1975)
3. Davidson, R.L., Gerald, P.S. : Improved techniques for the induction of mammalian cell hybridization by polyethylene glycol. Somat. Cell. Genet. 2, 165 (1976)
4. Davidson, R.L., O'Malley, K.A., Wheeler, T.B. : Polyethylene-glycol induced mammalian cell hybridization : effect of polyethylene-glycol molecular weight and cocentration. Somat.Cell.Genet. 2, 271(1976)
5. Cotton, R.G.H., Milstein,C. : Fusion of two immunoglobulin producing myeloma cells. Nature 244, 42 (1973)
6. Potter, M. : Immunoglobulin producing tumors and myeloma proteins of mice. Phys.Rew. 52, 631 (1972)
7. Horibata, K., Harris, A.W. : Mouse myelomas and lymphomas in culture Exptl. Cell.Res. 61, 77 (1970)
8. Potter,M., Humphrey, J.G., Walters, J.L. : Growth of primary plasmacytomas in the mineral oil-conditioned peritoneal environment. J. Natl.Cancer.Inst. 49, 305 (1972)
9. Hiramoto, R., Ghanka, V.K., McGee, J.R., Schrobenloher, R., Hamlin, N.M. : Use of dextran conjugated columns for the isolation of large quantities of MOPC 104E IgM. Immunochemistry 9, 1251 (1972)
10.Goetzl, E.J., Metzger, H. : Affinity labeling of a mouse myeloma protein which binds nitrophenyl ligands. Biochemistry 9, 1267(1970)
11.Porter, R.P. : The hydrolysis of rabbit γglobulin and antibodies with crystalline papain. Biochem.J.73, 119 (1959)
12.Shapiro, A.L., Vinuella, E., Maizel, J.V.Jr : Molecular weight estimation of polypeptide chains by electrophoresis in SDS-polyacrylamide gels. Biochem. Biophys. Res. Commun. 28, 815 (1967)
13.Cambiaso, C.L., Goffinet, A., Vaerman, J.P.,Heremans, J.F.: Glutaraldehyde activated aminohexyl derivative of Sepharose 4B as a new versatile immunoadsorbent. Immunochemistry 12, 273 (1975)

14. Spring, S.B., Nisonoff, A. : Allotypic markers on Fab fragments of mouse immunoglobulins. J. of Immunol. 113, 470 (1974)
15. Sirisinha, S., Eisen, H.N. : Autoimmune-like antibodies to the ligand binding sites of myeloma proteins. Proc.Natl.Acad.Sci.U.S.68, 3130 (1971)
16. Sakato, N., Eisen, N. : Antibodies to idiotypes of isologaus immunoglobulins. J.Exp.Med. 141, 1411 (1975)
17. Greenwood, F.C., Hunter, W.M., Glover, J.S. : The preparation of 132I-labelled human growth hormone of high specific radioactivity. Biochem.J. 89, 114 (1963)
18. Kuettner, M.G., Wang, A.L., Nisonoff, A. : Idiotypic specificity as a potential genetic marker for the variable regions of mouse immunoglobulin polypeptide chains. J.Exp.Med. 135, 579 (1972)
19. Landucci-Tosi, S., Mage, R.G. : A method for typing rabbit sera for A_{14} and A_{15} allotypes with cross-linked antisera. J. Immunol. 105, 1046 (1970)
20. Pardee, A.B., Jacob, F., Monod, J. : The genetic control and cytoplasmic expression of "Inducibility" in the synthesis of β-galactosidase by E.coli. J.Mol.Biol. 1, 165 (1959)
21. Cotton, R.G.H., Secher, D.S., Milstein, C. : Somatic Mutation and the origin of antibody diversity - Clonal variability of the immunoglobulin produced by MOPC21 cells in culture. Eur.J.Immunol. 3, 135 (1973)
22. Lemke, H., Hammerling, G.J., Hohmann, C., Rajewski, K. : Hybrid cell lines secreting monoclonal antibodies specific for major histocompatibility antigens of the mouse. Nature 271, 249 (1978)
23. Gefter, M.L., Margulies, D.H., Scharff, M.D. : A simple method for polyethylene-glycol-promoted hybridization of mouse myeloma cells. Somat.Cell.Genet. 3, 231 (1977)
24. Köhler, G., Milstein, C. : Derivation of specific antibody-producing tissue culture and tumor lines by cell fusion. Eur.J.Immunol. 6, 511 (1976)
25. Schwaber, J. : Human lymphocyte-mouse myeloma somatic cell hybrids: selective hybrid formation. Somat. Cell. Genet. 3, 295 (1977)
26. Cazenave, P.A. : Idiotypic-anti-idiotypic regulation of antibody synthesis in rabbits. Proc.Natl.Acad.Sci. U.S. 74, 5122 (1977)
27. Urbain, J., Wikler, M., Franssen, J.D., Collignon, C. : Idiotypic regulation of the immune system by the induction of antibodies against anti-idiotypic antibodies. Proc.Natl.Acad.Sci. U.S. 74, 5126 (1977)

A Hybridoma Secreting IgM Anti-Igl^b Allotype

R. DiPauli, W.C. Raschke

A. Introduction

The fusion of a myeloma and splenic plasma cell can yield a hybrid
which continues to secrete antibody, and includes the specific antibody being made by the plasma cell (1). This finding has opened the
way for producing continuous cloned cell lines as a source of monoclonal antibodies with known specificity. In this paper we summarize our experience in the generation of a hybrid cell line producing
antibody of the IgM class specific for the mouse Ig-1^b allotypic
marker antigen.

B. Results and Discussion

BALB/cke mice were hyperimmunized with the CBPC-101 myeloma protein
which is an immunoglobulin of the IgG2a class and carries the C_H
allotype marker Ig-1^b derived from C57BL/6 mice. Spleen cells were
obtained from the hyperimmune BALB/c mice 3 days after an intravenous injection of CBPC-101 protein and fused with the BALB/c
myeloma MPC-11 which was resistant to thioguanine and ouabain. The
fusion method of Gefter et al. (2) was used. Briefly, spleen cells
and myeloma cells were mixed at a 10:1 ratio (10^7:10^6 cells/tube)
and centrifuged at 200xg for 3 min. The cell pellet was suspended
in 0.2 ml 35% (w/v) polyethylene glycol (MW 1500 manufactured by
BDH Chemicals, Ltd., batch #6022810). After 2 min. at 37°C the tubes
were centrifuged at 600xg for 3 min. and left for a further 3 min.
at 37°C, after which the contents of the tube were diluted with
5 ml Dulbecco's Modified Eagle's minimal essential medium (DME)
containing no FCS, the cells were washed once in this medium before
final resuspension and transfer to a petri dish for overnight incubation. Next morning cells were transferred to HAT medium (3) and
distributed into microtiter wells (0.2 ml/well) such that there
were 2x10^5 total cells/well.

Of the 758 wells analyzed for growth 489 were positive (65%) by
day 20. Supernatants from positive wells were assayed for the presence of anti-Ig-1^b antibody by a hemagglutination assay (4) which
uses SRBC coated with a sub-agglutinating amount of Ig-1^b anti-SRBC
antibody. There were 45 wells showing specific anti-Ig-1^b activity
in the supernatants (control assays used SRBC alone or Ig-1^a coated
SRBC). Growing cells from the 45 HA positive wells were transferred
to 1 ml well cultures and then to 5 ml petri dishes. Within 1 month
after fusion many cultures began to lose activity and by the time
they were recloned, in all but 10 cases activity could not be recovered. These remaining 10 independent clones were injected into
syngeneic normal BALB/c mice (about 5x10^6 cells/mouse) and in each

case gave progressing tumors. Analysis of the sera from these animals showed that only clone 45 remained strongly positive with a titer of 1/2560 in the HA assay, and this antibody was completely inactivated by pretreatment with 2-mercaptoethanol, indicating the presence of IgM.

The products of clone 45 were analyzed in SDS-PAGE after labelling cells in vitro with ^{35}S-methionine (Figure 1).

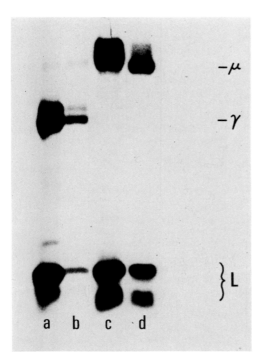

Fig. 1. Analysis of the intracellular products on SDS-polyacrylamide gel after reduction, a) MPC-11 intracellular immunoglobulin, b) MPC-11 secreted product, c) hybrid clone 45 intracellular and d) secreted product.

There was a clear band running in the position expected for μ and light chains but no band where the MPC-11 γ chain should be. There were 2 bands in the light chain region, one similar to MPC-11, the other ran in a position expected of a light chain 10% smaller than MPC-11. The reason for this large apparent size disparity in light chains is not known at present. However, there is every indication that clone 45 has lost the capacity to produce MPC-11 heavy chains, whether this is due to chromosome loss or not cannot be determined from the available data.

Interestingly, the clone 45 myeloma sera does not give a precipitation reaction in agar although it does show a partial inhibition pattern. From this we infer that the determinant recognized by the clone 45 antibody is represented only once on the IgG molecule or that if represented twice, as might be expected since each IgG has 2 H-chains, antibody binding to one determinant renders the other determinant inaccessible to another antibody.

One further piece of evidence that clone 45 is indeed producing an antibody directed at the CBPC-101 protein comes from the fact that labelled supernatant of 45 clone can be coprecipitated when incubated with CBPC-101 and a BALB/c anti-Ig-1^b serum. When instead of CBPC-101 a BALB/c immunoglobulin was used, no precipitation was found. The lack of inhibition of 45 for the anti-Ig-1^b suggest that the allotypic determinants are heterogeneous.

Acknowledgements. We wish to thank M. Cohn for his interest and support, R. Langman for many helpful discussions and for reading the manuscript critically. This work was supported by Grant Number RO1 CA19754 awarded by the National Cancer Institute to M. Cohn, and CA21531 awarded by the National Cancer Institute to W.C.R. and by a fellowship from the Deutsche Forschungsgemeinschaft to R.DP. W.C.R. also wishes to thank Mrs. Jean James for partial support of this work.

References

1. Köhler, G., Milstein, C.: Continuous cultures of fused cells secreting antibodies of predefined specificity. Nature 256, 495-497 (1975).
2. Gefter, M.L., Margulies, D.H., Scharff, M.D.: A simple method for polyethylene-promoted hybridization of mouse myeloma cells. Somatic Cell Genetics 3, 231-236 (1977).
3. Littlefield, J.W.: Selection of hybrids from mating of fibro-blasts in vitro and their presumed recombinants. Science 145, 709-710 (1964).
4. Kolb, C., DiPauli, R., Weiler, E.: Induction of IgG in young nude mice by Lipid-A or thymus grafts. J. Exp. Med. 144, 1031-1036 (1976).

Monoclonal Antibodies Against Developing Chick Brain and Muscle

D. Gottlieb, J. Greve

Introduction
============

The nervous system is characterized by a large repertoire of
functionally distinct cell types which form highly specific con-
nections with each other. In order to understand the morphogenesis
of the brain at a molecular level it will be necessary to have
specific markers for interacting cell types. The attractiveness of
the hybridoma methodology for addressing this problem rests on two
features. First, pure antibodies against cell surface determinants
can be generated without the need of purifying those determinants
or even the cells upon which they are found. Secondly, once
appropriate clones of antibody producing cells are isolated, large
quantities of antibody can be generated for use in detection,
isolation and functional studies of the antigens *in vivo* and *in
vitro*. With these considerations in mind we have initiated a
program of preparing monoclonal antibodies against the surface of
developing chick brain and muscle cells.

Materials and Methods
=====================

Viable dissociated brain cells from 8 and 9 day old chick embryos
were prepared according to (1) except that trypsin was omitted from
the dissociating medium. Muscle cells were cultured according to
(2). Balb/c mice were injected with 10^7 cells i.v. and boosted one
month later with 10^8 cells injected i.p. Spleens were removed
three days after the boost and used to prepare cells which were
fused to drug marked myeloma cells by the method of (3). Myeloma
lines 45.6 TG1·7 (4), P3-X63·Ag8, and P3-NS1/1-Ag4-1 (5) were used
in this study. Anti-brain and anti-muscle cell antibodies were
detected by direct and indirect binding assays to suspensions of
brain cells on monolayer cultures of muscle cells. In the case of
brain, 3×10^6 cells were used per assay. For muscle assays, cultures plated with 5×10^5 cells one week before assay were used. For
direct binding assays approximately 6×10^5 cpm of iodinated monoclonal antibody were added per sample, incubated for 30 minutes at
4°C and washed twice with 10% CMF-serum. The same first step with
unlabeled antibody was used for indirect binding assays followed by
a second incubation with 2×10^6 cpm ^{125}I rabbit anti-mouse IgG
and/or goat anti-mouse IgM and subsequent washing. Tumors of
hybrid lines were made by injecting Balb/c mice with approximately
10^7 cells subcutaneously. Isoelectric focusing of immunoglobulins
was performed as in (6). Iodination of immunoglobulins was performed by the method described in (7).

Results

Hybrids producing antibodies against the surface of developing brain
and muscle cells may be obtained.

Fusions were performed by the method described in "Materials and
Methods". A typical result using brain cells as an antigen and
another using cultured muscle is shown in Figure 1. In the case

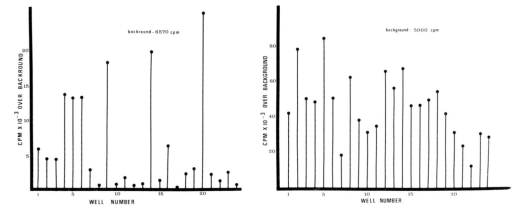

Figure 1: Left Panel. Supernatants from wells from fusion between
spleen from mouse immunized with brain cells are assayed for mouse
immunoglobulin directed against the surface of 8D dissociated brain
cells. Ten days after fusion, 100 λ samples of supernatant from
each well were added to assay tubes containing 3×10^6 brain cells
and the amount of bound antibody determined by indirect binding
assay. The input of ^{125}I rabbit anti-mouse IgG was 2×10^6 cpm per
tube. Background refers to the counts bound to brain cells exposed
to HAT medium.
Right Panel. An anti-muscle fusion is analyzed in a manner anal-
ogous to that described in Panel A. The test cells are muscle
cells grown in collagen coated tissue culture wells. Each test
well was seeded with 5×10^5 muscle myoblast cells which were then
kept *in vitro* for 7 days before performing the binding assay. At
this time cultures consisted predominantly of fused myotubes.
Occasional mononucleate cells were seen some of which are presumed
to be fibroblasts. 500 λ of fusion supernatant culture medium was
added to muscle cells in test wells. 2×10^6 cpm of ^{125}I rabbit
anti-mouse IgG was used in each assay well.

of the brain cells, approximately 1/3 of the wells produce detectable antibodies directed against the cell surface. Virtually all
of the wells in the anti-muscle fusion produce anti-surface antibody. Thus it is clear that hybrids against these two cell surfaces can be readily obtained. Since so many hybrids against brain
and muscle cell surfaces are generated in these experiments it will
be important to determine if these antibodies are directed against
many determinants on the cell surface or if, alternatively, the
surface of these embryonic cells are dominated by a few potent
antigenic determinants.

Evidence that the binding activity represents mouse immunoglobulin.

Figure 2 shows that the amount of second antibody bound is linearly related to the amount of added first antibody. This saturatable binding is consistent with the presence of a limited number of antigenic sites on the cell surface. Figure 2 also shows that the binding activity can be blocked by anti-mouse IgG but not anti-mouse IgM.

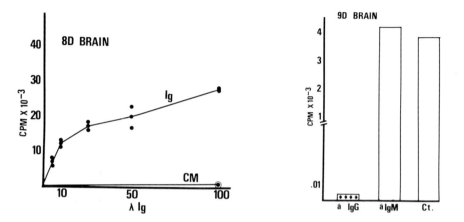

Figure 2. Characteristics of assay for detecting bound antibody on dispersed brain cells.
Left Panel. Antibody binding as a function of concentration of added supernatant fluid. Abscissa: Volume of supernatant fluid added to each standard binding assay tube containing 3×10^6 dispersed 8D brain cells. Ordinate total counts of second antibody bound. Ig indicates tubes receiving supernatant from culture Br.1. CM indicates tubes receiving medium conditioned by the corresponding parent myeloma 45.6 TG 1.7.
Right Panel. Antibody binding to brain cells is blocked by anti-mouse IgG. The binding of supernatant from culture Br.1 is tested with either 3 µg/ml anti-mouse Ig (āIgG) 3 µg/ml goat anti-mouse IgM (āIgM) or saline (Ct.) added to the assay tube during the first binding step. The radioactive antibody in this experiment contained a 1:1 mixture by weight of rabbit anti-mouse IgG and goat anti-mouse IgM. Only the anti IgG antibody appears to be an effective inhibitor.

In experiments not shown supernatants were chromatographed on Biogel A-0.5M. All binding activity co-migrated with IgG.

Isoelectric focusing patterns of monoclonal antibodies against the muscle cell surface.

Hybrids in 3 out of 24 wells producing anti-muscle antibody were cloned in soft agar. Clones were grown in the presence of 8 µCi/ml ^{35}S-methionine for 12 hours. Supernatants were harvested and analyzed by isoelectric focusing (6), with the results shown in Figure 4. The patterns of bands were relatively simple as expected from monoclonal antibodies. Patterns of clones from any one well were not found in the other two wells. The five clones of the first well gave four distinct patterns. The four clones from the second and third wells each gave two distinct patterns. The most likely explanation is that these clones originated from independent fusion events. However, we cannot exclude the possibility that the diversity seen in IEF patterns represents segregation after a single fusion even.

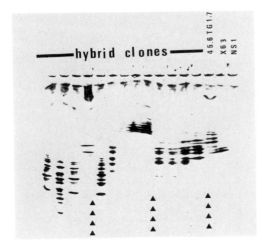

Figure 3. Secreted proteins of independently arising clones of anti-muscle hybrids have distinguishable isoelectric focusing bands. Arrows mark lanes with clones derived from each of three separate wells from the original fusion. Radioactive supernatants from the parent myeloma cells used in this study were focused in separate lanes on the right hand side of the gel.

Antibody B5 binds preferentially to brain cells. Table I shows the relative distribution of binding sites on a number of embryonic tissues for antibody B5. The tissue distribution of binding sites for B5 was determined by direct binding assays in which the plateau level of binding was measured.

Table I   Relative Saturation Binding Levels of B5 to Embryonic Tissues

	Brain	Heart	Liver	Gut	Muscle
cpm bound at saturation	8990	687	845	575	1745

B5 hybrid cells secreting anti-brain antibody were used to initiate tumors in Balb/c mice. The γ globulin fraction of serum from tumor bearing animals was obtained by ammonium sulfate fractionation and iodinated directly. Saturation binding curves were determined for 3×10^6 dissociated cells of each type and the value of counts above background bound at saturation tabulated above.

Conclusions

The present results allow several major conclusions. Hybrid clones secreting antibody against the surface of developing chick brain and muscle cell can be readily obtained. The detection of binding of secreted antibody to target cell surfaces can be readily accomplished by direct and indirect methods. In at least one case, B5, a distinctive tissue distribution of antigenic sites was noted. We feel that monoclonal antibodies will be useful tools for studying the function of many elements of the surface of developing brain and muscle cells.

Acknowledgements. Supported by N.I.H. Grant #NS12867. We thank Drs. J. Davie, B. Clevinger and D. Hansburg for advice.

Drs. C. Milstein and M. Sharff generously supplied drug marked myeloma lines.

References

1. Gottlieb, D.I., Rock, K. and Glaser, L.: A gradient of adhesive specificity in developing avian retina. Proc. Nat. Acad. Sc. 73, 410-414 (1976).

2. Fischbach, G.: Synapse formation between dissociated nerve and muscle cells in low density cell culture. Dev. Bio., 28, 407 (1972).

3. Galfre, G., Howe, S.C., Milstein, C. Butcher, G.W. and Howard, J.C.: Antibodies to major histocompatibility antigens produced by hybrid cell lines. Nature, 266, 550-552 (1977).

4. Marguiles, D., Kuehl, M., and Scharff, M.: Somatic cell hybridization of mouse myeloma cells. Cell, 8, 405-415 (1976).

5. Kohler, G. and Milstein, C.: Derivation of specific antibody tissue culture and tumor lines by cell fusion. Eur. J. Immunol., 6, 511-519 (1976).

6. Briles, D. and Davie, J.: Detection of isoelectric focused antibody by autoradiography and hemolysis of antigen-coated erythrocytes. J. Immunol. Methods, 8, 363-372 (1975).

7. Greenwood, F.C., Hunter, W.M. and Glover, J.S.: The preparation of ^{131}I-labeled human growth hormones of high specific radioactivity. Biochem. J., 89, 114-123 (1963).

Monoclonal Xenogeneic Antibodies to Mouse Leukocyte Antigens: Identification of Macrophage-Specific and Other Differentiation Antigens

T. Springer, G. Galfre, D. Secher, C. Milstein

A. Introduction

The identification and study of cell surface molecules which have specific immune functions and are markers of differentiated white blood cell subpopulations is of great interest in immunology. Antibodies are versatile probes with which to study these molecules. However, it is difficult to obtain highly specific antibodies recognizing individual cell surface molecules, because cell surfaces are complex mosaics of many different immunogenic glycoproteins and glycolipids. A general approach to this problem stems from the experiments of Köhler and Milstein (1). They fused myeloma cells and spleen cells from mice immunized with sheep red blood cells to derive continuous hybrid cell lines secreting antibodies to sheep red blood cells. Such lines can be manipulated in culture so that the multispecific response to a complex immunogen can be resolved into a set of monospecific responses by cloning. Hybrids have been obtained which secrete antibodies to rat major histo-compatibility antigens (2), rat cell surface xenoantigens (3), mouse IgD allotypes (4), and mouse H-2K antigens (5).

In this report, the application of this approach to the xenogeneic identification of cell surface antigens of the mouse is explored. Xeno- rather than allo-immunization was used because the difference between homologous proteins from different species is greater than between the polymorphic variants. Antibodies are thus elicited to a much wider range of molecules. We describe the identification of novel mouse cell surface differentiation antigens, and screening procedures which have been developed to allow detection of antigens present on as little as five percent of the immunizing cells.

I. Immunization and Fusion

DA strain rats received i.p. and s.c. injections of 10 x 10^6 nylon wool (6) and Ficoll-Isopaque (7) purified B10 spleen cells on days 1 and 15. 40 x 10^6 of the same cells mixed with 60 x 10^6 concanavalin-A-stimulated spleen cells were injected in the tail vein on day 31. Three days later the spleen cells were fused with P3-NSI/1-Ag4-1 (NSI) non-secretor myeloma line as previously described (3). The fused cells were distributed into 96 Linbro 2 ml wells and hybrid lines grew up in all. After 10 days and weekly thereafter supernatants from the hybrid cells were tested for binding to glutaraldehyde-fixed spleen cells with ^{125}I-rabbit F(ab')$_2$ anti-rat IgG or anti-rat Fab reagents (3). These reagents were absorbed with mouse IgG and IgM to prevent cross-reaction with surface Ig of the mouse spleen target cells. In the initial

screening at least 92 of 94 cultures were positive in the binding
assay. From the Poisson distribution it can be estimated that each
culture contained on the average about four independently-fused,
successful, specific antibody secreting cells. More clones were
positive for binding than for complement-mediated cytotoxicity,
and the former assay was therefore adopted for routine use. After
further growth over a seven week period, faster growing clones
became dominant and some cultures became negative.

II. Screening of Clones

The original 96 2 ml cultures are designated Ml/l-Ml/96. Clones
contain a second number, and if recloned, a third. Positive cultures were cloned in soft agar (1). About 48 clones were generally
picked, grown in liquid culture, and tested for activity in the
binding assay. Frequently, many positive clones were isolated from
a single culture. Because clones from the same culture might differ
in their antibody specificities, it became an important problem to
distinguish between identical and different clones. One approach
was to test ^{125}I-anti-Ig binding capacity at constant supernatant
concentration with serial spleen cell dilutions. This method distinguishes between antibodies recognizing antigens which differ in
their concentrations, since as cells become limiting, binding is

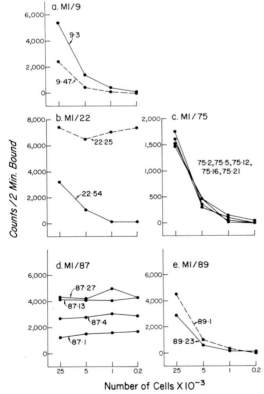

Figure 1. Binding Activity of
Ml Clone Supernatants Studied
by Spleen Cell Titrations.
Supernatants from clones which
had been transferred to liquid
culture medium (5μℓ) were mixed
with the indicated number of
glutaraldehyde-fixed (3) spleen
cells (in 5μℓ) in V well microtiter plates. After shaking
45 min. at 4°, 5μℓ of 5% filler
RBC (sheep here, ox or out-
dated human O in later experiments) were added and samples
were washed twice by adding
200μℓ buffered saline containing
2% bovine serum albumin, (BSA-
saline), centrifuging for 5
min. at 200g, and aspirating
the supernatant. The cells
were suspended in 5μℓ (∿30,000
cpm) of mouse Ig absorbed ^{125}I
rabbit F(ab')$_2$ anti-rat IgG
or Fab (prepared by iodinating
the appropriate antibodies while
bound to rat IgG-Sepharose (8))
in BSA-saline. After shaking
45 min. at 4°, cells were washed
3 times as above and transferred
to tubes for γ-counting. The
background (subtracted) was de-
termined for each spleen cell concentration using a rat anti-rat
histocompatibility antigen clone supernatant (4) as a control.

46

proportional to antigen concentration on the target cells (Fig. 1).
Here it can be seen that M1/9.3 and M1/9.47 are different·from each
other, M1/89.1 and M1/89.23 also differ, but all the M1/75 clones
appear identical. This test also allowed the detection at an early
stage of an interesting artifact. Binding activity of M1/22.25 (but
not M1/22.54) and M1/87 clones appeared unaffected by spleen cell
dilution. As will be shown later, the observed binding is to the
Forssman antigen on the sheep red blood cells used as filler cells
in the binding assay to prevent washing losses. In another set of
experiments (not shown), supernatants from different clones were
titrated by serial six-fold dilutions followed by the standard ^{125}I-
anti-Ig binding assay. In some cases, especially when 15x concentrated supernatants were used, saturation binding was achieved.
Clones which plateau at different levels of bound ^{125}I-anti-Ig can
be distinguished in this manner.

Supernatant titrations were used to guide cloning and recloning.
Once multiple identical clones were identified, one or two clones
with highest antibody titer were usually subcloned. Again, the subclone supernatants were titrated and the strongest subclones saved.
The importance of this procedure in isolating stable subclones is
emphasized by the examples of M1/69 and M1/87 clones. While M1/69.
16 and M1/69.17 clones were initially both highly positive, one
month later the titer of M1/69.16 was 1,000 fold greater than M1/69.
17. Ten M1/69.16 subclones were derived and the two most active were
saved. One, M1/69.16.2, has lost the myeloma κ chain. Seven M1/87
clones were positive. Clone M1/87.27 had five-fold more supernatant
activity than M1/87.1. Later, two subclones were randomly isolated
from each and analyzed. Subclones M1/87.27.5 and M1/87.27.7 were
both active, while subclones M1/87.1.2 and M1/87.1.5 are both inactive and have lost their specific heavy chains, but retain the
specific light chain. The observation that heavy chain loss variants
can overcome antibody secreting cells suggests that they may have a
selective growth advantage and emphasizes the importance of recloning
and screening antibody titer.

Of about 500 picked clones, 47 were positive in the binding assay.
By elimination of duplicate clones, the number was reduced to 10.
The characteristics of the monoclonal antibodies and the antigens
they recognize are summarized in Table 1. Two different types of

Table 1. Summary of M1 Clone Characteristics

Clone	Tissue Distribution	^{125}I-Antigen Precipitated	Stability at 120°C	Aggluti-nation MRBC	SRBC	Heavy Chain Class
M1/9.3	White blood cells	210,000 MW	−	−	−	γ
M1/89.18	White blood cells	210,000 MW	−	−	−	γ
M1/70.15	Macrophages	190,000 MW 105,000 MW	−	−	−	γ
M1/75.21	Complex	None	+	+	−	γ
M1/22.54	Complex	None	+	+	−	γ
M1/89.1	Complex	None	+	+	−	γ
M1/9.47	Complex	None	+	+	−	γ
M1/69.16	Complex	None	+	+	−	γ
M1/22.25	Forssman	None	+	−	+	μ
M1/87.27	Forssman	None	+	−	+	μ

clones were isolated from each of three cultures. The differences
between Ml/9.3 and Ml/9.47, between Ml/89.18 and Ml/89.1, and between Ml/22.54 and Ml/22.25 are confirmed by many criteria.

Four classes of antigen are recognized. Ml/9.3 and Ml/89.18 precipitate a heat labile antigen of 210,000 M.W. from detergent solubilized, ^{125}I-labeled spleen cells. Ml/70.15 precipitates two
polypeptides of 190,000 and 105,000 M.W. and the antigenic determinant is heat labile. Ml/75.21, Ml/22.54, Ml/9.47, and Ml/69.16
recognize a molecule on mouse RBC and other tissues which is stable
at 120°C and is not labeled by ^{125}I. These characteristics suggest
it is a glycolipid. Ml/22.25 and Ml/87.27 recognize a heat-stable
antigen which is not labeled by ^{125}I and is found on sheep red
blood cells as well as mouse tissues. Mice and sheep are Forssman
positive while rats are Forssman negative. Absorption of Ml/22.25
and Ml/87.27 activity by guinea pig kidney but not ox red blood
cells suggests these clones react with the Forssman antigen.

III. Screening for Differentiation Antigens using Tumor Cell Panels

Tumor lines are useful models for studying differentiation antigens,
since each line is usually a clone of cells arrested in a particular
stage of differentiation. Therefore, the monoclonal antibodies were
used to quantitatively compare the amounts of antigen expressed on
several tumor cell lines (Table 2, see legend for details). Ml/89.18
binds to a number of white blood cell tumor lines but not to the NS-
1 myeloma or red blood cells. Ml/69.16 shows a different pattern,
reacting with red blood cells, thymocytes, T lymphomas and an
Abelson lymphoma but not a myeloma or a macrophage-like line.
Fluorescent labeling experiments (12) show Ml/69.16 is expressed on
mature B but not T lymphocytes and thus the T lymphomas resemble thymocytes but not mature peripheral T lymphocytes in this respect.
Ml/69.16 typing may be of interest to investigators doing T-T fusions
who are looking for cultured lines with mature T lymphocyte characteristics. A macrophage-like line, P 388D$_1$, is the only line positive

Table 2. Antigen Titers on Tumor Lines and Normal Cells

	Ml/89.18	Ml/69.16	Ml/70.15
Red blood cells	<0.2	150	<0.02
Spleen (Ficoll-Isopaque purified)	125	67	(3)
Thymocytes	130	60	<0.8
SIA T lymphoma (9)	650	2,500	<4
S49 T lymphoma (9)	770	1,700	<4
BW5147 T lymphoma (9)	2,000	270	<0.8
R8CL7 Abelson lymphoma (10)	500	570	<4
NS-I myeloma (3)	<6	(5)	<0.2
P 388D$_1$ macrophage-like line (11)	250	(3)	480

The binding assay (see legend to Fig. 1) was conducted with serial
5-fold dilutions of glutaraldehyde-fixed tumor cells, beginning
at 1 to 10 x 10^7 cells/ml. Antigen titer = 10^9 x (cells/ml giving
half-maximal ^{125}I-anti-Ig binding)$^{-1}$. ( ): extrapolated to half-
maximal binding.

for Ml/70.15, and expresses 100-fold greater quantities of this
antigen than spleen white cells. Testing with the tumor panel thus
shows that Ml/89.18, Ml/69.16, and Ml/70.15 recognize differentiation
antigens and that lymphoma, myeloma, and macrophage-like lines differ
in the mosaic of antigens they express.

Further testing has shown that Ml/70.15 recognizes an antigen found
on macrophages and their precursors but not on lymphocytes (12).
This antigen is expressed in low amounts on only 5-8% of the
cells in spleen. Binding by Ml/70.15 antibody to spleen cells in
the standard assay was only 2 times background, which is just
marginally detectable. It is interesting that the Forssman clones,
which also gave only barely detectable binding to mouse spleen cells,
give strong binding to teratocarcinomas but not to many other types of
tumor cell lines tested (13). The ability of the myeloma hybrid
procedure to produce monoclonal antibodies recognizing antigens
present in low amounts or on minor subpopulations of the immunizing
cells, has been observed repeatedly. Because Ml/70.15, Ml/22.25,
and Ml/87.27 binding to spleen cells was barely above the threshold
for detection, we suspect that other interesting clones may have
been missed because the screening procedure was not sensitive enough.
We thus propose initial screening with tumor cell lines. Each line
expresses a different mosaic of surface antigens and the problem
of minor subpopulations could be avoided.

B. Concluding Remarks

The cell surface antigens of the mouse are better characterized than
any other species and at least 20 surface markers have been identified, mainly by alloantisera. Of the four types of antigens recognized by the antibodies from the ten clones isolated in this
study, specific antisera to only the Forssman antigen had been
previously prepared. So far our experiments to produce monoclonal
antibodies to differentiation antigens have been based on screening
hybrid myeloma cultures with a similar cell population to the one
used in the immunization. No serious attempt to select for specific
target antigens was made. In spite of this simple approach, monoclonal reagents to three novel types of mouse white blood cell
antigens, not previously available as specific allo- or xenoantisera,
have been obtained in the present study. This not only represents
a significant extension of our catalogue of cell surface reagents,
but implies that a large number remain yet to be identified. Thus,
obtaining clones of hybrid cells secreting xenoantibodies is a
powerful technique for identifying novel surface antigens, even in
a well studied animal such as the mouse.

References

1.   Köhler, G. and Milstein, C.: Derivation of specific antibody-
      producing tissue culture and tumor lines by cell fusion.
      Eur. J. Immunol. 6, 511 (1976).
2.   Galfre, G., Howe, S.C., Milstein, C., Butcher, G.W., Howard,
      J.C.: Antibodies to major histocompatibility antigens produced
      by hybrid cell lines. Nature 266, 550 (1977).
3.   Williams, A.F., Galfre, G., and Milstein, C: Analysis of cell

surfaces by xenogeneic myeloma-hybrid antibodies: differentiation antigens of rat lymphocytes. Cell <u>12</u>, 663 (1977).

4. Pearson, T., Galfre, G., Ziegler, A., and Milstein, C.: A myeloma hybrid producing antibody specific for an allotypic determinant on "IgD-like" molecules of the mouse. Eur. J. Immunol. <u>7</u>, 684 (1977).

5. Lemke, H., Hämmerling, G.J., Höhmann, C., and Rajewsky, K.: Hybrid lines secreting monoclonal antibody specific for major histocompatibility antigens of the mouse. Nature <u>271</u>, 249 (1977).

6. Julius, M.H., Simpson, E., and Herzenberg, L.A.: A rapid method for the isolation of functional thymus-derived murine lymphocytes. Eur. J. Immunol. <u>3</u>, 645 (1973).

7. Davidson, W.F., and Parish, C.R.: Removing reds and deads from lymphoid cell suspensions. J. Imm. Methods <u>7</u>, 291 (1975).

8. Herzenberg, L.A., and Herzenberg, L.A. in Weir, D.M. (Ed). Handbook of Experimental Immunology, 3rd Edit., Blackwell, Oxford p. 12.1 (1978).

9. Trowbridge, I.S., and Hyman, R.: Thy-1 variants of mouse lymphomas: biochemical characterization of the genetic defect. Cell <u>6</u>, 279 (1975).

10. Raschke, W.C., Ralph, P., Watson, J., Sklar, M., and Coon, H.: Oncogenic transformation of murine lymphoid cells by *in vitro* infection with Abelson leukemia virus. J. Nat. Cancer Inst. <u>54</u>, 1249 (1975).

11. Dawe, C.J., and Potter, M.: Morphologic and biologic progression of a lymphoid neoplasm of the mouse *in vivo* and *in vitro*. Am. J. Pathol. <u>33</u>, 603 (1957).

12. Springer, T., Galfre, G., Secher, D.S., and Milstein, C.: Monoclonal xenogeneic antibodies to murine cell surface antigens: identification of novel leukocyte differentiation antigens. Eur. J. Immunol. submitted.

13. Stern, P., Willison, K., Lennox, E., Galfre, G., Milstein, C., Secher, D., Ziegler, A., and Springer, T.: Monoclonal antibodies as probes for differentiation and tumor associated antigens: a Forssman specificity on teratocarcinomas and early mouse embryos. Cell submitted.

50

Monoclonal Antibody Detecting a Stage-Specific Embryonic Antigen (SSEA-1) on Preimplantation Mouse Embryos and Teratocarcinoma Cells

B.B. Knowles, D.P. Aden, D. Solter

A.   Introduction

Syngeneic and xenogeneic immunizations with teratocarcinoma stem cells (1, 2, 3) and mouse embryos (4) have resulted in antisera which after appropriate absorption react with teratocarcinoma stem cells, germ cells and embryos at various stages of development as well as some adult tissues. Such sera contain antibodies to multiple antigenic determinants, which precludes precise definition of embryo stage-specific antigens.˙ In order to circumvent this difficulty and to study the stage-specific molecules in a methodical fashion, we are producing monoclonal antibodies (5) reactive with teratocarcinoma cells and embryos. Production and characterization of one such antibody is described here.

B.   Production and Characterization of Monoclonal Reagent

BALB/c mice were immunized weekly with irradiated F9 teratocarcinoma cells (10^7 cells/mouse), tail bled and their sera tested for reactivity with F9 cells after absorption with BALB/c spleen cells. After six immunizations, the cytotoxic titer was 1:500. The mice were then immunized once more, their spleens removed 4 days later and fused with P3-X63-Ag 8 myeloma cells as described (5). Supernatants from wells with growing colonies were tested using a radioactive binding assay (RIA) on F9 cells. In this preliminary screening rabbit IgG anti-mouse IgG (heavy and light chain) labeled with ^{125}I was used as second reagent. Supernatants from 14 separate colonies were tested and one was positive. This colony was subsequently cloned; six clones were positive (6000 cpm/5 x 10^5 cells) using supernatant at dilution 1:2; one intermediate (1600 cpm/5 x 10^5 cells) and one negative. Clones were grown, one of them expanded to mass culture and injected intra-peritoneally into BALB/c mice. Three weeks later when ascites was well developed, sera and ascites fluids were collected from hybridoma-bearing mice and tested on F9 cells in RIA. Serum from one mouse was used subsequently in all assays, though all sera and ascites fluids gave comparable results. Supernatants from the original colony and clones were tested with mouse immunoglobulin class specific reagents in Ouchterlony gel diffusion. Precipitation lines of identity were formed with anti- γ1 (secreted by the parental myeloma line) and with anti-μ chain reagents. No reactivity with anti-λ light chain reagents was found. The presence of μ and K chains were confirmed by SDS gel electrophoresis. In all subsequent RIAs rabbit IgG anti-mouse IgM was used as the tracer.

1. Reactivity of Monoclonal Antibody with Cell Lines

The RIA titration curve on F9 cells of serum from a hybridoma-bearing mouse is shown on Fig. 1. All other cell lines were tested at three

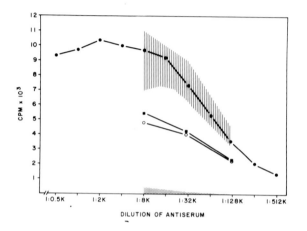

DILUTION OF ANTISERUM

dilutions on the descending portion of the binding curve. Cell lines which react in the same way as F9 (shaded area) include several mouse teratocarcinoma cell lines derived from spontaneous or experimental teratocarcinomas of 129, BALB/c, AKR and LT mouse strains and a cell line derived from a human teratocarcinoma. Two independently derived human teratocarcinoma cell lines gave intermediate reactions ([]-[], o-o). Nonreactive cells (stippled area) include several differentiated cell lines derived from the same mouse and human teratocarcinomas tested, human transformed cell lines and mouse transformed cell lines from 129, C57BL/6, BALB/c, AKR, and C3H inbred mouse strains. Since monoclonal antibody reacts with teratocarcinoma stem cells regardless of the species or strain of origin, and since it does not react with several nonteratocarcinoma cell lines derived from 129 mouse strains we can conclude that the parental spleen cell was stimulated to respond to a stem cell-specific antigen. Results obtained in radioactive binding assay were confirmed using cytotoxicity, indirect immunofluorescence and quantitative absorption assays (data not given).

2. Reactivity with Early Mouse Embryos

The same serum was tested using RIA, complement-mediated cytotoxicity and indirect immunofluorescence assays on preimplantation mouse embryos. Binding of antibody is first detected at late 8-cell stage; all earlier stages including unfertilized eggs are nonreactive. Binding increases on morulae, decreases on trophectoderm of the blastocyst and disappears completely when blastocysts are cultured for 24 hrs. Inner cell masses isolated immunosurgically (6) from blastocysts are reactive and binding increases during their growth *in vitro*. Complement-mediated cytotoxicity and indirect immunofluorescence give essentially the same results. There is no lysis or fluorescence of embryos up to the 8-cell stage. At the 8-cell stage some embryos (10-20%) are partially or completely lysed at dilutions from 1:10-1:128,000. Observation after indirect immunofluorescence reveals that not all of the embryos are reactive with the antisera; frequently only a few fluorescent blastomeres are seen. Although the percentage of reactive embryos increases (80-90%) at later embryonic stages, all of the embryos are never stained or lysed, regardless of the developmental stage or the antibody concentration.

C. Concluding Remarks

Our results indicate that it is possible to produce monoclonal antibody against an antigen (SSEA-1) common to both mouse and human teratocarcinoma stem cells and to early mouse embryos of a given developmental stage. The pattern of reactivity of the monoclonal reagent with embryos of various developmental stages does not correspond to that of any of the antisera described previously (1-4) although it might represent part of the reactivity found in any or all of them. Using the same immunization schedule, we are now trying to isolate clones producing different, stage-specific monoclonal antibodies. Such an approach should give us some insight into the number of stage-specific antigens. Moreover, the availability of these pure reagents will lead to the isolation and characterization of these molecules and the assessment of their role in development and differentiation.

References

1. Jacob, F.: Mouse teratocarcinoma and embryonic antigens. Immunol. Rev. 33, 3-32 (1977)
2. Edidin, M., Gooding, L.R.: Teratoma-defined and transplantation antigens in early mouse embryos. In: Teratomas and Differentiation. Sherman, M.I., Solter, D. (eds.). New York: Academic Press, 1975, pp. 109-121
3. Stern, P.L., Martin, G.R., Evans, M.J.: Cell surface antigens of clonal teratocarcinoma cells at various stages of differentiation. Cell 6, 455-465 (1975)
4. Wiley, L.M., Calarco, P.G.: The effect of anti-embryo sera and their localization on the cell surface during mouse preimplantation development. Develop. Biol. 47, 407-418 (1975)
5. Galfre, G., Howe, S.C., Milstein, C., Butcher, G.W., Howard, J.C.: Antibodies to major histocompatibility antigens produced by hybrid cell lines. Nature 266, 550-552 (1977)
6. Solter, D., Knowles, B.B.: Immunosurgery of the mouse blastocyst. Proc. Natl. Acad. Sci. U.S. 72, 5099-5102 (1975)
7. Supported by grants from the A.C.S. IM-88 and PDT-26, the NIH CA-18470, CA-10815, CA-17546, an RCDA-AI-00053 (BBK), and the N.F.M. of D. 1-301.

Monoclonal Anti-Rat MHC (H-1) Alloantibodies

J.C. Howard, G.W. Butcher, G. Galfre, C. Milstein

A. Introduction: Cell Fusion and Cloning

We wish to report the isolation of 6 clones of hybrid myelomas secreting rat antibodies with specificity for alloantigens of the rat major
histocompatibility complex H-1 (AgB). An account of the isolation of
the first such clone including details of the immunization and fusion
techniques has been published (1). Briefly two AO (H-1^w, AgB2) rats
were hyperimmunized against DA (H-1^a, AgB4) lymphoid cells and finally boosted with an intravenous injection of DA cells 2.5 days or 4
days before fusion. Spleen cells were fused with P3-X63-Ag8 BALB/c
mouse myeloma cells in the presence of Polyethylene Glycol (PEG) 1500.
We used the intravenous route to boost before fusing to favour localization of the antibody response in the spleen (2). After fusion,
aliquots initially containing 2×10^6 spleen cells were grown in the
wells of 6 x 4 Linbro trays in HAT medium for two weeks. Active growth
of presumptive hybrids was seen in all wells. Supernatants obtained
at this stage were assayed for specific antibody by ^{51}Cr release from
AO and DA RBC in a complement dependent lytic assay. 106/113 wells
contained significant lytic activity against DA RBC while no autoantibody activity against AO RBC was detected.

After the successful derivation of the clone R3/13 (1) three strongly
lytic cultures, R2/10, R2/15 and R3/47 were selected for cloning. Uncloned R2/10 supernatant had been noted previously to be far more lytic than R3/13 clone supernatants, which characteristically release
about 50% of the ^{51}Cr from DA RBC with a plateau at this level (1).
Supernatants from 194 clones of the R2/10 hybrid were assayed for lytic activity on DA RBC. Only a single clone (R2/10P) with lytic activity was found and, surprisingly in view of the strong lytic activity
of the uncloned hybrid supernatant, had the 'partial' lytic activity
characteristic of the cloned R3/13 (1). It seemed possible that the
lytic activity of the uncloned hybrid resulted from synergy between
the antibody product of the partially lytic clone R2/10P and a non-
lytic 'silent' clone. Accordingly two pools of R2/10 clone supernatants were prepared, one lacking the 'partial' lytic supernatant
R2/10P, the other containing it. The pool lacking R2/10P did not lyse
DA RBC, while the other pool showed the strong lytic activity of the
uncloned hybrid R2/10. When individual clone supernatants from the
'negative' group were tested for synergistic lysis of RBC in the presence of R2/10P, 27/97 were found to be positive. 'Synergistic' clone
supernatants (but not the others) were shown to contain an antibody
against DA RBC by an indirect binding assay as described (1). One
such clone, purified and named R2/10S. Two partially lytic antibody
secreting clones R2/15P and R2/15S were isolated by a combination of
limiting dilutions and agar cloning from the hybrid R2/15. As with
R2/10, neither clone had the lytic activity of the uncloned hybrid on
DA RBC, while a combination of the two was strongly lytic. The uncloned hybrid R3/47 lost its lytic activity before cloning but the
supernatant could be shown still to contain an antibody by its abili-

54

ty to contribute synergistic lytic activity to R/10P. A clone with this synergistic lytic activity was subsequently isolated by agar cloning and named R3/47.

A total of 6 active clones was therefore isolated from 4 uncloned hybrids. Analysis of the earliest HAT supernatant of R3/13 demonstrated lytic activity far higher than that shown by the isolated clone from R3/13. It may be inferred that a second synergistic antibody was also present in this hybrid, but was lost before cloning. Similar considerations apply to R3/47, when a second active clone was also presumably lost before cloning. It therefore seems likely that multiple active hybrids were present in each well shortly after fusion. If it is assumed that a maximum of one hybrid in 10 was secreting specific antibody in these two fusions, the fusion efficiency must have been $1/10^5$ cells or higher. The extraordinary amount of activity recovered from this allogeneic immunization in rats contrasts with the difficulty reported in isolating hybrids secreting anti-H-2 antibody in alloimmunizations between mice (3).

The 6 clones isolated from the R2 and R3 fusions have been repeatedly recloned in agar, and have been grown for long periods *in vitro*. Initially R2/10P, R2/10S and R2/15P were relatively unstable with few active clones being recovered from recloning, but eventually lines were established which have been stable for several months. By SDS-polyacrylamide gelelectrophoresis of ^{14}C-lysine labelled products all 6 clones secrete a new γ chain and light chain presumably of rat origin. Variants lacking X63 mouse chains have been selected and the following clones are now available: R2/10P-HL; R2/15S-HL; R3/13-HL; R3/47-HL; R2/10S-HLK; R2/10P-HL; R2/15P-HLGK.

B. Competitive Inhibition of Binding by Monoclonal Antibodies

The topographical relationship between the targets of the 6 monoclonal antibodies described in this report has been studied by competitive inhibition of binding. All 6 monoclonal antibodies have been analysed for their ability to inhibit the binding of subplateau concentrations of 4 internally labelled antibodies (R2/10P-HLGK-^{3}H-Lys., R2/15S-HL-^{3}H-Lys., R3/13-HL-^{3}H-Lys and R3/47-HL-^{3}H-Lys) to DA RBC. The results are summarized in Table 1.

Table 1. Competitive inhibition of binding to DA RBC of 4 monoclonal anti DA alloantibodies internally labelled with ^{3}H-Lysine.

	% inhibition of binding of ^{3}H-Lysine-labelled monoclonal antibody			
Competitive inhibitor	R2/10P-HLGK	R3/13-HL	R2/15S-HL	R3/47-HL
R2/10P-HLGK	88.0	92.1	3.5	13.7
R2/15P-HLGK	93.3	97.2	2.2	39.5
R3/13-HL	95.5	98.9	3.7	-13.6
R2/10S-HLK	-42.7	16.8	3.8	-10.5
R2/15S-HL	- 3.9	11.2	97.0	86.4
R3/47-HL	4.5	11.0	18.6	88.1
R3/13-HK	0	- 5.9	- 1.8	10.8

Competitive supernatants were assayed at a final concentration of 1/6. R3/13-HK is a variant of R3/13 which secrets only the rat heavy chain (H) and the mouse light chain (K). The supernatant has no detectable antibody activity on DA targets and is used as a negative control.

The three antibodies R2/1OP, R2/15P and R3/13 show complete mutual
inhibition and may be said roughly to define a 'site' on the alloantigenic target. R3/47 and R2/15S clearly have topographically related
targets that are different from the site defined by the R2/1OP group.
The asymmetrical inhibitory behaviour of these two antibodies is
probably due to a low affinity of binding by R3/47, making it a relatively poor competitor with R2/15S. A relationship between the R2/15S
site and the R2/1OP group site is suggested by the partial (35%) inhibition of binding of R3/47 by R2/15P. Partial inhibition of R3/47
by R2/15P is seen as a plateau at this level over several dilutions
of R2/15P and may therefore indicate heterogeneity in the target for
R3/47, one site adjacent to the R2/15P target, and one relatively
distant from it. In any event the relationship between the two targets strongly indicates that the R2/1OP group and the R2/15S group
are carried at least in part on the same molecules. The site for
R2/1OS is not yet defined relative to the sites for the R2/1OP and
R2/15S groups. However, a relationship of some kind between R2/1OP
and R2/1OS is suggested by the markedly enhanced binding of R2/1OP in
the presence of R2/1OS. This reproducible result resembles a similar
finding in an analysis of the targets for 2 mouse anti-Ig1^b allotype
monoclonal antibodies reported by Oi *et al.* in this volume.

By co-precipitation from NP40 lysates of ^{125}I-labelled DA lymphocytes,
the targets for the antibodies R2/1OP, R2/15P, R3/13 and R2/15S have
been shown to be carried on the same molecules (unpublished results
with C.P. Milstein), confirming the assignment already made by competitive inhibition analysis. The molecular locations of the targets
for R2/1OS and R3/47 have not yet been determined.

C. Synergistic Lysis by Monoclonal Antibodies

It appears that individual monoclonal antibodies do not lyse 100% of
a population of DA lymphocytes or erythrocytes while appropriate mixtures of two monoclonals will do so. The lysis of cells by a monoclonal antibody is extremely sensitive to the mean antigen concentration on the cells. This can be demonstrated by comparing the lytic
activity of monoclonal antibodies on cell populations from DA and
from (DA x AO)F$_1$ in which the mean antigen concentration is reduced
by about 50% (Table 3). The fact that increasing antibody concentration gave a plateau significantly lower than 100% lysis indicates
heterogeneity of antigen concentration among erythrocytes or lymphocytes and/or heterogeneity of cell susceptibility to lysis. The former heterogeneity has been demonstrated directly in the fluorescence
activated cell sorter (FACS)(data not shown). The dramatic effect of
a mere 50% reduction in mean antigen concentration on the lytic activity of a monoclonal IgG antibody accounts for our previously reported
reduction in plateau from nearly complete lysis of DA to about 50%
lysis of (DA x HO)F$_1$ lymph node cells by R3/13 (1). At that time we
discussed the possibility that the expression of the monoclonal antibody target might be selectively reduced in a proportion of F$_1$ lymphocytes. There is no suggestion of any such effect when the binding of
R3/13 to DA and (DA x HO)F$_1$ lymphnode cells is assayed in the FACS.

Synergistic lysis by a mixture of two monoclonal antibodies occurs
only when the two antibodies do not compete for a single determinant.
This principle is illustrated with R2/1OP, R3/13 and R2/15S in Table
2.

Table 2. Synergistic lysis of DA RBC by non-competitive monoclonal antibodies.

^{51}Cr release as % input using monoclonal antibody mixtures.

	R2/10P(1/8)	R3/13(1/8)	R2/15S(1/64)
R2/10P-HLGK (1/8)	28.6	38.8	101.8
R3/13-HLK (1/8)		51.5	105.8
R2/15S-HL (1/64)			31.5

R2/10P and R3/13 are mutually competitive antibodies: R2/15S is competitive with neither R2/10P nor R3/13 (see Table 1). Synergistic lysis occurs only with non-competitive mixtures. Spontaneous release in this experiment was 2.1%.

R2/10P and R3/13 are competitive and do not synergize with each other; R2/15S is competitive with neither R2/10P nor R3/13 and synergizes with both.

Synergistic lysis by non-competitive monoclonal antibodies could be due either to binding of the two antibodies to sites randomly dispersed relative to each other, or to sites on a single molecule or in an aggregate and thus non-randomly dispersed. If the two sites are randomly dispersed relative to each other, then a mixture of two antibodies should be no more effective in lysis than an equivalent increase in antigenic sites followed by lysis with a single antibody. If the sites are adjacent then a mixture of two antibodies should be more efficient than predicted by the increase in antigenic concentration. Such effects can be investigated using cells from homozygous and heterozygous animals of different genotype. Erythrocytes were obtained from rats of DA, (DA x AO)F_1, PVG.1R and (PVG.1R x AO)F_1 origin chosen to vary the mean antigen density on RBC over a range of about 5 fold. The PVG.1R strain carries a recombinant MHC haplotype H-1^{ac1} in which the segment determining the targets for R2/15P and R2/15S is of DA (H-1^a) origin. We have observed that the density of these and other H-1^a determinants is reduced on cells from the PVG.1R strain to about 40% of its level on the DA strain. The two F_1 hybrids with AO provided further mean antigen density variation. In this experiment relative levels of monoclonal antibody bound to the 4 erythrocyte populations were measured from an indirect binding assay; complement-dependent lysis of erythrocytes was then assayed by ^{51}Cr release for the two monoclonal antibodies alone and as a mixture. The results show (Table 3) that the mixture of two monoclonal antibodies is far more efficient in lysis than is either antibody alone, even with half as many antibody molecules bound.

Table 3. Synergistic lysis of RBC by a mixture of non-competitive monoclonal antibodies is more efficient than would be expected if the targets for the two antibodies were randomly dispersed on the membrane.

	Target RBC : Strain of Origin			
	DA	(DAxAO)F_1	PVG.1R	(PVG.1RxAO)F_1
R2/15P-HLGK alone				
Antibody Bound (DA=1.0)	1.0	0.42	0.29	0.21
% Specific ^{51}Cr Release	44.0	3.5	3.3	1.9
R2/15S-HL alone				
Antibody Bound (DA=1.0)	1.0	0.56	0.41	0.28
% Specific ^{51}Cr Release	52.8	3.6	1.9	<1.0
R2/15P & R2/15S mixed				
Antibody Bound (DA=2.0)	2.0	0.98	0.70	0.49
% Specific ^{51}Cr Release	107.2	101.4	103.6	71.9

All assays were performed at a final concentration of each monoclonal antibody of 1/32. Antibody Bound for individual monoclonal antibodies was measured from an indirect binding assay using Rabbit ^{125}I-F(ab')$_2$ anti rat Fab. The level of binding to DA RBC was arbitrarily set to 1.0 for each antibody. Antibody Bound with the mixture of monoclonals was calculated as the sum of each bound alone.

Specific ^{51}Cr release was calculated as

$$\frac{\text{Experimental Release - Spontaneous Release}}{100- \text{ Spontaneous Release}} \%$$

Spontaneous release was 1.8% of input in this experiment.

We therefore conclude that the targets for the two synergistic antibodies are adjacent. This result is compatible with the evidence from co-precipitation experiments that the R2/15P and R2/15S targets are carried on the same molecules. The results have implications for the mechanism of complement activation mediated by IgG molecules. They confirm the old idea (4) that interaction between neighbouring IgG antibody molecules is essential in the fixation of complement or at least in initiating the activation cascade. A similar conclusion was reached recently in a study of antibodies bound to different sites on membrane immunoglobulin in a guinea pig lymphoma (5).

D. Genetic Control of Monoclonal Antibody Targets

The targets for all 6 monoclonal antibodies are specified by the A region of the H-1 complex. In an indirect binding assay binding was positive on the two H-1^a congenic lines PVG-H-1^a (DA) and LEW.1A, and negative on PVG(H-1^c,AgB5) and LEW (H-1^l,AgB1). Binding was also positive on the two H-1 recombinant congenic lines PVG.1R and LEW.1AR1 which carry the (K+D-like) A region of H-1^a, and the (I-like) B region of H-1^c and H-1^w respectively (6,7). Three classes of cross-reactions could be distinguished when the 6 monoclonal anti H-1^a antibodies were assayed for binding to RBC from strains carrying other H-1 haplotypes.
a) Co-ordinate Expression. All six monoclonal antibodies bound to RBC from the strain BD VII (H-1^e). The ratio of binding levels of the 6 antibodies was similar to that seen on DA, and the end points of the titration curves were as seen on DA. However the absolute binding levels were reduced to about 20% of the DA level. The observation is compatible with diminished expression on BD VII RBC of the same antigenic molecule(s) expressed by DA. By conventional serological tests H-1^a and H-1^e are strongly cross-reactive haplotypes distinguished by a weak private specificity (8).
b) Strong cross reactions. The monoclonal antibody R2/15S bound strongly to RBC from rats carrying the haplotypes H-1^d (AS2 and LEW.1D) and H-1^f (AS2 and LEW.1F). In level and titre the binding was comparable to that shown by R2/15S on H-1^a RBC. The other monoclonal anti H-1^a antibodies either did not bind at all to these haplotypes or bound with very weak cross-reactions (see below). Since the determinants for R2/10P, R2/15P, R3/13 and R3/47 are probably on the same molecule or molecules as R2/15S in the H-1^a haplotype, the isolated strong cross-reactions of R2/15S are consistent with common polymorphic structures shared by H-1^d, H-1^f, and H-1^a molecules.
c) Weak cross reactions. Weak binding reactions were noted with several monoclonal antibodies on a variety of H-1 haplotypes. These reactions, which were characterised by low binding and the absence of any plateau, appear consistent with low affinity cross reactions to non-identical polymorphic determinants.

58

The level of expression of H-1^a specificities detected on RBC by monoclonal antibodies appears to be a variable under complex genetic control. At least two genetic influences on the level of expression were detected, associated respectively with the non-MHC background, and with the MHC B region, a region which expresses no known antigenic specificities on RBC in rats (9). These effects are illustrated by the plateau binding levels of R2/15P on RBC from two groups of strains in Table 4. The low binding level on PVG.1R was exploited in the analysis of synergistic lysis reported above. The genetic specification of the control of the level of H-1 antigen expression is currently under investigation.

Table 4. Variation in apparent determinant density for monoclonal anti-H-1A^a antibody R2/15P-HLGK on RBC influenced by the non-MHC background and by the MHC B region.

Strain	H-1 Haplotype H-1A	H-1B	Non-MHC Background	Bound F(ab')$_2$ anti Fab ^{125}I counts per 3 min.
DA	a	a	DA	2636
PVG-H-1^a(DA)	a	a	PVG	1758
PVG.1R	a	c	PVG	698
AVN	a	a	AVN	2669
LEW.1A(AVN)	a	a	LEW	1536
LEW.1AR1	a	w	LEW	921

R2/15P at a final concentration of 1/16. Data taken from titrations of R2/15P on each target. Counts represent plateau binding in each case. Binding in the absence of R2/15P was <170 counts per 3 min. on all targets.

E. Summary

Some properties of 6 monoclonal rat alloantibodies prepared from fusions of hyperimmune rat spleen cells with mouse P3-X63-Ag8 myeloma cells are described. All the antibodies are IgG and have specificity for targets on a molecule or molecules determined by the K/D-like A region of the H-1^a (AgB4) MHC haplotype. Analysis of the topography of the antigenic molecule or molecules on the RBC membrane by competitive inhibition of binding indicates the presence of at least two alloantigenic sites per molecule. A study of the lytic properties of the monoclonal antibodies for RBC strongly suggests that adjacency between bound IgG molecules is a necessary and possibly sufficient condition for complement activation. Analysis of binding of monoclonal alloantibodies to cells carrying other haplotypes than H-1^a indicates three distinct forms of cross-reaction, while a similar analysis of binding to RBC from H-1^a congenic and recombinant strains indicates complex genetic control of the level of expression of the determinant-carrying molecules.

Acknowledgements. We thank Diana Suckling, Shirley Howe and Bruce Wright for technical assistance, and Jenny Williams for typing the manuscript. Dr. A.F. Williams, M.R.C. Cellular Immunology Unit, Sir William Dunn, School of Pathology, University of Oxford kindly provided the rabbit F(ab')$_2$ anti-Fab used in this study. RBC from a variety of rat strains were kindly provided by Dr. E. Gunther, Max-Planck-Institut für Immunbiologie, Freiburg, F.R.G.

This work was supported in part by U.S.P.H.S. grant no AI 13162.

References

1. Galfre, E., Howe, S.C., Milstein, C., Butcher, G.W., Howard, J.C.: Antibodies to major histocompatibility antigens produced by hybrid cell lines. Nature 266, 550-552 (1977)
2. Rowley, D.A.: The effect of splenectomy on the formation of circulating antibody in the adult male albino rat. J. Immunol. 64, 289-295 (1950)
3. Lemke, H., Hammerling, G.J., Hohmann, C., Rajewsky, K.: Hybrid cell lines secreting monoclonal antibody specific for major histocompatibility antigens of the mouse. Nature 271, 249-251 (1978)
4. Humphrey, J.H., and Dourmashkin, R.R.: Electron microscope studies of immune cell lysis. Ciba Found. Symp. Complement. Churchill, London (1975) p. 175
5. Elliot, E.V., Pindar, A., Stevenson, F.K., Stevenson, G.T.: Synergistic cytotoxic effects of antibodies directed against different cell surface determinants. Immunology 34, 405-409 (1978)
6. Butcher, G.W. and Howard, J.C.: A recombinant in the major histocompatibility complex of the rat. Nature 266, 362-364 (1977)
7. Stark, O., Gunther, E., Kohoutova, M., Vojcik, L.: Genetic recombination in the major histocompatibility complex (H-1,Ag-B) of the rat. Immunogenetics 5, 183-187 (1977)
8. Gunther, E., Stark, O.: Personal communication
9. Davies, H. ff. S.D., and Butcher, G.W.: Kidney alloantigens determined by two regions of the rat major histocompatibility complex. Immunogenetics (in press)

Selective Suppression of Reactivity to Rat Histocompatibility Antigens by Hybridoma Antibodies

T.J. McKearn, M. Sarmiento, A. Weiss, F.P. Stuart, F.W. Fitch

A. Introduction

Alloimmune serum can selectively suppress humoral and cell-mediated immunity towards the histocompatibility antigens of inbred rats. In the case of renal allograft enhancement, the suppression induced by one injection of alloimmune serum often lasts for the life of the adult rat and displays striking immunological specificity (1,2). Due in part to the variability in biological effects of various pools of alloimmune serum, much controversy currently exists regarding the immunoglobulin class, alloantigen specificity and idiotypic repertoire of those alloantibodies which are presumed responsible for the suppressive effects of an effective alloantiserum (3,4). Hybridoma technology offers an unique opportunity to dissect the component immunoglobulins of an immunized individual and study the biological effects of these hybridoma antibodies.

I. Materials and Methods

Lewis (Ag-B^1), Brown Norway (AgB^3), ACI (AgB^4) inbred rats and their respective F$_1$ hybrids were purchased from Microbiological Associates (Bethesda, Md.). Female Lewis rats were immunized towards Brown Norway (BN) alloantigens by two intravenous injections of 5 x 10^7 viable BN spleen cells one month apart. Spleen cells were taken from the immunized Lewis rat three days after the second antigen injection. A single cell suspension of spleen cells was prepared in high glucose Dulbecco's Modified Eagle's Medium (DMEM) (Gibco #H-21) using a loose-fitting tissue grinder. Suspended cells were washed once with DMEM and adjusted to 2.5 x 10^7 cells/ml and then separated on a Ficoll-Hypaque gradient as described by Davidson and Parish (5). Cells which remained at the gradient interface were washed twice with DMEM and adjusted to 12.5 x 10^6 cells/ml. P3-X63-Ag8 cells (kindly provided by Dr. C. Milstein) were washed twice with DMEM and adjusted to 5 x 10^6 cells/ml. One ml of P3-X63-Ag8 cells and 4 ml of rat spleen cells were mixed in a 60 mm Petri dish (Falcon 3002) and these dishes placed in centrifuge carriers for microtiter plates (Cooke). The dishes were spun for three minutes at 250 x g at room temperature. This results in the formation of an adherent monolayer of cells on the bottom of the dish. Supernatant was aspirated and the dishes gently flooded with 50% polyethylene glycol (PEG 1500, Fisher) in DMEM. After 30 seconds at room temperature, the plate was flooded with 5 ml of DMEM and the supernatant aspirated. This wash procedure was repeated once. The plates are then filled with DMEM containing 20% fetal calf serum and the plates were incubated overnight at 37° in a humidified 5% CO$_2$ environment. The following day, cells were harvested from the plates by gentle rinsing with fresh DMEM followed by scraping with a rubber policeman. The cells were centrifuged at 50 x g for 5 minutes and the supernatant aspirated. The cells were resuspended to approximately 30 ml with DMEM containing 20% fetal calf serum and the following additives: 1 x 10^{-4}M hypoxanthine, 4 x 10^{-7}M aminopterin, 3 x 10^{-6}M thymidine. L-arginine (0.116 gm/l), L-glutamine (0.216 gm/l), folic acid (0.006 gm/l) L-asparagine (0.036 gm/l), sodium bicarbonate (2.0 gm/l) and sodium pyruvate (0.11 gm/l). The cells were distributed into 288 microtiter wells of 96 well flat bottom dishes (Costar) and kept in a 5% CO$_2$ humidified environment. After 6-7 days, each well received an additional 100 ul of DMEM with 20% fetal calf serum. Supernatants were withdrawn for assay

when the pH of the medium began to acidify and clusters of proliferating hybrids
were grossly apparent within the wells. With the P3-X63-Ag8 parent, this occurred between 10 and 24 days after fusion.

Supernatant fluids were assayed for cytotoxic antibodies against BN alloantigens
by a modification of the technique of Weiss & Fitch (6). In brief, 50 ul of
fluid was mixed with 10^4 ^{51}Cr labelled ConA blasts derived from BN spleen cells
in round bottom 96 well plates (Linbro). After 30 to 45 minutes of incubation,
the plates were centrifuged at 300 x g for 5 minutes and the supernatants removed. One hundred microliters of agar absorbed rabbit complement (diluted 1:8
with DMEM) were then added and the plates incubated for an additional 30 to 45
minutes at 37°. Plates are then centrifuged and aliquots of supernatant removed
for counting.

Supernatant fluids were also assayed for hemagglutinating antibodies by an indirect hemagglutination assay; 50 microliters of serial dilutions of supernatant
were mixed in 96 well-V-bottom trays (Linbro) with an equal volume of 0.1% (v/v)
BN erythrocytes in saline. After 30 minutes incubation at room temperature, 50
microliters of a 1:250 dilution of goat anti-rat Ig were added and the plates
incubated at 37° for 30 minutes. Wells were scored for agglutination by pellet-
ting the erythrocytes at 300 x g for one minute and then tipping the trays at
approximately 60° using a special rack. Cells which are agglutinated remain as
a pellet while those which are not agglutinated run from the pellet.

Hybridoma cells were cloned by limiting dilution in 96 well cluster dishes
(Costar) which had previously been seeded with 10^6 irradiated (1200 R.) Lewis
rat thymocytes. The wells were seeded at a density of 0.5 to 1.0 cell/well and
the wells microscopically scored for presence of one or more clusters of proliferating cells 5 or 6 days after cloning.

Renal allografts were performed as previously described (7). Recipient animals
underwent bilateral nephrectomy at the time of renal allograft transplantation
and therefore serial blood urea nitrogen determinations can serve as a measure
of allograft function.

II. Results

Figure 1 shows a representative plot of the type and amount of ^{51}Cr release
seen upon screening 140 supernatants containing proliferating hybrids derived
in selective H-A-T medium by fusing spleen from a Lewis rat immunized to BN
antigens to the mouse myeloma mutant P3-X63-Ag8.

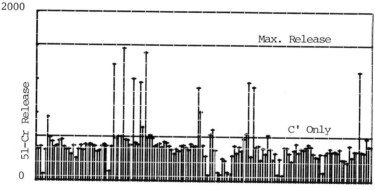

Figure 1

62

The specificity of lysis was ascertained by measuring the amount of lysis seen against a panel of ^{51}Cr labelled ConA blasts (Table 1).

Table 1

	[51]Cr CPM release		
	Lewis ConA Blasts	BN ConA Blasts	ACI ConA Blasts
Maximum Release	1532	1263	1285
Normal Serum +C^1	406	375	390
	Hybridoma Supernatant +C^1		
D4, 41	409	777	394
D4, 44	321	607	394
D4, 46	394	1162	844
D4, 68	373	810	327
D4, 69	272	673	402
D4, 91	364	795	663
D4, 135	343	1085	416

Several of the supernatants possessed cytotoxicity towards both BN and ACI targets, while none of the supernatants were cytotoxic when tested against Lewis targets. All of these cytotoxic supernatants possessed binding activity towards BN erythrocytes, implying recognition of a determinant common to erythrocytes and ConA blasts. In capping experiments using ConA blasts, hybridoma supernatants and "conventional" anti-AgB alloantisera showed co-capping between all supernatants and the anti-AgB alloantisera, thereby confirming the nature of the determinants recognized by the hybridoma antibodies.

Following cloning, three hybridoma cell lines were expanded and the immunoglobulin class of the rat immunoglobulins determined (Table 2).

Table 2

Hybridoma	Heavy Chain	Light Chain	Cytotoxicity Titer
D4,37	μ	(lacks Kappa allotype marker)	1:1,000,000
D4,68	γ	Kappa	1:32
D4,69	γ	Kappa	1:16

The biological effects of cloned hybridoma supernatants were assessed in two in vivo assays. The effect of hybridoma supernatant on subsequent humoral immune responses was tested by injecting Lewis rats with one ml of hybridoma supernatant (containing 60-80 μgm of rat immunoglobulin). One day later these animals were injected intravenously with 5 x 10[7] Lewis X Brown Norway F_1 spleen cells. Serial blood samples were obtained by tail vein puncture and hemagglutinating antibodies measured (Table 3).

Table 3

Supernatant Injected	Anti BN Hemagglutinin Titer			Anti ACI Hemagglutinin Titer		
	Day 6	Day 8	Day 10	Day 6	Day 8	Day 10
P3-X63-Ag8	10.3+1.6	7.3+1.4	8.0+1.1	10.7+0.6	11.3+0.6	10.3+0.6
D4, 68	2.1+1.5	1.0+1.2	0.4+0.7	11.8+0.5	12.0+0.8	11.9+0.6
D4, 37	9.2+1.4	8.6+1.1	7.2+1.2	N.D.		
D4, 69	10.3+0.8	9.8+1.0	8.5+1.4	N.D.		

Those Lewis animals who received one ml of D4,68 supernatant showed marked suppression of their anti BN antibody response. Furthermore, this suppression of antibody responses was specific in that Lewis rats injected with D4,68 responded normally to challenge with ACI alloantigens.

The effect of hybridoma antibodies on cell-mediated immune responses was tested by injecting Lewis recipient animals with one ml of hybridoma supernatant on day −10 and 5 x 10^7 LBN F, spleen cells on day −9 relative to transplantation of an LBN renal allograft on day 0. Those animals who received hybridoma supernatants prior to renal allograft placement showed marked suppression of renal allograft rejection (Figure 2).

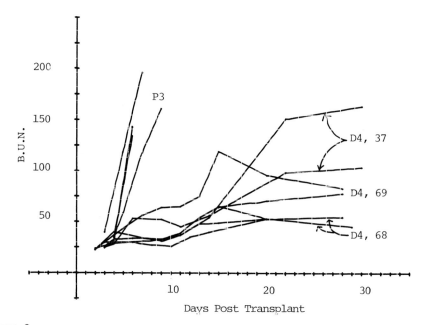

Figure 2

B. Discussion

Hybridoma cell lines can be derived by fusion of alloimmunized rat spleen cells
to mouse myeloma mutants. Cloned lines of such hybridoma cells secreted up to
80 µgm/ml of rat immunoglobulins. Hybridomas secreting antibody reactive with
rat Ag-B alloantigens can be used to selectively suppress humoral and cell-
mediated immunity directed against those antigens.

References

1. Stuart, F.P., D.M. Scollard, T.J. McKearn and F.W. Fitch Cellular & Humoral
    Immunity after Allogeneic Renal Transplantation in the Rat. V Appearance
    of Anti-idiotypic Antibody and its Relationship to Cellular Immunity after
    Treatment with Donor Spleen Cells and Alloantibody. Transpl. 22, 455-466
    (1976).

2. French, M.E. and J.R. Batchelor Immunological Enhancement of Rat Kidney
    Grafts. Lancet 2, 1103-1106 (1969).

3. Tilney, N.L., J. Bancewicz, W. Rowinski, J. Notis-McConarty, A. Finnegan,
    and D. Booth Enhancement of Cardiac Allografts in Rats. Comparison of
    host responses to different treatment protocols Transpl. 25, 1-6 (1978).

4. Soulillou, J-P, C.B. Carpenter, A.J.F. d'Apice, and T.B. Strom. The Role
    of Nonclassical, Fc Receptor-Associated Ag-B Antigens (Ia) in Rat Allograft
    Enhancement J. exp. Med. 143, 405-421 (1976).

5. Davidson, W.F. and C.R. Parish, A Procedure for Removing Red Cells and Dead
    Cells from Lymphoid Cell Suspensions J. Immun. Meth. 7, 291-299 (1975).

6. Weiss, A. and F.W. Fitch Macrophages Suppress CTL Generation in Rat Mixed
    Leukocyte Cultures J. Immunol. 119, 510-516 (1977).

7. Stuart, F.P., T. Saitoh and F.W. Fitch, Rejection of Renal Allografts:
    Specific Immunological Suppression Science 160, 1463-1465 (1968).

Monoclonal Antibodies to Human Lymphocyte Membrane Antigens

M.M. Trucco, J.W. Stocker, R. Ceppellini

A.   Introduction

Recent reports have shown that the original technique of Köhler et al. (1) can be successfully applied for obtaining specific antibodies against single determinants of complex antigens, through cloning of immunoglobulin secreting hybridomas (references in this volume).   Here we discuss the variety of antibodies obtained with this technique after primary immunization of mice with human lymphocytes.

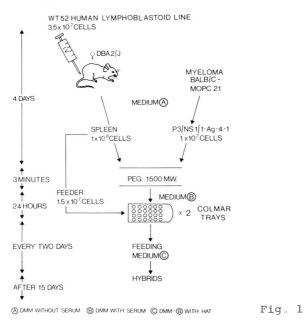

IMMUNIZATION AND HYBRIDIZATION

Fig. 1

Ⓐ DMM WITHOUT SERUM   Ⓑ DMM WITH SERUM   Ⓒ DMM−Ⓑ WITH HAT

B.   Materials and Methods

They are summarized in Fig. 1.   The myeloma used, the 8-Azaguanin resistant variant of the P3/NS1 line, was kindly provided by Dr. C. Milstein.   It is a producer of κ light chains only, which in general are secreted together with the antibody immunoglobulins of the derived hybridomas (1).   The human lymphoblastoid line (HLL) used for immunization, WT52 (2), as well as the other HLL used as targets for screening, were obtained by EB virus infection of HLA homozygous cells from the Torino-Basel panel (3).   Daudi, a Burkitt lymphoma line lacking surface $\beta2$ microglobulin ($\beta2$m) and HLA:ABC, was obtained from Prof. W. Bodmer.   Molt 4, a human line of T origin, was obtained from Dr. S. Ferrone.   One of the rabbit $F(ab')_2$ anti-mouse Ig preparations, used for the indirect radiobinding assay was a gift of Dr. A.S. Williams.

66

Presence of hybrids was looked for after 15 days, when no living cells could be detected in the control wells (3 with splenocytes and 3 with myeloma cells only). Supernatants of growing cultures were tested against the specific target (WT52) for: a) complement dependent lysis (Cdl) with a variety of techniques (fluorochromasia, ^{51}Cr release, NIH dye exclusion test); b) indirect immunofluorescence (IIF) with anti-mouse Ig reagents; c) indirect radiobinding (IRB) using a ^{125}I labeled rabbit anti-mouse Ig (4). If positive, the corresponding cultures were expanded and subsequently cloned in soft agarose.

C. Results and Discussion

All the 42 wells in which the hybridization cell mixture was distributed, produced cultures growing in HAT selective medium. Complement dependent cytotoxic activity against the specific WT52 target was found in the supernatnats of 7 cups. These cultures were expanded while the others (Cdl negative) were stored in liquid nitrogen. After expansion, only 5 of the 7 retained cytoxic activity (cultures nos. 10, 16, 19, 24, 26). Later, 8 of the 35 Cdl-negative cultures, recovered from storage, were tested against WT52 by IRB: 3 supernatants (cultures nos. 5, 6, 34) were positive. Thus it can be estimated that about 50% of the primary growing cultures produced antibodies against the target.

Clones from the "positive" populations were derived in soft agarose. The class and type and the monoclonal nature of the antibodies produced by clone cultures were analyzed by SDS-PAGE and IEF, respectively. All supernatants analyzed appeared to be IgG, κ (Fig. 2). This was confirmed by specific reactions with anti-mouse γ and κ but not with anti-μ and λ. The majority of clones express a second κ light chain, corresponding to the one secreted by the parental MOPC 21 myeloma. Some clones however, during the establishment of cultures, have lost it. Preliminary results after passage of the clones in histocompatible mice with production of ascites and solid tumors have shown that antibody titers in the order of 10^5 can be obtained. None of these antibodies reacted against mouse cells or against human or sheep erythrocytes.

Table 1: Patterns of reactivity of representative clones

	Human Lymphoblastoid Lines*				PBL**	
Cultures:	WT 52	DAUDI	MOLT 4	Other HLL	Cdl	IIF
No. 16	+	-	+	11/11	100	95
24a	+	-	+	11/11	100	95
26	+	-	+	11/11	100	95
No. 34	+	-	n.d.	11/11	0	95
No. 19	+	+	-	9/11	0	1-4
24b	+	-	-	7/11	0	1-7
5	+	-	n.d.	7/11	0	1-10
6	+	-	n.d.	7/11	0	1-10
No. 10	+	-	+	9/11	0-100	30-95

* Positive line (HLL) over total tested in cytotoxicity or (nos. 5, 6, and 34) in radiobinding.

** Peripheral blood lymphocytes (PBL).

Table 1 shows the specificities of the better studied clones. Three of the initial cultures, nos. 16, 24, 26 (and derived clones) were soon suspected to be producers of anti-human β2m. In fact they were active against all 12 HLL tested, except for Daudi, and were cytotoxic for all peripheral blood lymphocytes (PBL) from 30 unrelated donors (Cd1 titer about 1:100). This specificity was substantiated by the following evidence: a) preincubation of supernatants with highly purified preparations of human β2m removed all cytotoxic activity; b) radiolabeled β2m was precipitated by these supernatants in the presence of ammonium sulfate at 50% saturation; c) affinity chromatography of supernatants on a β2m-Sepharose column was able to remove the cytotoxic activity, which was then recovered by elution with 0.1 M acetic acid in 0.9% NaCl; d) monoclonal antibodies were displaced from the β2m antigen antibody complex by an avid turkey anti-human β2m antiserum (5). As expected for monoclonal antibodies, they do not give any line of precipitation with the specific antigen in the Outcherlony test. Functional studies of these antibodies (lymphocyte stimulation, block of MLC) are under way.

Clones from non-cytotoxic culture no. 34 showed, using the IIF and IRB techniques, a pattern of reactions against the HLL and PBL panels, identical to the one given by anti-β2m. In fact they were non-reactive only with Daudi. Their activity, however, was not neutralized by preincubation with purified human β2m. While further studies are in progress, it is tentatively suggested that no. 34 reacts with structures common to a variety of HLA antigens. In fact previous coating of the cells with turkey anti-β2m inhibited the uptake of no. 34 as well as the uptake of specific anti-HLA sera (6).

Monoclonal supernatants of cultures nos. 19 and 24b (a different family of clones from primary culture no. 24), which are Cd1-positive for the WT52 target, as well as nos. 5 and 6 (Cd1-negative, IIF- and IRB-positive) can be grouped together for a number of similarities: a) they react with some but not all HLL (their patterns of reactivity against the HLL panel are, however, not identical). b) No significant Cd1 is observed over non-enriched preparations of PBL. With IIF they react, however, with a small percentage of total PBL of any donor tested. By double staining with fluoresceine-labeled anti-human Ig, which detects surface immunoglobulins, and rhodamin-labeled anti-mouse Ig, it was ascertained that all supernatant-positive cells are Ig$^+$ (while not all Ig$^+$ are supernatant-positive). It seems therefore that these antibodies detect a subclass (or subclasses) of B lymphocytes or a special surface structure expressed during the B cell cycle. Double coating with nos. 19 and 24b gave, in IIF, estimates of positive cells intermediate between the values obtained with single coating and the sum of the two. The two B subpopulations (or functional phenotypes) seem therefore to be distinct but partly overlapping. However, it is noteworthy that among the panel of EB virus-transformed HLL (even after single cell cloning of the line), some are positive while others are negative. Thus the heterogeneity found among Ig$^+$ PBL is reflected on HLL. No obvious correlation was observed with the EB virus-dependent cell surface antigens. Moreover, B lymphocytes from several umbilical cord bloods showed the same frequency of positive cells as in adults, thus suggesting that the corresponding antigens are not acquired after birth, e.g., through infection. They seem also to be unrelated to HLA and β2m. Their immunochemical characterization is in progress. Binding at saturation according to Williams et al. (4) allows approximate estimates of 500,000 molecules per cell for β2m, as detected by no. 26/5 supernatant, and of 150,000 molecules for no. 19/4 determinants. Finally, no. 10 culture (and derived clones) shows an antibody specificity (not yet fully characterized) against a structure present on both B and T cells.

The distributions of these antigens in lymphatic leukemia cells as well as in tissues other than lymphoid are being studied.

The part of the work done in Torino was supported by CNR: Centro di Studi per l'immunogenetica e l'istocompatibilitá, Torino, Italy.

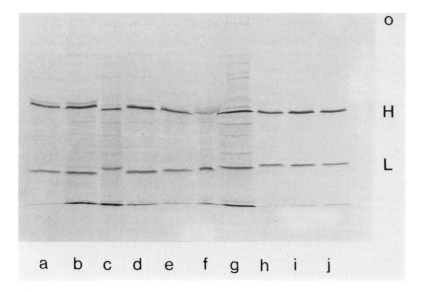

Fig. 2   Supernatants from clones: no. 19 (a, b), no. 26 (d, e, f), no. 34 (h, i, j), and from cell line X63 (c, g).

References

1. Köhler, G., Howe, S.C., Milstein, C.: Fusion between immunoglobulin secreting and non-secreting myeloma cell lines. Eur. J. Immunol. 6, 292-295 (1976).
2. Trucco, M., Galfré, G., De Marchi, M., Varetto, O., Carbonara, A.O.: "tb9", A new HLA-D Specificity defined by two homozygous typing cells. Tissue Antigens 10, 343-344 (1977).
3. Carbonara, A.O., Miggiano, V.C., De Marchi, M., Trucco, M., Galfré, G., Ceppellini, R.: Collaborative participation to the Sixth International Workshop. II: The LD data. Histocompatibility testing, Copenhagen: Munksgaard, 1975, p. 494.
4. Williams, A.F., Galfré, G., Milstein, C.: Analysis of cell surface by xenogeneic myeloma hybrid antibodies. Cell 12, 663-673 (1977).
5. Bernoco, B., Bernoco, M., Ceppellini, R., van Leeuwen, A., van Rood, J.J.: B cell antigens of the HLA system: a simple serotyping technique based on non-cytotoxic anti-β2 microglobulin reagents. Tissue Antigens 8, 253-260 (1976).
6. Poulik, M.D., Bernoco, M., Bernoco, D., Ceppellini, R.: Aggregation of HLA antigens on the lymphocyte surface induced by antiserum to β2-microglobulin. Science 182, 1352-1355 (1976).

Expression of Murine IgM, IgD and Ia Molecules on Hybrids of Murine LPS Blasts with a Syrian Hamster B Lymphoma

W.C. Raschke

A. Introduction

The production of hybrids with characteristics of mouse B lymphocytes will be of considerable interest since only a few mouse B lymphomas with high surface immunoglobulin levels and low secretion have been identified and no routine procedure is available for generating them. The ultimate goal of a B lymphoma fusion with a B lymphocyte is the production of a hybrid which represents a cloned, long-lived antigen sensitive cell with its programs for immune induction and paralysis intact and functional. Based on the success of antigen specific monoclonal antibody production in myeloma hybrids, the search for the appropriate B cell hybrid holds considerable promise. In this report, a hamster B lymphoma is shown to fuse with low frequency with mouse lymphocytes to yield hybrids with properties characteristic of mouse B lymphocytes.

B. Methods

The GD36A.Ag.1 line originated as a tumor induced by intravenous injection of Simian virus 40 into a weanling Syrian hamster (1). The tumor cell is characterized by low levels of cytoplasmic and secreted immunoglobulin (7S IgG_2) and a high level of membrane immunoglobulin (2). The cell line was obtained from Dr. G. Diamandopoulos (Harvard Medical School, Boston, Mass.) and the 8-azaguanine resistance marker introduced at the Salk Institute.

Lipopolysaccharide (LPS) blasts were obtained with only slight modification of the procedure of Andersson et al. (3). Spleen cells were seeded at 10^6 per ml in RPMI 1640 medium with the following supplements (final concentrations given): 2 mM glutamine, 10 mM HEPES, 50 μM 2-mercaptoethanol, 15 μg/ml *E. coli* O111:B4 LPS, and 10% fetal bovine serum (FBS). The culture medium was replaced every other day and cells were harvested on day 7.

MPC11TG1.7Oua.3 was obtained as the 45.6TG1.7 clone from Dr. M. Scharff (Albert Einstein College of Medicine, Bronx, New York) and the ouabain resistance marker was introduced at the Salk Institute. The line secretes IgG_{2b}.

Several fusion procedures have been used with similar results. The procedure now used routinely utilizes 40% polyethylene glycol (PEG) in Dulbecco's Modified Eagles (DME) and 10^7 cells of each cell population. To a pellet of the two cell populations 1 ml of 40% PEG is added. The cells are gently mixed at room temperature and after two minutes 1 ml of DME is added with further gentle mixing. · At two minute intervals 2 ml, 4 ml, 8 ml and 16 ml of DME are added. The cells are then pelletted at 500xg for 5 minutes, the supernatant removed, and 2 ml DME with 20% FBS added without disturbing the pellet. After 1 hour in a 37°C CO$_2$ incubator the cells are sus-pended and plated in 10 ml DME with 10% FBS overnight. The next day the cells are collected and diluted to 100 ml with DME and 10% FBS (10^5 cells/ml based on the initial number of cells of one parent). HAT ingredients are added and the cells are distributed in Falcon 3040 microwell trays at 0.2 ml per well.

Rabbit anti-mouse immunoglobulin serum was prepared by multiple in-jections with DEAE-purified mouse gamma globulin. Rabbit anti-μ and anti-δ sera were prepared by Dr. R. Coffman as described (4). Anti-κ serum was provided by Dr. W. Geckeler of this institute. Rabbit anti-GD36 immunoglobulin was the gift of Dr. J. Coe (2). Rabbit anti-mouse lymphocyte serum was obtained by repeated intra-venous injections of BALB/c thymocytes. Anti-H-2^d serum was pre-pared in C57B1/6 mice by repeated intraperitoneal injections of BALB/c spleen cells.

Specific immune precipitation of ^{125}I-labeled immunoglobulin and Ia molecules from detergent extracts of surface labeled cells were per-formed by "sandwich" precipitation with goat anti-rabbit immuno-globulin as previously described (5) or by the use of *Staphylococcus aureus* (6). The labeled molecules were solubilized from the washed precipitates or bacteria by boiling in sample buffer containing SDS and 2-mercaptoethanol and the samples were analyzed by polyacrylamide slab gel electrophoresis (5) or by the two dimensional resolving system, isoelectric focusing followed by polyacrylamide slab gel electrophoresis, described by Garrels and Gibson (7). Labeled mole-cules were visualized by autoradiography of dried gels. Quantitation of bands on X-ray film was obtained by densitometer scan.

The half-life of cellular immunoglobulin was measured by labeling 10^6 cells for 4 hr with 250 μCi ^{35}S-methionine in 1 ml of complete medium. Aliquots were removed at hourly intervals and tested for labeled immunoglobulin by immune precipitation.

To determine DNA content in parental cells and hybrids, cells were fixed, stained with mithramycin and analyzed by flow microfluori-metry (8).

C. Results

The GD36 cell line has properties characteristic of a B lymphocyte. Biosynthetic labeling of the cell reveals a low level of cell asso-ciated immunoglobulin compared with plasmacytoma lines and a negli-gible amount of immunoglobulin secreted. By surface labeling, how-ever, a large amount of the label can be demonstrated in immunoglobulin indicating that much of the cellular immunoglobulin is located on the cell surface.

The first fusion attempt with GD36A.Ag.1 with LPS blasts yielded a single hybrid, GCL2, which will be the subject of most of this paper.

The subsequent seven attempts with similar cell populations failed to yield viable hybrids. Recently, however, hybrids have been obtained with an approximate frequency of 1-2 per 10^7 input GD36A.Ag.1 cells. The reason for the poor fusion frequency of GD36A.Ag.1 in the early experiments is not known. Similar fusion experiments with MPC11 and LPS blasts have a frequency of 1-5 hybrids per 10^4 input myeloma cells. Hybrids of GD36A.Ag.1 with spleen cells have just recently been obtained and show a fusion frequency about 4 fold less than that found with LPS blasts. The GD36-spleen cell fusions have not yet been analyzed for immunoglobulin nor chromosome content.

The immunoglobulin synthesized by the hybrid GCL2.1 was analyzed by biosynthetic and surface labeling procedures. The results of surface labeling (Figure 1) show both IgM and IgD present on the surface of the hybrid.

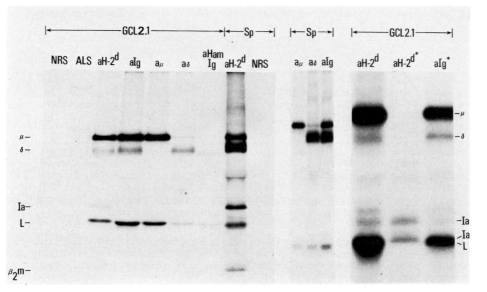

Fig. 1. Surface immunoglobulin and Ia molecules on GCL2.1 and BALB/c spleen cells labeled with ^{125}I. In the two columns labeled with an asterisk *S. aureus* was used while all other columns represent sandwich precipitations.

The percent of δ chain of the total heavy chains on the hybrid cell surface is 23% (77%μ) compared with 56% (44%μ) on spleen cells and 5% (93%μ) on LPS blasts. No mixed molecules containing μ and δ are present on the hybrid as indicated by the lack of δ chain precipitated by the anti-μ serum. The small amount of label present in the μ region precipitated by anti-δ serum in both the hybrid and spleen cells indicates either a small anti-μ component in the anti-δ serum or a true δ-like component with a higher molecular weight. The hamster γ2 immunoglobulin expression is severely depressed in the hybrid and is barely detectable (Figure 1). A 4 hr biosynthetic labeling of GCL2.1 cells yielded no detectable immunoglobulin released into the supernatant while both IgM and IgD were detectable in the cell extract (not shown). A study of the turnover rate of the cellular immuno-

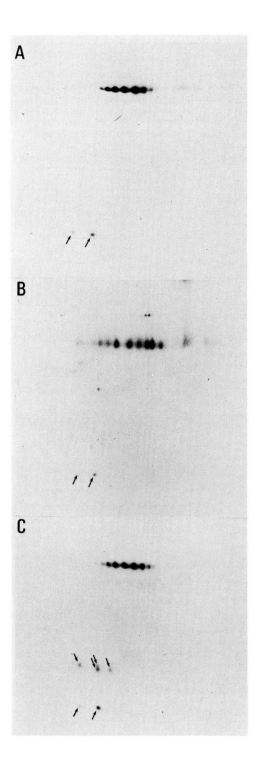

Fig. 2. Autoradiogram of two dimensional gels of IgM (A), IgD (B), and molecules recognized by anti-H-2^d plus anti-immunoglobulin sera (C). The light chains are indicated by upward arrows and the Ia molecules by downward arrows. pH 3.5-10 ampholytes were used in the first dimension giving a pH range of 4.5-7.0 (left to right) in the portion of the gels shown. The second dimension is a 10% acrylamide gel.

globulin of the GCL2.1 hybrid, while not precisely determined, revealed a half-life greater than 10 hr compared with 3-5 hr for that
of the myeloma, MPC11.

To determine whether the IgM and IgD had different light chains, the
reduced immunoglobulins were subjected to analysis by two dimensional
gels (Figure 2). Two light chains, both κ, are found associated with
both IgM and IgD. Although one light chain component is present in
larger quantity than the other, the ratio between them is approximately constant in both the IgM and IgD fractions. At present, the
possibility that one light chain is a biosynthetic or proteolytic
modification of the other cannot be excluded. The multiple spots of
IgM and IgD reflect the different degrees of sialic acid content in
these molecules.

Precipitation of an extract with ^{125}I-surface labeled GCL2.1 cells
with anti-H-2^d serum reveals the presence of 33,000 and 25,000 dalton Ia
molecules (Figure 1). Two dimensional gel analysis shows the 33,000
dalton material is composed of at least four molecular components
(Figure 2).

The stability of GCL2 was tested by recloning after approximately
three months in culture. Eight clones were selected and the surface
immunoglobulin analyzed. Figure 3 shows that two clones have markedly

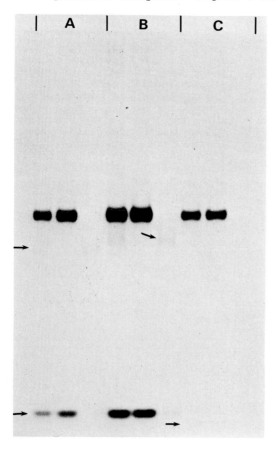

Fig. 3. Surface immuno-
globulin of three clones
of GCL2. GCL2.6 (A) gives
the pattern shown by the
majority of the clones
tested while GCL2.3 (B)
and GCL2.5 (C) show
altered products as
shown. The three wells
(left to right) for each
line represent precipi-
tations with anti-mouse
immunoglobulin serum,
anti-μ and anti-δ.

different immunoglobulins from the others as detected by polyacrylamide
slab gel electrophoresis. The δ chain of GCL2.3 is shifted to a
slightly higher apparent molecular weight while the light chain of
GCL2.5 has a lower apparent molecular weight. In subsequent recloning
the ratio of μ to δ has also been found to change with a decrease in
the relative δ content being most common.

Evidence that GCL2 is a fusion between GD36A.Ag.1 and a mouse cell,
probably a B lymphocyte, is implicit in the finding of both mouse
and hamster immunoglobulin synthesis by the cloned line. Mithramycin
staining of the DNA of the hybrid, the GD36 line and mouse thymocytes
followed by quantitation with a flow microfluorimeter reveals the
mouse thymocytes to have 40 chomosomes (the standard), the GD36A.Ag.1
line to have approximately 63 and GCL2.1 to have approximately 98.
The hybrid thus appears to be a fusion of one GD36 cell with one
mouse cell with some chromosome loss.

Table 1. Hybrids of LPS Blasts

	Total	New Ig	μ	μδ	δ	γ	L
MPC11	74	61	56	0	0	4	1
GD36	10	10	0	8	2	0	0

A comparison of the fusion products of MPC11 and GD36 with LPS blasts
is shown in Table 1. The MPC11 hybrids which show new immunoglobulin
synthesis all secrete the new immunoglobulin. Futhermore, the ratio
of IgM to other classes roughly reflects the high content of IgM found
in LPS blasts. Out of ten hybrids of LPS blasts with GD36, each shows
expression of IgD and two synthesize IgD only. The immunoglobulin
from seven of these hybrids are shown in Figure 4.

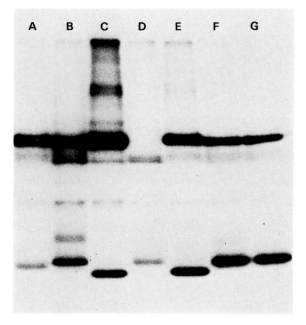

Fig. 4. Surface
immunoglobulin of
seven hybrids of
GD36A.Ag.1 with
LPS blasts.

D. Discussion

In a population of LPS blasts over 90% of the total surface immuno-globulin is IgM. Upon fusion with hamster B lymphoma, GD36, the LPS blasts yield hybrids which express IgM and IgD or IgD alone. Several possible explanations are currently being considered. (A) The GD36 cells preferentially fuse with only the subpopulation of the LPS blasts which express IgD. (B) The GD36 parent provides an element(s) which derepresses IgD expression in the mouse genome. (C) The GD36 parent provides splicing nucleases necessary for the expression of the mouse δ chain. None of the possibilities have yet been ruled out.

Mouse B lymphocytes fused with GD36 yield hybrids which appear to represent B lymphocytes halted at a stage of maturation considered to be that of the mature antigen sensitive cell; that is, IgM$^+$, IgD$^+$, and Ia$^+$. Studies are currently underway to isolate hybrids with surface immunoglobulin specific for the immunizing antigen and to determine whether such B lymphocyte hybrids will allow questions concerning the biochemical nature of induction and paralysis to be answered.

Acknowledgements. I would like to thank Diana Barritt and Scott Lindner for excellent technical assistance. This work was supported by Grant Number CA21531 awarded the National Cancer Institute to W.C.R. and Grant Number RO1 CA19754 awarded by the National Cancer Institute to Dr. Mel Cohn. Partial support of this work was pro-vided by Mrs. Jean James.

References

1. Diamandopoulos, G.T.: Induction of lymphocytic leukemia, lympho-sarcoma, reticulum cell sarcoma, and osteogenic sarcoma in the Syrian golden hamster by oncogenic DNA Simian virus 40. J. Natl. Cancer Inst. 50, 1347-1365 (1973).
2. Coe, J.E.: Immunoglobulin synthesis by an SV-40-induced hamster lymphoma. Immunology 31, 495-502 (1976).
3. Andersson, J., Coutinho, A., Lernhardt, W., Melchers, F.: Clonal growth and maturation to immunoglobulin secretion *in vitro* of every growth-inducible B lymphocyte. Cell 10, 27-34 (1977).
4. Coffman, R., Cohn, M.: The class of surface immunoglobulin on virgin and memory B lymphocytes. J. Immunol. 118, 1806-1815 (1977).
5. Dennert, G., Raschke, W.C.: Continuously proliferating allo-specific T cells, lifespan and antigen receptors. Eur. J. Immunol. 7, 352-359 (1977).
6. Kessler, S.W.: Rapid isolation of antigens from cells with a Staphylococcal protein A-antibody adsorbent: Parameters of the interaction of antibody-antigen complexes with protein A. J. Immunol. 115, 1617-1624 (1975).
7. Garrels, J.I., Gibson, W.: Identification and characterization of multiple forms of actin. Cell 9, 793-805 (1976).
8. Crissman, H.A., Tobey, R.A.: Cell-cycle analysis in 20 minutes. Science 184, 1297-1298 (1974).

Hybrid Plasmacytoma Production: Fusions with Adult Spleen Cells, Monoclonal Spleen Fragments, Neonatal Spleen Cells and Human Spleen Cells

R.H. Kennett, K.A. Denis, A.S. Tung, N.R. Klinman

A.   Introduction

Development by Milstein et al. (1,2,3) of methods for producing hybrid cell lines synthesizing antibody against antigens of choice should facilitate progress in many areas of biology and immunochemistry. In general it will make possible advances in at least two areas: 1) analysis of the mouse B-cell repertoire and 2) production of monoclonal antibodies to be used as reagents. We will describe here the methods we have used to make progress in both of these areas.

I.   Analysis of the Mouse B-Cell Repertoire

Hybrid cells making antibody against the haptenic groups 2,4-dinitro-phenol (DNP) and 2,4,6-trinitrophenol (TNP) have been made by fusion of cell lines with adult spleen cells from mice injected with hapten-carrier conjugates (1,2). Such hybrid lines may provide a source of monoclonal antibody which could be used for amino acid sequence analysis and potentially a source of nucleic acid for isolation and analysis of the DNA sequences from which the antibodies are synthe-sized (4). This provides an opportunity for production of hybrids with B cells that are from one or more of the early predominant clonotypes that have been characterized in inbred strains of mice. For example there appear to be three predominant clones which produce antibody in response to DNP in Balb/C mice during the first few days of neonatal development (5). By making hybrid cells synthesizing antibodies of these early clonotypes as well as later clonotypes one could perhaps ascertain by sequencing their proteins and nucleic acids the genetic relationship between these early and late clonotypes. This of course should bring us nearer to understanding the mechanism by which antibody diversity is generated.

Approaching such questions in this way depends on the ability to produce hybrids with B-cells expressing early clonotypes as well as those expressing clonotypes of older neonates and adult mice. To isolate hybrids representing early clonotypes it seemed that the most convenient source of cells for hybridization would be neonatal spleen cells and monoclonal spleen fragments derived from neonatal B-cells in which antibody class, specificity, and clonotype or isoelectric point (pI) had been characterized. We will present below methods for producing hybrids with B-cells from these sources.

II.   Production of Monoclonal Antibodies as Reagents

Undoubtedly hybrid cells will be a source of antibodies that will be used in many techniques in which antibodies have been useful previously.

77

In addition such antibodies will be applied to many new techniques made possible because of the higher levels of discrimination, lower backgrounds, and constantly available supplies resulting from production of antibodies in this way.

One area in which this system will be particularly useful is in the production of antibodies against human antigens. Antibodies against human antigens can be made in other species, however such antibodies often show a high background of heterospecific antibodies. Alternatively investigators have often depended on the production of antibodies by multiparous women resulting in weak antisera which are usually directed against more than a single polymorphic specificity.

We have been particularly interested in the production of antibodies against human differentiation antigens and tumor specific antigens so that these antibodies can be used for detecting the expression of antigens on mouse x human hybrids and for isolation and characterization of these antigens. We will discuss the progress made in producing hybridomas secreting antibodies against human cell surface antigens.

By extending the hybridization procedures to interspecific hybridizations one has the potential for making hybrids synthesizing antibody of other species which are directed against specific antigens. We describe here methods for making hybrids between mouse plasmacytomas and human lymphocytes including human spleenic lymphocytes from a patient with Hodgkin's disease in which spleen cells were taken from a spleen biopsy containing nodules of Hodgkin's tumor.

B. Materials and Methods

I. Plasmacytoma Cell Lines

P3/X63-Ag8 is a Balb/C plasmacytoma derived from MOPC-21 by C. Milstein (6). 45.6TG1.7 is a Balb/C line derived by M. Scharff from MPC-11 (7). Both of these lines lack the enzyme hypoxanthine phosphoribosyl transferase (HPRT) (E.C. 2.4.2.8) and are thus killed in HAT (hypoxanthine, aminopterin, thymidine) selective medium (8).

II. Media

The parental lines are maintained in stationary suspension cultures in Dulbecco's Modified Eagle's Medium (DMEM) with high glucose (4.5 g/L) supplemented with 10-20% fetal calf serum in 8-10% CO_2 at cell concentrations between 10^5-10^6/ml.

After fusion the hybrids are grown and cloned in HY medium: DMEM, 10% NCTC 109 medium (Microbiological Associates), 20% fetal calf serum, (GIBCO, Microbiological Associates), 0.2 units bovine insulin/ml (Sigma), 0.45 mM pyruvate, 1 mM oxaloacetate, and antibiotics of choice. Littlefield's concentrations of thymidine (1.6×10^{-5}M) and hypoxanthine (1×10^{-4}M) are added in addition to the concentrations of these two bases which are already present in NCTC 109. To make HY-HAT medium aminopterin is added as indicated below.

78

III. Immunizations and Preparation of Spleen Cells

The optimal time for obtaining hybrids producing antibody against a
given antigenic determinant may of course vary with the antigen and
the route and timing of immunization as well as the time between the
last boast and the removal of the spleen cells. We have found that
a primary intraperitoneal injection followed by an intravenous
injection is sufficient to produce as many hybrids as can be conven-
iently screened and grown up for analysis. We have also produced
hybrids from hyperimmunized animals which have been given injections
of hapten-carrier conjugates or tumor cells over a period of a few
months and there has been no significant effect on the number of
hybrids obtained.

Animals that were injected with cell culture supernatant were given
a primary immunization of concentrated supernatant in complete Freunds
adjuvant (1/1) followed by an intravenous injection of 0.1 ml of the
supernatant concentrate. The supernatant was obtained from a human
B-lymphoblastoid line grown for four days in RPMI 1640 with 0.5%
human AB serum. The supernatant was collected and concentrated in
an Amicon pressure cell with a filter which retains molecules of
larger than 10,100 M.W.

To produce antibodies against human cells mice were given a primary
intraperitoneal injection of 10^7 cells in 0.2 ml saline. A second
injection of $2-5 \times 10^6$ cells was given intravenously three days before
the spleen cells were removed for fusion.

Cells precoated with an anti-human antiserum before injection were
incubated at $4^{\circ}C$ in 1 ml of diluted antiserum (1/100 in RPMI 1640)
for one hour, pelleted and resuspended in 0.2 ml of medium for
injection. The anti-human serum used was a from hyperimmune mice
given repeated intraperitoneal injections of a human B-lymphoblastoid
line.

On the third day after the last intravenous injection the spleen is
removed and placed in a 60 mm petri dish in DMEM with 20% fetal calf
serum (DMEM-S20) at room temperature. We perfuse the cells from the
spleen by using two syringes with 26 gauge needles. Several holes
are poked in the spleen and medium forced gently through the spleen
with one syringe while it is held in place with the other. Care
must be taken that the dispensed cells are not drawn up into the
needles and forced through the small orifice. The spleen cells are
removed to a conical centrifuge tube and pelleted at 1000 rpm in an
IEC-MS2 centrifuge (radius = 35 cm.). The supernatant is removed;
the pellet tapped loose, and then suspended in 5 ml of ice cold
0.17 NH_4Cl for 10 minutes. Chilled DMEM-S20 is added and the cells
pelleted. They are suspended in 10 ml DMEM-S20 and the cell number
and viability determined by phase microscopy or trypan blue exclusion.
We obtain $5 - 10 \times 10^7$ spleenic lymphocytes per spleen with this
method. The cells are usually greater than 95% viable.

IV. Fusion with Adult Spleen Cells

The following is an adaptation of the method used by Gefter et al. (9)

for fusion of mouse plasmacytomas.

The 30% PEG solution is made by heating sterile PEG 1000 (Baker, MW. 950-1050) and DMEM without serum (DMEM-SO) to 41°C and mixing 3 ml of PEG with 7 ml of DMEM-SO by repeatedly drawing the mixture into a warmed pipette to assure complete mixing. The pH of the mixture should be 7.4 - 7.6. After mixing the solution is maintained at 37°C until use. Care must also be taken to maintain the pH of the DMEM-SO used for washing at 7.4 - 7.6.

It is important that the plasmacytoma line be used when it has a good viability with a high mitotic index. A culture which is growing well in mid-log phase will produce more hybrids than one that is taken as it approaches a cell density or other culture conditions in which the growth rate is reduced.

The cells from a single spleen are mixed with 10^7 of the plasmacytoma line and they are pelleted together in a round bottomed tube (Falcon 2001). Wash the cells once in 10 ml of DMEM-SO and remove all the medium. Loosen the pellet by tapping and add 0.2 ml of 30% PEG 1000. The cells are exposed to PEG for eight minutes. During this time the cells are centrifuged at 1000 rpm for 3-6 minutes. The longer centrifugation generally produces more hybrids. At the end of 8 minutes 5 ml of DMEM-SO is added gently so that the PEG is diluted and the pellet is suspended. Five more ml of medium are added and the cells are pelleted at 1000 rpm. The cell pellet is suspended in 30 ml of HY medium and this is distributed into 6 microplates (Linbro FB 96 TC) with 1 drop ($\sim$50 µl) per well. The next day 1 drop of HY medium with 2 X aminopterin (8 x 10^{-7}M) is added to each well. Two drops of HY medium is added 6-7 days later and clones usually appear macroscopically between 10-20 days. HAT may be added immediately after the fusion with no obvious effect on the number of hybrids obtained and the hybrids may be maintained in HAT after the first week if desired.

V.   Fusion with Neonatal Spleen Cells

A single cell suspension of neonatal spleen cells from Balb/c mice less than three days of age was depleted of red cells by a ten minute exposure to 0.17 M NH_4Cl on ice. These cells were then resuspended in DMEM-S20 at a density of 2 x 10^6/ml. An equal volume of LPS (Difco Bacto LPS W E. Coli 055:B5) in DMEM-S20 (40 µg/ml) was then added to make a final concentration of 1 x 10^6 cells/ml in 20 µg/ml of LPS. Fusions with 1 x 10^7 spleen cells and 1 x 10^6 myeloma cells were then done at various time periods after LPS stimulation according to the methods described above. The supernatants from wells showing growth after 14-21 days are assayed for IgM production as described below.

VI.   Fusion with Monoclonal Spleen Fragments

The generation of antibody producing clones in fragment culture is extensively described in numerous publications (10,11,12). Briefly, limited numbers of spleen cells (2-4 x 10^6) from mice less than three days of age are injected intravenously into lethally irradiated (1300 r.) hemocyanin primed adult mice. After 16-18 hours, the recipient spleen now containing neonatal cells is removed and chopped into approximately 50 fragments. These are individually placed into microtiter wells and

80

stimulated with DNP-hemocyanin for three days. The culture fluid is
changed every three days and collected on day nine for detection of
anti-DNP antibody production by radioimmunoassay as outlined below.

Fragments producing anti-DNP antibody are teased and perfused into a
single cell suspension with 26 gauge needles. The cells from each
fragment are then mixed with $1-2 \times 10^5$ myeloma cells in a 12 x 75 mm
sterile tube (Falcon 2003) and pelleted. After washing with serum free
medium 0.1 ml 30% PEG is added and the fusion procedure outlined above
is followed. After the final wash the cells are resuspended in 0.3 -
0.5 ml of HY medium and distributed into 5 or 6 microwells. Aminopterin
is added the next day.

The supernatants from wells showing growth after 14-21 days is assayed
for anti-DNP activity as below.

VII. Fusion with Human Lymphocytes

Human lymphocytes can be fused with the two mouse plasmacytomas mentioned
above. We have done fusions using 10^7 of the mouse lines and $10^7 - 10^8$
human lymphocytes. The fusion procedure is as indicated above. We have
found that without an added feeder layer the number of mouse x human
hybrids is relatively low but this can be increased by addition of
1000-2000 irradiated (4500 r) human fetal fibroblasts to each microwell.
They may be added the day before the fusion or added to the fusion
cells and dispensed with them.

VIII. Collection of Supernatants for Assays

After clones have appeared macroscopically the medium is changed by
removing most of the medium and adding fresh medium. This should be
done at least twice to remove residual antibody that has been made by
unfused spleen cells. After the second change the medium is allowed
to remain for at least 4 days and then collected for assays of activity
and specificity.

The problem of false positives arising from the residual antibody
synthesized by unfused spleen cells seems to be less of a problem when
the antigen is a hapten then when the antigen is a complex assortment
of determinants such as a human cell line.

IX. Assay Procedures

1. Radioimmunoassay for Anti-Hapten Antibody and for Immunoglobulin
   Chain Production

The details of this assay are described elsewhere (10). Briefly,
culture fluids are added to polyvinyl chloride microtiter plates
(Cooke Laboratory Products Div., Dynatech Laboratories Inc., Alexandria,
Virginia) precoated with DNP-BSA and blocking protein. After washing,
bound specific antibody is detected by the addition of ^{125}I labeled
anti-mouse Fab or class specific reagent. As little as 0.5 ng of

anti-DNP antibody can be detected.

For the detection of IgM production in the LPS stimulated neonatal spleen hybrids, several modifications of the above procedure are made. The polyvinyl chloride plates are initially coated with rabbit anti-mouse Fab (40 ng/well) and then blocking protein. Final detection of bound immunoglobulin from the culture fluid is made by the use of ^{125}I anti-mouse IgM.

2. Radioimmunoassay using Plasma Membranes or Proteins Bound to Filter Paper Discs

Antibodies against membrane components or soluble enzymes can be detected by covalently binding these antigens to cyanogen bromide activated filter paper discs. We prepare membranes by the method of Crumpton (29) and have used either the purified plasma membrane or the crude membrane fraction coupled to filters. The preparation of the filters and the assay procedure are described in this volume by L. Manson et al.

3. Radioimmunoassay with Cell Pellets

This is a modification of A. Williams (13) binding assay introduced to us by K. Bechtol. All antibody reagents were centrifuged to pellet aggregates that would pellet with the cells. The target cells are washed in 0.1% bovine serum albumin (BSA) in HEPES (15 mm) buffered medium or balanced salt solution. Twenty five μl of antiserum or supernatant to be tested is put into each tube and 5-10 x 10^5 cells in 50 μl added. All antisera and reagents are kept on melting ice. The cells are incubated with antibody for one hour and then 0.6 ml of 0.1% BSA is added. The tubes are centrifuged for five minutes at 2500 RPM, the supernatant is carefully removed and the wash is repeated. ^{125}I labeled anti mouse immunoglobulin (Fab or class specific) or Staph. aureus protein A (Pharmacia) (14) is diluted in 0.5% BSA to 200,000 cpm/100 μl. 100 μl of the labeled reagent is added to each tube and these are incubated on ice for one hour. The cells are again washed twice in 0.1% BSA and the pellet counted in a gamma counter. Samples are done in duplicate.

4. Cytotoxicity Assays

The details of the microcytotoxicity assay have been reported previously (15). Cell death is detected by trypan blue exclusion or by cyto-fluorochromasia. Rabbit serum is used as a complement source and when necessary is absorbed by target cells to remove heterospecific antibodies (15).

X. Passage of Clones at Early Stages

One should choose the wells that contain the desired antibody activity as soon after the fusion as possible to avoid the necessity of passaging unwanted cells. At this early stage of growth the hybrids must be watched carefully to avoid losing them due to overcrowding or attempting to grow them at a density of viable cells which is too low.

In general the cells do better at first in the microwells of 96 well plates then in the larger 16 mm or 50 mm wells. The cells grow best in distinct clumps or patches. When the initial colonies are grown to cover ~1/4 the bottom of the original well we usually transfer it to a 16 mm well or to 4-6 microwells. If the cells begin to do poorly in larger wells they can often be saved by putting them back in microwells and waiting for macroscopic clumps to become visible again. The cells should be cloned as soon as possible to avoid overgrowth by non-producers and to assure the isolation of a subline from an early stage of growth which will remain stable with respect to antibody production.

XI. Cloning in Semi-Solid Agarose

We clone the hybrids at as early a stage as possible in semi-solid agarose over a human fibroblast feeder layer (16). $5 \times 10^4 - 10^5$ fibroblasts are plated per 60 mm plate. A layer of 4.5 ml HY medium plus 0.5 ml 2.4% Seakem Agarose (Microbiological Associates) in 0.9% NaCl which is mixed at $39-40^{\circ}$C is layered over the plate of fibroblasts from which all the medium has been removed. A second 5 ml layer of the same medium-agarose mixture to which the cells to be cloned (100-1000 in ~0.1 ml) have been added is pipetted on top of the first layer. Clones appear in 10-14 days. These clones are picked with a paster pipette into 2 drops of HY medium in a microwell. These clones grow more rapidly and are less fastidious than the cells in the initial wells. We usually pick 12-20 clones of each line, check them for antibody activity and specificity and save 4 positive clones which are grown up, and samples frozen. One of these clones is then grown up to obtain antibody and sometimes passaged in pristane primed mice (17). These mice receive 0.5 ml Pristane (Pfaltz and Baur Inc., Stamford, Conn.) 2-4 weeks prior to the intraperitoneal injection of $10^4 - 10^6$ cells. We have injected the contents of a microwell into a pristane primed mouse and obtained ascites containing antibody and viable hybrids after two weeks.

XII. Freezing Clones

To assure recovery and retention of producing hybrids one should freeze several ampoules of cells from the antibody producing hybridomas. We pellet the cells from a rapidly dividing healthy culture and suspend the pellet ($5-10 \times 10^6$ cells/ml) in cold 5% Dimethylsulfoxide (DMSO) in fetal calf serum (15). The cells are put into freezing ampoules (Wheaton) in a -70°C freezer and transferred to liquid nitrogen the next day. For reconstitution cells are thawed quickly, washed free of DMSO and suspended in HY medium.

C. Results and Discussion

I. Anti-Hapten Antibodies

1. Adult spleen cells

A Balb/C mouse hyperimmunized with DNP-hemocyanin (DNP-Hy) was given a final intravenous injection and the spleen cells fused to P3/X63-Ag8. Growth appeared in 80% of the wells in 6 microplates. 170 wells were assayed for anti-DNP activity. 38 (22%) exhibited anti-DNP activity. 33 (87%) of the positive clones produced antibody of the γ class and 5 (13%) of μ class. This proportion of γ and μ is consistant with that seen in the class analysis of spleen fragment cultures derived from adult Balb/C cells (18). Several of these lines have been cloned and still produce anti-DNP antibody. One has been passaged in mice and ascites fluid containing 2-5 mg/ml of anti-DNP antibody has been obtained.

Another Balb/C mouse was given two intraperitoneal injections of NP-Hy, the first in complete Freunds adjuvant and the second in saline. The cells were fused four days after the second injection and of 78 clones tested 7 had anti-NP activity. One clone has been kept in culture for four months and appears stable.

2.   Neonatal Spleen Cells

In order to isolate hybridomas producing the anti-DNP antibody that is produced by one of the three predominant neonatal anti-DNP clonotypes it is desirable to be able to fuse neonatal cells at a time when they are expressing these clonotypes. One potentially useful method is to fuse neonatal cells and to screen hybrids for anti-DNP activity. This procedure would of course apply to the isolation of neonatal clonotypes directed against any particular antigenic determinant.

Our first three attempts to fuse neonatal spleen cells (3 day old Balb/C mice) with P3/X63-Ag8 resulted in the production of no hybrids, so we attempted a similar fusion after stimulation with LPS.

Table 1 indicates that neonatal spleen cells stimulated with LPS and then fused to either P3/X63-Ag8 or 45.6TG1.7 do produce hybrid cells. The hybrids produced were screened for IgM production because as defined in the spleen culture system neonatal clonotypes are of the IgM class. One of the IgM producers does make antibody which reacts with the hapten NP as well as DNP.

Table 1.   LPS stimulation neonatal spleen fusion

Time after stimulation		#hybrids	#IgM producing
1	hour	1	
5-6	hours	31	3
24	hours	7	1
48	hours	0	

3.   Monoclonal Spleen Fragments

Attempts to obtain hybridomas producing specific neonatal clonotypes could be facilitated if one were able to identify a clone of cells producing a desired clonotype and to fuse those cells to produce a hybridoma making the specific preselected antibody.

The results in Table 2 indicate that it is possible to fuse monoclonal spleen fragments and to derive hybridomas producing antibody with the same specificity as that detected in the spleen fragment supernatant. It is our experience that the line 45.6TG1.7 has worked well in this procedure and the line P3/X63Ag-8 has not been successful in parallel experiments. It is clear, therefore, that the fusion conditions are adequate to produce hybrids with the dispersed fragments.

Table 2. Spleen Fragment Fusions

Fragments fused	#Hybrids	#Anti-DNP
26	12	1 IgM
11	8	4 IgM
7	1	
TOTALS:44	21	5 IgM

Spleen fragments producing monoclonal anti-DNP antibody derived from neonatal B-cells fused on day 10-12 after _in vitro_ DNP stimulation.

II. Antibodies Against Human Cell Surface Antigens

Production of hybridomas making antibodies against human cell surface antigens can be approached generally in two ways: 1) The cells expressing the chosen antigen(s) can be injected and the hybrids screened for production of antibody against the specific antigen of interest, or 2) An immunization procedure which restricts the anti-human response to a selected sub-set of antigens can be used. Since there is apparently preferential hybrid formation with cells that have been stimulated with antigen, immunization protocols which restrict stimulation to a specific sub-set of cell surface antigens will decrease the number of clones derived from a fusion and increase the relative frequency of antibodies against the antigen(s) of interest. There are several methods that have been used to restrict the response against human cell surface antigens (Table 3) which can be applied to hybridoma production. We have begun to use some of these methods and will report some of the results obtained.

Table 3. Methods for restricting heterospecific responses to human
cell surface antigens

Method	Reference
Immunization with purified antigen-applicable to well characterized antigens such as HLA, β_2 microglobulin.	(19)
Immunization with cells coated with antibodies made against another human cell type.	(20)
Immunization with cell culture supernatant.	(21)
Immunization with mouse x human hybrids having only a few human chromosomes and thus expressing a limited number of human antigens.	(22)
Identification of monoclonal spleen fragments making antibody against human alloantigens. As shown here these fragments can be used for hybridoma production.	(23)
Immunization of primates with human cells. Hybridization of the primate spleen or lymph node cells may thus result in production of primate antibodies against specific human antigens.	(24)

1. Immunization with Cell Culture Supernatant

We had found previously that immunization of mice with concentrated
culture supernatant from a B-lymphoblastoid line resulted in production
of antisera reacting with human B-cells, B- and null cell leukemias,
monocytes, and 20-30% of peripheral blood lymphocytes from every
individual tested, but not with T-cells, fibroblasts, or tumor cells
from various other tissue types (21).

Cells from a C57Bl/6J mouse immunized with concentrated B-lymphoblastoid
line culture supernatant were fused with P3/X63-Ag8. Growth appeared
in 80% of the microwells within two weeks. 50% of these wells had
cytotoxic activity against a human B lymphoblastoid line, 8866 (25).
84% of these did not react with the human T cell line MOLT4 (26). We
cloned and grew 30 of the B-cell specific - lines and 5 of the lines
reacting with both the B- and T-cell lines. Figure 1 shows the binding
of the antibody in mid-log phase supernatant from one of the lines,
B1C3, which is detected with ^{125}I labelled Staph. aureus protein A.

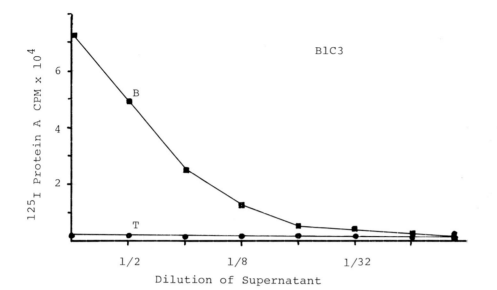

Figure 1. Binding of antibody from the supernatant of hybridoma B1C3
to a human B-cell line and T-cell line as detected by ^{125}I protein A.

We have assayed the reaction of the supernatants from these cloned
hybridomas with a B-lymphoblastoid line using ^{125}I protein A, anti
Fab, anti μ, anti γ and cytotoxicity. As expected the most general
assay reagent is anti Fab. The combination of protein A and
cytotoxicity will detect many but not all of the antibodies produced
against cell surface antigens. As reported previously some supernatants which initially have cytotoxic activity are not cytotoxic
after the hybrid is cloned but do retain binding activity. This
probably results from the original supernatant containing antibody
from two different clones which act synergistically to produce lysis
but that after cloning and production of a single type of antibody
only binding activity can be detected (3).

Most of the supernatants tested show the pattern of reactivity shown
by the antiserum from the mouse indicated above.

2.   Immunizations with Antibody Coated Cells

M. Greaves and co-workers (22) have reported that if leukemia cells
are coated with a high titered antiserum against normal human cells
and then injected into a rabbit that an antiserum can be produced
which reacts specifically with the leukemia cells and contains little
background against normal human cells. We have repeated the procedure
using mice with the same results.

If as this indicates one can restrict responses to antigens that have not been reacted with antibody then injecting such coated cells into mice before hybridoma production should reduce the number of clones that must be screened in order to obtain a clone making antibody against a specific type of antigen.

In one fusion experiment we coated a B-lymphoblastoid line GM1530 from a cystic fibrosis patient (Institute for Medical Research, Camden, N.J.) with a hyperimmune antiserum against the B-lymphoblastoid line 8866. One immunization of 10^7 coated GM1530 was followed by an intravenous injection of the cells not coated with antiserum. Hybrids appeared in every well (576) of 6 microplates. When 200 of these were screened only one reacted with GM1530 and not 8866.

In a second experiment the mouse was given both injections with anti-serum coated cells and the number of myeloma cells was reduced from 10^7 to 5×10^6. This fusion resulted in 40 clones in 576 wells and 15 of these react with GM1530 and not 8866. Whether these are against polymorphic differences between the two cell lines such as HLA or whether they react with GM1530 because it is a cystic fibrosis cell line is now being determined.

Similar experiments with acute lymphocytic leukemia (ALL) cells and chronic lymphocytic leukemia (CLL) cells have yielded hybridomas making antibodies that react specifically with the leukemia cells. Whether they react only with the injected cells or also with leukemias of the same or different types remains to be seen.

III. Hybrids between Mouse Myelomas and Human Lymphocytes

Using the indicated fusion procedure we have hybridized P3/X63-Ag8 with human peripheral blood lymphocytes, human CLL cells, and human spleen cells obtained from a Hodgkin's disease biopsy. These hybrids can be used for screening antibodies against the corresponding types of human cells as indicated above and can also be used both as a source of these antigens for immunoprecipitation and for determining which chromosomes contain the genes for the antigens detected (22,27). Many of these hybrids produce human immunoglobulin chains (28) and are thus potential sources of human antibody directed against specific antigens.

We have also obtained, by the same methods, mouse x monkey spleen cell hybrids. The monkeys were immunized with human leukemia cells and the hybrids are thus a potential source of either anti-leukemia, specific or anti-HLA antibodies.

D. Conclusions

It is clear that the production of hybridomas for analysis of the mouse B cell repertoire and as a source of antibodies against human cell surface antigens is pratical and should be exploited. It is possible to obtain hybrids by fusing myeloma lines with adult spleen cells, neonatal cells or monoclonal spleen fragments.

In order to obtain antibody against particular antigens that are present in a complex collection of antigens it is helpful to restrict the response and to set up the cultures so that there are not too many wells containing multiple clones which in effect makes it difficult to determine specificity.

The assay of choice, depends, of course, on the projected use of the antibody. In the case of cell surface antigens the combination of cytotoxicity and ^{125}I protein A binding detects most but not all of the antibodies detected by an anti-Fab reagent.

In order to obtain stable clones it is important to clone at as early a stage as possible. When this has been done we have not failed to obtain clones which remain stable. In the one case where the activity of a clone was lost we then took another clone derived from the same original well and this clone has grown in culture for several months, been passaged through mice, and is still producing antibody.

Table 4 shows the various hybridomas that have been produced in our laboratories and the collaborators involved in their production and analysis. Many of the producing clones have been stable for more than six months and by this time there are several ampoules of cells from early stages that have been frozen for future use by ourselves and other investigators who will find these antibodies useful.

Table 4. Fusions with Mouse Myelomas from which we have Isolated Cloned Antibody Producing Lines

Antigen	Collaborators
DNP Adult spleen cells NP Neonatal spleen cells + LPS PC Monoclonal spleen fragments GLA5	K. Denis, N. Klinman, A. Tung P. Stashenko, C. Merryman
Mouse Testes Mouse Lymphoma L5178Y Ag-B Human Alkaline Phosphatase Ricin Human Neuroblastoma Human ALL (Null) Human CLL (B) Human B Cell (concentrated culture supernatant)	K. Bechtol, S. Brown L. Manson D. Gasser C. Slaughter, M.Cancro R. Polin F. Gilbert M. Greaves P. Nowell

We would like to thank L. Sherman for introducing Scharff's PEG fusion technique to us and H. Davis and J. Haas for expert technical assistance. R. Kennett is supported by grants from USPHS, CA 18930, NSF PCM76-82997, and the Cystic Fibrosis Foundation. N. Klinman is supported by USPHS, AI08778.

References

1. Kohler, G., Milstein, C.: Derivation of specific antibody-producing tissue culture and tumor lines by cell fusion. Eur. J. Immunol. 6, 511-519 (1976).
2. Kohler, G. and Milstein, C.: Continuous cultures of fused cells secreting antibody of predefined specificity. Nature 256, 495-497 (1975).
3. Galfre, G., Howe, S.C., Milstein, C., Butcher, G.W., Howard, J.C.: Antibodies to major histocompatibility antigens produced by hybrid cell lines. Nature 266, 550-552 (1977).
4. Toneqawa, S., Brack, C., Hozumi, N., Matthyssens, G., Schuller, R.: Dynamics of immunoglobulin genes. Immunological Rev. 36, 73-94 (1977).
5. Klinman, N.R., Sigal, W.H., Metcalf, E.S., Pierce, S.K., Gearhart, P.J.: The interplay of evolution and environment in B-cell diversification. Cold Spring Harbor Symp. Quant. Biol. 41, 165-173 (1977).
6. Svasti, J., Milstein, C.: The complete amino acid sequence of a mouse κ light chain. Biochem. J. 128, 427-444 (1972).
7. Margulies, D.H., Kuehl, N.M., Scharff, M.D.: Somatic cell hybridization of mouse myeloma cells. Cell 8, 405-415 (1976).
8. Littlefield, J.W.: Selection of hybrids from matings of fibroblasts in vitro and their presumed recombinants. Science 145, 709-710 (1964).
9. Gefter, M.L., Margulies, D.H., Scharff, M.D.: A simple method for polyethylene glycol promoted hybridization of mouse myeloma cells. Somatic Cell Genetics 3, 231-236 (1977).
10. Klinman, N.R.: The mechanism of antigenic stimulation of primary and secondary clonal precursor cells. J. Exp. Med. 136, 241-260 (1972).
11. Klinman, N.R., Press, J.L.: The B cell specificity repertoire: Its relationship to definable subpopulations. Transplant. Rev. 24, 49-83 (1975).
12. Klinman, N.R., Metcalf, E.S., Sigal, N.H.: The functional analysis of B cell muturation at the clonotype level. In: Development of Host Defenses. Dayton, D. (ed.), New York: Academic Press, 1976, pp. 93-106.
13. Williams, A.F.: Differentiation antigens of the lymphocyte cell surface. Contemp. Topics Mol. Immunol. 6, 83-116 (1977).
14. Welsh, K.I., Dorval, G., Wigzell, H.: Rapid quantitation of membrane antigens. Nature 254, 67-69 (1975).
15. Kennett, R.H., Fairbrother, T., Hampshire, B., Bodmer, N.F.: The specificity of naturally ocurring heterophile antibodies in normal rebbit serum. Tissue Antigens 8, 21-28 (1976).
16. Sato, K., Slesinski, R., Littlefield, J.: Chemical mutagenesis at the phosphoribosyltransferase locus in cultured human lymphoblasts Proc. Natl. Acad. Sci. 69, 1244-1248 (1972).
17. Potter, M.: Immunoglobulin-producing tumors and myeloma proteins of mice. Physiol. Rev. 52, 631-719 (1972).
18. Press, J.L., Klinman, N.R.: Frequency of hapten-specific B cells in neonatal and adult murine spleens. Europ. J. Immunol. 4, 155-159 (1974).
19. Goodfellow, P., Barnstable, C., Jones, E., Bodmer, W., Crumpton, M., Snary, D.: Production of specific antisera to human B lymphocytes. Tissue Antigens 7, 105-117 (1976).
20. Brown, G., Capellaro, D., Greaves, M.: J. Natl. Canc. Instit. 55, 1281-1289 (1975).

21. Kennett, R.: Unpublished observations, in preparation.
22. Buck, D., Bodmer, W.F.: The human species antigen on chromosome 11. Birth Defects: Original Article Series 11, 87-89 (1975).
23. Lampson, L., Levy, R., Grumet, F.C., Ness, D., Pious, D.: Production of in vitro murine antibody to a human histocompatibility alloantigen. Nature 271, 461-462 (1978).
24. Metzgar, R.S., Mohanakumar, T., Miller, D.S.: Antigens specific for human lymphocytic and myeloid leukemia cells: Detection by nonhuman primate antiserums. Science 178, 986-988 (1972).
25. Pious, D., Bodmer, J., Bodmer, W.F.: Antigenic expression and cross reactions in HLA variants of lymphoid cell lines. Tissue Antigens 4, 247-256 (1974).
26. Minowada, J., Ohnuma, T., Moore, G.E.: Brief communication rosette-forming human lymphoid cell lines I. Establishment and evidence for origin of thymus-derived lymphocytes. J. Nat. Cancer Inst. 49, 891-895 (1972).
27. McKusick, V.A., Ruddle, F.H.: The status of the gene map of the human chromosomes. Science 196, 390-405 (1977).
28. Schwaber, J.: Human lymphocyte-mouse myeloma somatic cell hybrids: selective hybrid formation. Somatic Cell Genetics 3, 295-302, (1977).
29. Crumpton, M.J., Snary, D.: Preparation and properties of lymphocyte plasma membranes. Contemp. Topics in Mol. Immunol. 3, 27 (1974).

Fusion of in vitro Immunized Lymphoid Cells with X63Ag8

H. Hengartner, A.L. Luzzati, M. Schreier

A. Introduction

Most hybridomas derived from fusion of myeloma cells with immunized
or hyperimmune spleen cells do not secrete antibody of the expected
specificity. This is related to the fact that only a small fraction
of the spleen cells which are able to fuse with the myeloma cells
secrete specific antibodies (1). The fraction of specific PFC in a
spleen immunized with SRBC is of the order of 0.1%. By appropriate
in vitro immunization this low fraction can be increased by one to
two orders of magnitude. This enrichment is due to both selective
survival of antigen stimulated cells in culture and escape from *in
vivo* regulatory mechanisms.

We fused *in vitro* immunized mouse spleen cells at different times
after initiation of *in vitro* culture with mouse myeloma cells X63Ag8
asking the following questions. a) Do *in vitro* immunized spleen
cells fuse at all? b) Do some cells fuse preferentially? c) Does the
higher fraction of specific PFC after *in vitro* stimulation give rise
to a higher fraction of specific hybridomas?

An extensive collection of myeloma cell lines exist only in rodents.
We have fused human and frog lymphocytes with mouse myeloma in order
to see whether such hybridomas would secrete non-murine immunoglobulins or exhibit other non-murine lymphoid markers.

B. Methods

I. *in vitro* immunization of murine spleen cells

Two modifications of the culture system of Mishell and Dutton were
used as described in detail elsewhere (2,3). System A contains a
slightly modified MEM, supplemented with 10% selected FBS, penicillin, streptomycin and L-glutamine. Hydrocortisone acetate (Nutrit-
ional Biochem. Corp.) (10^{-8}M) was added as indicated in the Tables.
System B contains RPMI 1640 medium, supplemented with 20% selected
FBS, 5×10^{-5}M, 2-mercaptoethanol, L-glutamine, penicillin and
streptomycin.

Normal or immune spleen cells were cultured in tissue culture flasks
(Falcon No. 3013) containing 5 ml medium, 2.5 to 5×10^{7} spleen
cells and 5×10^{7} SRBC. Cultures were incubated with gentle agitation in an atmosphere of 10% CO_2, 8% O_2 and 82% N_2.

II. *in vitro* immunization of human PBL

9×10^5 human PBL were cultured in a volume of 0.2 ml in Falcon II (3040)   plates. The culture medium, a modification of MEM, was supplemented with supernatant of the marmoset cell line B95/8 as the source of EBV. SRBC were used as antigen. The details of the method were described recently (4).

III. Cell fusion

Sendai virus induced fusion of murine spleen cells with X63Ag8 was essentially done as described by Köhler and Milstein (5). 5×10^7 spleen cells were fused with X63Ag8 in 1 ml Earles BSS containing 1000 HAU of β-propione lactone treated Sendai virus (6). After fusion, the cells were washed twice in culture medium (DMEM supplemented with 15 or 20% horse serum, 5×10^{-5} 2 ME, 1 mM L-glutamine, penicillin and streptomycin) and distributed in 48 1 ml aliquots into tissue culture wells (Costar 3524). Mouse peritoneal macrophages (10^5 cells in 0.1 ml medium) were added as indicated in the Results. 24 hr after fusion, 1 ml HAT supplemented medium was added to each well. On each of the following three days, later in 2-3 day intervals, 1 ml medium was removed and replaced with fresh medium.

IV. PEG induced fusion

All interspecies hybridizations described in this paper were done with PEG 1500 according to the procedure of Galfre et al. (7). 10^5 peritoneal macrophages were added to each culture well.

V. Preparation of peritoneal macrophages

Peritoneal macrophages are recovered from the peritoneal cavity of normal BALB/c mice as previously described (2). Briefly, the peritoneal cavity was flushed with 4-5 ml cold 0.34 M sucrose and the fluid was withdrawn with a syringe fitted to a 18G needle. After centrifugation at 600 g for 10 min, the pellet was suspended in culture medium. The recovery of macrophages is about 2 to 3×10^6 per mouse.

VI. Tranformation of human PBL with EBV, establishment and cloning of human B-cell lines

After 10 days of *in vitro* culture in the presence of antigen and EBV (see above), the human lymphoid cells of two 0.2 ml culture wells were harvested and transfered to one well of a Costar tissue culture plate (Costar No. 3524). These cultures contained a confluent feeder layer of 3300 R X-irradiated normal human fibroblasts. The original volume of 1 ml was increased to 2 ml after one week. The cultures were fed every 3 to 4 days by replacing 1 ml culture supernatant with fresh RPMI 1640 medium, supplemented with 10% FBS, L-glutamine, penicillin and streptomycin. After two to three weeks, multiple clusters of proliferating cells appeared in all culture wells. The cells were grown up to large numbers in tissue culture bottles. For cloning, 1 to 20 of these cells were plated in a

volume of .2 ml in Falcon II microtiter plates. Each flat bottom
well contained an X-irradiated feeder layer of human fibroblasts.
Cultures were fed twice a week by removal and addition of 0.1 ml
medium. Cultures containing clusters of viable cells were transferred to 1 ml cultures and finally into culture bottles.

VII. Analysis of secreted products by SDS-PAGE

2×10^5 hybridoma cells or 2×10^5 transformed human PBL were cultured
overnight in .2 ml medium containing ^{14}C-leucine. The supernatant
was analysed by 10% SDS-PAGE under reducing conditions as described
by Cowan et al. (8).

B. Results

I. Hybridization of *in vitro* immunized murine spleen cells with the mouse myeloma cell line X63Ag8

Table 1 shows that murine spleen cells cultured for 48 hr and 96 hr
in the presence of SRBC fuse well with myeloma cells and that the
class distribution of secreted immunoglobulin chains of the established hybridomas is very simlar in both cases. The fusion of 10^7
myeloma cells with 5×10^7 spleen cells containing 25 PFC/10^6 cells
(after 48 hr of *in vitro* culture) did not yield any specific hybrids
whereas fusion with 5×10^7 cells containing 15000 PFC/10^6 cells (after 96 hr of *in vitro* culture) gave 20 specific hybridomas. This is
almost one order of magnitude higher than what we usually obtained
in experiments using *in vivo* immunized spleen cells.

Table 1. Immunization *in vitro*: 10^7 normal spleen cells/ml C57Bl/6J
+ 10^7 SRBC/ml in RPMI 1640, 20% supportive FBS, 5.10^{-5}M ME, P/S, Gln

Exp	*in vitro* immunization			Fusion					
	Time in culture w. SRBC	Recovery of viable cells	PFC/10^6 cells	Positive wells	Spec. hybr.	P3 only	Secretion products		
							P3+		
							μ	γ	L
R2	48 hr	45%	25	36/48	0	15/32	15/32	0/32	2/32
R4	96 hr	27%	15000 (1.5%)	48/48	20	24/42	18/42	0/42	0/32

These results establish that *in vitro* immunized murine spleen cells
fuse at least as efficiently as freshly prepared murine spleen cells
and that the higher fraction of specific PFC elicited *in vitro* is
expressed in a higher number of specific antibody secreting hybridomas. *in vitro* immunization gives rise mostly to direct PFC under
standard culture conditions. To study the question whether IgM or
IgG secreting cells fuse preferentially, we chose *in vitro* culture
conditions which increased indirect PFC. This is shown in Table 2.

94

Upon addition of 10^{-8}M hydrocortisone 50% of the PFC are of the indirect type as compared to 15% in the control culture (expt. T2 and T3, respectively).

Table 2. Immunization *in vitro*: 10^7 hyperimmune spleen cells (C57Bl/6J)/ml + 10^7 SRBC/ml in MEM 10% FBS, P/S, Gln

Exp	Hydro-cort. added to cult.	Recov. viable cells	PFC/10^6 cells (day 5) dir.	ind.	Pos. wells	αSRBC spec. hybrids dir.	indir.	Secretion products P3 only	P3+ μ	γ	L
T2	10^{-8}M	12%	11200 (50%)	11000 (50%)	48/48	4	8	15/31	14/31	0/31	2/31
T3	no	16%	14600 (85%)	2700 (15%)	48/48	7	3	23/42	16/42	2/42	1/42

* 10% SDS-PAGE under reducing conditions

The fusion of 5 x 10^7 cells of these cultures with 10^7 myeloma cells revealed 4 hybridomas producing direct and 8 producing indirect plaques in the first case and 7 hybridomas producing direct and 3 producing indirect plaques in the second case. These results are consistent with the hypothesis that there is no fusion preference between IgM or IgG anti-SRBC specific antibodies secreting plasma cells and the mouse myeloma cell line X63Ag8. The majority of hybridomas secrete an additional heavy chain of the μ-class as analyzed by SDS-PAGE. This is a general phenomenon observed also in hybridomas derived from fusions between *in vivo* immunized spleen cells and X63Ag8.

Since the experiments shown so far have established that the higher fraction of specific PFC, as obtained by *in vitro* immunization yields higher numbers of specific hybrids, we have attempted to further exploit this approach. To obtain a higher fraction of specific PFC it is necessary to use lower spleen cell density during *in vitro* immunization. Under these conditions we have obtained up to 10% specific PFC. Such an experiment is shown in Table 3. Spleen cells, derived from the same spleen cell pool, were immunized at 10^7 (expt. T3) and 5 x 10^6 spleen cells/ml (expt. T5). The fraction of specific PFC is about twofold higher in expt. T5. Spleen cells cultured at the higher cell concentration as described under T2 and T3 fused at high frequency, whereas by fusing spleen cells cultured at lower cell concentration (expt. T5), only 5 hybridomas, none of them showing anti-SRBC specificity, could be established. During the first week after fusion we observed no major difference in the cultures derived from the 3 different conditions. However, around day 10 after fusion, clusters of viable cells started to die off in expt. T5. The most striking difference between experiment T3 and T5 is the persistance of cell debris under the conditions of expt. T5 and the almost total lack of adherent cells in these cultures. This suggests that macrophages are required for the estab-

Table 3. *in vitro* immunization: hyperimmune spleen cells (C57Bl/6J) in MEM, 10% FBS, Gln, P/S

	in vitro immunization					Fusion	
Exp	Spleen cell concen. in culture	Recovery of viable cells	PFC/10^6 cells (day 5)		Fraction of specific PFC	Positive wells	αSRBC spec. hybrids
			direct	indir.			
T3	10^7/ml	16%	14600 (85%)	2700 (15%)	1.7%	48/48	10
T5	$5x10^6$/ml	25%	26000 (85%)	4700 (15%)	3.1%	5/48	0

lishment of somatic cell hybrids. Macrophages presumably have some beneficial effect on the hybridomas either by clearing cultures from debris or by releasing growth supportive mediators or nutrients. Very often clusters of hybridomas develop around groups of adherent cells.

II. Effect of macrophages on cloning efficiency and establishment of hybridomas

Very often it is important to clone hybridomas as early as possible after fusion. Whenever multiple clones arise in the same tissue culture well, specific hybridomas can be lost by overgrowth of non-specific clones. In these early stages of adaptation to tissue cul-ture conditions at low cell density, many hybrids die upon transfer to culture bottles or by further dilution in the culture well. These difficulties can be overcome by addition of peritoneal macro-phages. In fact, some of our hybrid cell lines have never been adapted to growth in the absence of macrophages.

The growth supportive effect of macrophages has been routinely and successfully used to cloning under limiting dilution conditions in liquid culture as well as for soft agar cloning. The cloning effic-iency under limiting dilution conditions, using one cell per ml in Costar trays or 1 cell per .2 ml in Microtiter plates, was regularly found to be 100% for many different B- and T-cell hybridomas. The number of macrophages required is 5 x 10^4 to 10^5 per ml. Similarly, cloning in 0.3% soft agar on top of a 0.5% agar feeder layer contain-ing macrophages increased the cloning efficiency by at least one or-der of magnitude. Some of our established T- and B-cell hybrids could not be cloned without macrophages.

III. Fusion of X63Ag8 with frog and human lymphoid cells

Frog spleen cells and *in vitro* immunized human peripheral blood lymphocytes (see below) were fused with the mouse myeloma cell line X63Ag8 and cultured in the absence or presence of 10^5 peritoneal macrophages per well. Both frog-mouse hybrids and human-mouse hybrids were only established in culture wells supplemented with

96

murine peritoneal cells. The human-mouse fusion will be described
in detail below. The established frog-mouse hybridomas reveal one
to three frog chromosomes, but so far none of them could be shown to
secrete frog immunoglobulin chains (Hengartner and Du Pasquier,
unpublished).

IV. Approaches to establish human specific antibody secreting cell lines

We have recently shown that human peripheral blood lymphocytes give
rise to specific antibody-secreting cells if Epstein-Barr-Virus
(EBV) and antigen are added to the culture. The response is antigen
specific (4). After 6 to 10 days of culture, 0.05 to .5% of the
recovered viable cells are specific PFC, directed against the immunizing antigen (SRBC, HRBC). We have attempted to establish permanent
cell lines secreting SRBC-specific human antibodies 1) by somatic
cell hybridization and 2) by transformation with EBV.

1. Fusion of *in vitro* immunized human PBL with mouse myeloma X63Ag8

3×10^7 human PBL, cultured for 8 to 10 days with SRBC and EBV were
fused with 6×10^6 X63Ag8 using polyethylenglycol 1500 as fusing
agent. The fused cells were distributed in 24 culture wells. In 80
to 90% of the wells hybrids grew in selective medium. To kill EBV-
transformed human PBL which arise in high frequency in these cultures,
the HAT-medium was supplemented with 5×10^{-5} M ouabain. This drug
is toxic for human, but not for mouse lymphoid cells at this concentration. None of the established human-mouse hybrids secreted
αSRBC-antibodies. Three weeks after fusion, 4 out of 49 hybridoma
cultures secreted human μ-chains in addition to the mouse myeloma
protein, as shown by analysis of radioactively labeled supernatants
of the cultures on reducing SDS-PAGE (Fig. 1) and on Ouchterlony
gel diffusion plates.

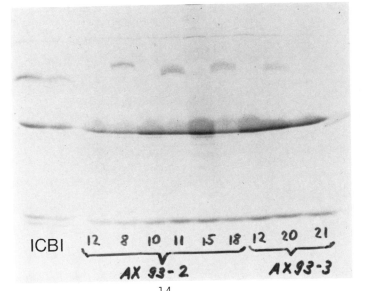

μ-chains

γ-chains

L-chains

Fig. 1. SDS-PAGE of ^{14}C-leucinine labeled secreted products of 2
mouse-mouse (1CB1) and 9 human-mouse (AX93-2 and AX93-3) B-cell
hybridomas.

After 3 months in culture and upon cloning in normal medium, the human-mouse hybrids no longer secreted human antibody and had lost all human chromosomes. The selective loss of human chromosomes is well known in human-mouse fibroblast hybrids. We can, however, not exclude that hybridomas containing human chromosomes were preferentially lost during cloning.

2.   Transformation of stimulated human PBL by EBV

The EBV required for the induction if an *in vitro* antibody response in human PBL regularly gave rise to transformed human B-cell lines. Culture wells containing 0.1 to 0.5% specific PFC on day 8-10 after initiation, were fed every 3 to 4 days, subcultured and screened for specific antibody secreting cells. Though all wells gave rise to continuous cell lines the number of PFC decreased continuously. Some cultures revealed specific PFC for a period of up to three months. The fraction of PFC, however, decreased to as low as one in 10^4. All attempts to clone SRBC-specific antibody secreting cells, by picking single cells from the center of a plaque or by limiting dilutions on irradiated human fibroblasts as a feeder layer, failed. Mouse peritoneal macrophages do not increase the low cloning efficiency of EBV-transformed PBL.

During this work we established numerous EBV-transformed human B-cell lines (Fig. 2) which were in part cloned under limiting dilution conditions. Numerous sublines derived from these transformed cultures secreted IgM with either κ or λ light chains. The amount of antibody secreted by these cells is equivalent to that of mouse myeloma cells in culture.

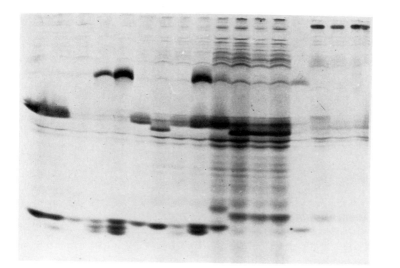

μ-chains

γ-chains

L-chains

Fig. 2.   SDS-PAGE of ^{14}C-leucine labeled supernatants of 17 EB virus transformed human cell lines.

All attempts to select HGPRT$^-$ and TK$^-$ mutants from these cell lines for further fusion experiments have failed. EBV-transformed cells survive and grow readily in 2×10^{-4}M azaguanine, 5-bromodeoxyridine thioguanine-supplemented medium, which is more than ten times the concentration required for the selection of mouse myeloma mutants.

Acknowledgments. We would like to acknowledge the expert technical assistance of C. Olsson, M. Steiner and R. Taslimi and the excellent secretarial assistance of K. Perret-Thurston.

References

1. Köhler, G. and Milstein, C.: Continuous cultures of fused cells secreting antibody of predefined specificity. Nature. 256, 495-497 (1975).
2. Schreier, M.H. and Nordin, A.A.: An evaluation of the immune response *in vitro*. Loor, F. and Roelants, G.E. (ed.). B and T cells in immune recognition. N.Y.: Wiley, 1977, p. 127-151.
3. Schreier, M.H. *in vitro* immunization of dissociated murine spleen cells. Lefkovits, I. and Pernis, B. (eds.). Research Methods in Immunology, N.Y.: Academic Press. In press.
4. Luzzati, A.L., Hengartner, H. and Schreier, M.H.: Induction of plaque-forming cells in cultured human lymphocytes by combined action of antigen and EB virus. Nature 269, 419-420 (1977).
5. Köhler, G. and Milstein, C. Derivation of specific antibody-producing tissue culture and tumor lines by cell fusion. Eur. J. Immunol. 6, 511-519 (1976).
6. Tees, R.: Preparation of Sendai virus for cell fusion. Lefkovits, I. and Pernis, B. (eds.) Research Methods in Immunology, N.Y.: Academic Press. In press.
7. Galfre, G., Howe, S.C., Milstein, C., Butcher, B.W. and Howard, J.C.: Antibodies to major histocompatibility antigens produced by hybrid cell lines. Nature 266, 550-552 (1977).
8. Cowan, N.J., Secher, D.S. and Milstein, C.: Intracellular immuno globulin chain synthesis in non-secreting variants of a mouse myeloma: Detection of inactive light-chain m-RNA. J. Mol. Biol. 90, 691-701 (1974).

Monoclonal Antibodies against Murine Cell Surface Antigens: Anti-H-2, Anti-Ia and Anti-T Cell Antibodies

G.J. Hämmerling, H. Lemke, U. Hämmerling, C. Höhmann, R. Wallich, K. Rajewsky

Introduction

During the course of antibody responses many different lymphocyte clones are activated. This results in the appearance of a mixture of different antibody species in the serum. Moreoever, different immunizations with the same antigen may yield different antibody populations. Recently, cell hybridization of myeloma cells with immune lymphocytes has been proven to be a powerful tool for the immortalization of individual antibody producing cells (1). These hybrid cell lines (hybridomas) which continuously secrete antibody of a desired specificity can be grown as tumors in mice and allow the isolation of large amounts of monoclonal antibodies.

To date we have produced hybridomas with specificity for a variety of antigens (2) such as the hapten 4-hydroxy-3-nitrophenacetyl (NP) (3), soluble proteins (chicken gamma globulin, hen egg lysozyme and the synthetic polypeptide (T,G)-A--L) and particulate antigens (Streptococcus Group A vaccine (4), sheep and horse red blood cells). A large number of other hybridomas has been established by other investigators (see this volume).

In this communication the derivation by the cell fusion technique of monoclonal antibodies against lymphocyte surface antigens is described, including anti-H-2, Thy-1, TL, Lyb-2 and xenoantibodies against murine lymphocytes. These monospecific antibodies represent standardized reagents and they will be useful for the exploration of the complexity of cell surfaces.

Methodology and Derivation of Alloantibody Secreting Hybridomas

The methodology can be summarized as follows (for further details see ref. 5): Spleen cells from immune mice were mixed with the BALB/c derived myeloma lines P3-X63-Ag8 or NS-1 (gift of Dr. C. Milstein (1)) at ratios of 10:1 to 2:1 and pelleted by centrifugation. The cells were resuspended in 1 ml polyethylene glycol (PEG 4000, 41.7% in Dulbecco's PBS containing 15% dimethylsulfoxide). After 60 seconds at 37° or 22°C the mixture was diluted by dropwise addition of 5-10 ml Dulbecco's PBS. Then the cells were centrifuged, resuspended in Littlefield's selective HAT (6) medium (RPMI 1640, 10% FCS, 2×10^{-5} M 2-mercaptoethanol plus HAT (6)) and aliquoted at a density of 0.2 - 1×10^{6} cells per well per 1.5 ml into 24 well tissue culture plates (Costar 3524). Since the myeloma cells are deficient for the enzyme hypoxanthine guanosyl phosphoribosyl transferase they are not able to

Table 1. Derivation of monoclonal alloantibodies by cell hybridization

Fusion	Recipient[a]	Cells for immun.	No. of immun.	Growth/ wells	positive[b] wells	Number of hybrid lines pos.after[c] subculture	pos.after cloning	Specificity of cloned hybrids
6	$(A/Thy-1.1 \times AKR/H-2^b)F_1$	ASL 1 T leukemia	8x	156/384	5	3	1	Thy-1.2
7	A.TH	A.TL Spleen	3x	207/384	5	3	1	Ia.7
9	AKR	B6/Ly-1.1 Thy	3x	148/336	6	4	3	Thy-1.2
10	$(C57BL/6 \times C3H/H-2^I)F_1$	I 29 B leukemia	6x	132/288	4	4	4	Lyb-2.1
12	$(C57BL/6 \times A/TL^-)F_1$	A/J Thy	6x	202/384	9	8	3	TL[d)
13	A.TH	A.TL Spleen	3x	40/288	2	2	1	Ia[d)
15	A.TL	A.TH Spleen	3x	200/288	2	2	0	
17	A.TH	A.TL Spleen	4x	580/596	4	3	ND	Ia[d)
H100	BALB/c	CBA Spleen	5x	200/240	5	3	3	$H-2^k$
H116	BALB/c	C3H Spleen	4x	257/312	8	8	2	$H-2^k$, $Ia-A^k$
B15	BALB/c	AKR Spleen	2x	125/192	1	1	1	$Ia-A^k$

(a) Cells from 2 spleens were fused to P3-X63-Ag8 cells 3-4 days after i.v. boost with 2-5 x 10^7 cells.

(b) Alloantibody activity measured by microcytotoxicity assay.

(c) Hybrid cells from positive wells were transferred into other wells and supernatants were assayed several days later for activity.

(d) Fine specificity not yet determined.

grow in HAT medium. Only hybrid cells survive, the growth of which can be observed after 10 to 20 days in many wells without further feeding.

It should be mentioned that hybrid cells seemed to grow faster when feeder cells were used. Either about 1×10^5 non-hybridized spleen cells were added per well or adherent spleen cells were used as feeders by plating 10^5 spleen cells per well and subsequent removal of non adherent cells prior to the fusion experiment.

In Table 1 several of the successful hybridization experiments are listed which resulted in hybridomas secreting alloantibodies. These were detected in culture supernatants by microcytotoxicity assays with selected rabbit complement or sometimes by cellular binding assays utilizing radiolabeled Staphylococcal protein A (5). However, it should be emphasized that there also exists a long list of fusions performed under comparable conditions which did not yield positive hybrids. Several aspects of Table 1 are worthwhile to be discussed. First, while it is apparent from previous studies (2, 3, 7) that antigen activated B cells seeem to hybridize preferentially it is not yet known which immunization schedule is optimal for the production of alloantigen specific hybridomas. The number of immunizations does not seem to be critical. Two injections of antigen appear to be sufficient (Fusion B15 and undocumented data). In general, mice were immunized at 2-3 week intervals i.p. or i.v. The last injection of cells was always intraveneously and spleen cells were harvested 2-4 days later. Preliminary evidence obtained in our laboratories indicates that fusions with spleen cells taken at later times after boost are less successful. However, it should be emphasized that it is unclear which cell type is optimal for fusion, e.g. whether B cells early after activation hybridize better than mature plasma cells.

It can be seen from Table 1 that a small fraction of the original hybrid cultures was found to produce alloantibodies. However, when these hybrids were subcultured by transfer into other dishes the loss of acvitiy was frequently observed. Possible explanations are that positive hybrids ceased to produce antibody or were overgrown by negative hybrid cells. Therefore it is important to seed the cells at a low density after fusion as a first cloning step. Even then the growth of several distinct colonies can frequently be observed in individual wells. Since it is possible that not all colonies in a positive well may produce antibody it is advisable to transfer the individual colonies into separate culture dishes. The usefulness of this procedure is demonstrated by the following examples: In fusion No. 10 a positive well consisted of 5 colonies; only 1 of them produced anti-Lyb-2 after separation. Hybrid culture H116-32 consisted of 3 colonies, 2 of which produced anti-Ia antibodies. In another case 12 out of 12 colonies were positive.

Another aspect of Table 1 is that attempts to clone positive hybrids in semi solid agar were not always successful. Again this could be due to rapid occurrence of negative variants and/or to selective growth of non producing hybridomas in agar. In many cases stability of cloned hybridomas may be a problem. For example, when hybridoma 6/68 (anti-Thy-1.2) was recloned 2 weeks after the first cloning in soft agar only 30% of subclones were positive. It should be noted in this context that Köhler and Milstein (7) have reported frequent loss of immunoglobulin chains in hybrids. On the other hand many hybridomas in our laboratory are producing large quantities of antibodies over a period of 2 years and thus seem to be stable. In any case frequent quality controls are advisable.

Serological analysis of monoclonal anti-H-2 and anti-Ia antibodies

The specificity of most monoclonal antibodies recorded in Table 1 was determined by the use of mice congenic for Thy-1, TL, H-2 and Ia, respectively. Some of the anti-H-2 (5) and anti-Ia antibodies were studied in more detail. For this purpose hybridoma cells were injected i.p. into BALB/c mice pretreated with pristane. This allowed the recovery of large quantities of ascites fluids containing several mg of antibody/ml. For typing of antibody B15-124 culture supernatant was employed while for antibody 7/24 serum from tumor bearing Swiss nu/nu mice was used. The cytotoxicity patterns of these antibodies against a variety of splenocytes from mice with different H-2 haplotypes are presented in Table 2. It can be seen that in ascites fluids respectable titers of up to $1/10^6$ can be obtained.

Since antibodies H100-5, H100-25, H100-30 and H116-22 are positive on B10.BR and negative on B10.D2 it is clear that they react with H-2^k antigens. The strong reaction with B10.A compared to the weak or absent reactions with C3H.OH and A.TL indicates that these antibodies react strongly with the K region of H-2^k. Cytotoxicity with spleen cells from other strains demonstrates that these antibodies react with public H-2 determinants. On the other hand antibodies H116-32, B15-124 and 7/24 appear to react with Ia antigens as indicated by the strong reactivity against A.TL compared to the weak or absent reaction with A.SW. Results obtained with the intra I region recombinant mice B10.A(4R) and B10.HTT suggest that H116-32 and B15-124 recognize Ia determinants controlled by the I-C region, quite like Ia.7 as suggested by the positive reaction with BALB/c.

The notion that these antibodies react with Ia determinants is further strengthened by the observation that they lyse only 50-60% of spleen cells, 95% of B cells and no T cells while the anti-H-2 antibodies lyse 100% of both B and T cells (See Fig. 1). However, the negative reaction of these anti-Ia antibodies with T cells needs to be confirmed with high titered ascites fluids.

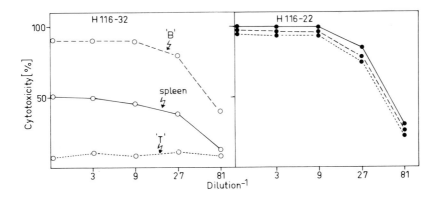

Fig. 1. Cytotoxicity of monoclonal anti-Ia and anti-H-2 antibodies against T and B cells. Culture supernatants were used for this test. CBA-T cells were obtained by nylon wool fractionation and B cells after treatment of spleen with anti-Thy-1.2 plus complement.

103

Table 2. Cytotoxic activity of alloantibodies obtained from cloned hybrid lines

| Strain | H-2 regions | | | | | | | | | Reciprocal 50% cytotoxic titer of monoclonal antibody[a] | | | | | | |
	K	A	B	J	E	C	S	G	D	H100-5	H100-27	H100-30	H116-22	H116-32	B15-124	7/24
B10.BR, CBA	k	k	k	k	k	k	k	k	k	90.000	40.000	100.000	1.000.000	20.000	240	1.200
B10.D2, BALB/c	d	d	d	d	d	d	d	d	d	<1	<1	<1	<1	<1	<1	1.200
B10.A	k	k	k	k	d	d	d	d	d	25.000	20.000	200.000	1.000.000	10.000	200	800
C3H.OH	d	d	d	d	d	d	d	d	k	<1	50[b]	50	<1	<1	<1	nd
A.TL	s	k	k	k	k	k	k	k	d	<1	<1	9.000	10.000	40.000	240	800
B10.A(4R)	k	k	b	b	b	b	b	b	b	nd	nd	nd	1.000.000	40.000	240	10
B10.HTT	s	s	s	k	k	k	k	k	d	nd	nd	nd	10.000	<1	<1	2.000
C57BL/6	b	b	b	b	b	b	b	b	b	<1	<1	400	2.500	<1[b]	<1	10
ASW,SJL	s	s	s	s	s	s	s	s	s	<1	<1	10.000	10.000	10[b]	<1	<1
B10.RIII	r	r	r	r	r	r	r	r	r	25.000	100.000	400	nd	nd	nd	nd
SWR	q	q	q	q	q	q	q	q	q	4.000	<1	50	nd	nd	nd	nd
A.CA	f	f	f	f	f	f	f	f	f	<1	<1	<1	nd	nd	nd	nd
Suggested specificity or subregion										H-2.11	H-2.25	H-2.5	H-2K	I-A	I-A	Ia.7
H and L chain class										Y2a, κ	Y2a, κ	Y2b, κ	nd	nd	nd	nd

(a) Ascites fluids of tumor bearing mice were used with the exception of B15-124 which represents culture supernatants and 7/24 which represents serum from tumor bearing Swiss nu/nu.

(b) Although cytotoxicity is low, considerable antibody activity was observed in a protein A binding test

nd = not determined

Since the hybridoma antibodies are monoclonal and therefore monospecific they define individual H-2 or Ia determinants, respectively. Comparison of the reactivity patterns depicted in Table 2 with the known H-2 and Ia specificities which were established with conventional antisera shows that the determinants detected by the monoclonal antibodies correlate largely although not in a perfect manner with known specificities (5). Thus, H100-5 appears to react with determinants H-2.11, H100-27 with H-2.25, H100-30 with H-2.5 and 7/24 with Ia.7. For correlation of the other antibodies more serological information is required. It should be noted that the strain distribution patterns (Table 2) are solely based on cytotoxic assays which may not be suitable for all target cells. Indeed, recent studies by H. Lemke (unpublished) indicate that in some instances considerable binding activity of monoclonal antibodies can be observed in a radiolabeled protein A assay although no or only weak cytolytic activity is observed (see Table 2). This phenomenon is presently under investigation.

A striking aspect of Table 2 are the enormous titer differences observed with ascites fluids on different strains. These differences which can be up to 40.000 fold as can be seen in the case of H100-30 when tested on B10.A, C3H.OH and SWR cells very likely reflect the variations of H-2 determinants on different H-2 haplotypes. In view of these titer differences the question can be raised whether it is possible to obtain monoclonal antibodies which are exclusively specific for a particular haplotype or whether these powerful reagents will always show weak crossreactions with some other H-2 determinants.

Monoclonal rat anti-mouse antibodies

Another approach to the production of hybridomas with specificity for murine lymphocyte surfaces was to immunize Wistar rats with strain DBA/2 derived T lymphoma cells and to subsequently hybridize their spleen cells with P3-X63-Ag8 cells. Hybrid cell lines of this type have also been produced by Springer et al. (this volume). The value of this approach is manifold. Monoclonal antibodies against known determinants such as Thy-1 or Lyt may be obtained, but also antibodies against unknown cell surface determinants or tumor specific antigens may be isolated.

In a first experiment 13 hybrid lines were obtained 2 of which produced antibodies which also reacted with normal C57BL/6 lymphocytes. The cytotoxic reaction of culture supernatants against purified T and B cells is presented in Fig. 2 indicating that hybridoma R3-1 reacts only with T but not with B lymphocytes. It thus parallels the reactivity of monoclonal anti-Thy-1.2 alloantibody No. 6/68. Since R3-1 reacts with T cells of all strains including AKR it seems to detect a determinant common to T cells of all strains. Blocking and cocapping experiments should reveal whether or not this determinant is located on the Thy-1 antigen. On the other hand, antibody R3-11 reacts with both B and T cells (Fig. 2). Reactivity has also been observed with the mastocytoma P815 and with P3-X63-Ag8. Thus, a determinant common to most or all murine lymphoid tissue is recognized. Both hybridomas grow as tumors in nude mice which yield antibody titers of 1:300.000 for R3-1 or 1:3.000 for R3-11, respectively.

In another experiment spleen cells from Wistar rats immune for C57BL/6 lymphocytes were used for hybridization. 136 out of 406 wells showed growth of hybrids. Of these 136 wells 51 displayed cytotoxic activity against C57BL/6 cells. These numbers demonstrate that in xenoimmuniza-

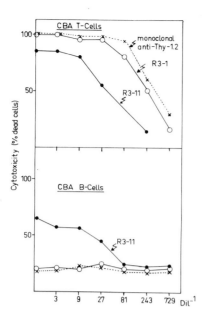

Fig. 2. Cytotoxicity of monoclonal rat anti-mouse antibodies derived from hybridizations of P3-X63-Ag8 with spleens from Wistar rats immune for the DBA/2 T lymphoma Eb. Culture supernatants were assayed against CBA splenic T and B cells (see Fig. 1). o——o Rat-mouse hybridoma R3-1; ●——● Rat-mouse hybridoma R3-11; x----x Hybridoma 6/68 (anti-Thy-1.2, see Table 1, Fusion No. 6).

tions a surprisingly high number of positive hybridomas can be isolated many of which may secrete useful antibodies.

Acknowledgements. We thank Ms. M. Palmen for expert technical help and Ms. A. Böhm for preparing the manuscript. This work was supported by the Deutsche Forschungsgemeinschaft.

1. Köhler, G., Milstein, C.: Continuous cultures of fused cells secreting antibody of predefined specificity. Nature 256, 495-497 (1975)
2. Hämmerling, G.J., Reth, M., Lemke, H., Hewitt, J., Melchers, I., Rajewsky, K.: Fusions of myelomas and T lymphomas with immunocytes. In XXV Colloquium "Protides of the Biological Fluids". Peeters, H. (ed.). Oxford: Pergamon Press, 1978, pp. 551-558
3. Reth, M., Hämmerling, G.J., Rajewsky, K.: Analysis of the repertoire of anti-NP antibodies in C57BL/6 mice by cell fusion. I. Characterization of antibody families in the primary and hyperimmune response. Eur. J. Immunol. in press
4. Rajewsky, K., von Hesberg, G., Lemke, H., Hämmerling, G.J.: The isolation of thirteen cloned hybrid cell lines secreting mouse strain A derived antibodies with specificity for group A streptococcal carbohydrate. Ann. Immunol. (Inst. Pasteur) 129 C, 389-400 (1978)
5. Lemke, H., Hämmerling, G.J., Höhmann, C., Rajewsky, K.: Hybrid cell lines secreting monoclonal antibody specific for major histocompatibility antigens of the mouse. Nature 271, 249-251 (1978)
6. Littlefield, J.W.: Selection of hybrids from matings of fibroblasts in vitro and their presumed recombinants. Science 145, 709-710 (1964)
7. Köhler, G., Milstein, C.: Derivation of specific antibody producing tissue culture and tumor lines by cell fusion. Eur. J. Immunol. 6, 511-519 (1976)

Mouse Myeloma — Spleen Cell Hybrids: Enhanced Hybridization Frequencies and Rapid Screening Procedures

L. Claflin, K. Williams

A. Introduction

One aspect of our research on the genetic control of antibody diversity has been to dissect the clonal repertoire to PC in mice, to determine the molecular heterogeneity among idiotypically related molecules within a strain and analyze structural and genetic, similarities among equivalent clonotypes from different strains (1). Recent advances in somatic cell hybridization techniques developed by Milstein (2) and Scharff (3) make this analysis possible. We here described specific methodology which we use in our laboratory to generate high yields of immunoglobulin-secreting hybrids and to simply and rapidly screen hybrids producing antibody of a given specificity.

B. Experimental Procedures and Results

The myeloma cell line used for fusion was 45.6TG1.7 (MPC 11) which was obtained from Dr. M. Scharff, Albert Einstein College of Medicine, Bronx, N.Y. Its phenotype is $\gamma 2b^+$, K^+, $6\text{-}TG^R$ (3). The cell line was grown in suspension in supplemented Dulbecco's modified Eagle's medium and 15% fetal calf serum (DME + FCS). Cloning medium consisted of DME + 20% FCS supplemented with 10% NCTC 109; HAT consisted of Cloning Medium plus 100µM hypoxanthine, 10µM aminopterin, and 30µM thymidine (3,4). Cells are cultured in a humid atmosphere of 5% CO_2 at 37°C. Spleen cells were obtained from mice immunized with 3 to 4 injections of *S. pneumoniae* (R36A) (5). Red cells were lysed with 0.17 M NH_4Cl prior to fusion.

Polyethylene glycol-mediated fusion followed a modification of the basic protocol described by Gefter, et al. (6). (a) Myeloma cells (10^7) and spleen cells (10^8) are processed separately in serum-free DME-HEPES, mixed in Falcon #2001 tubes, and centrifuged at 1200 rpm for 5 min in a Sorval RC3 fitted with an HLA-8 rotor. (b) The mixture of cells is suspended by swirling in 0.2ml of 30% PEG (mw 1000), then centrifuged at 700 rpm for 3 min. (c) After a total exposure time to 30% PEG of 8 min, 5ml of DME-HEPES is added with little suspension of the cell pellet. (d) After 1-2 minutes, the cell pellet is gently suspended by swirling intermittently for 3-4 minutes. The cells are centrifuged at 1200 rpm for 5 min, the medium is aspirated and replaced with 5ml Cloning Medium. (e) The cell pellet is left undisturbed for 5-7 min and then swirled and pipetted once. 2.5ml of cell suspension is dispensed into 2 100 x 20mm petri dishes containing 7.5ml Cloning Medium. (f) After 2 days growth, during which time the volume of medium is doubled twice, the cells are centrifuged at 1200 rpm for 10 min, resuspended in HAT medium to a density of 2-12 x 10^5 cells/ml, and 2 drops are seeded in each well of a 96 well Cluster Dish (Costar, Cambridge, Mass.). Up to 50 dishes for a potential of 2000 hybridomes can be obtained from 10^8 spleen cells. Each well is fed with 1 drop of HAT every 4-5 days. (g) Isolated clones are visible in 1 to 2 weeks, scored for anti-PC activity and selected clones are grown to mass culture in HT for detailed study. Between 20 and 40% of hybrids survive. Maintenance of close contact in

a cell pellet in steps (c) and (e) is a departure from Gefter's procedure and is critical for obtaining high yields of hybrids.

Because of the large number of 96 well dishes that can be generated in
a single experiment, simple direct screening procedures for the few
desirable hybrids are required. Solid phase radioimmunoassays for PC
binding and idiotypes were initially used which in large experiments
were time consuming and expensive. To circumvent these inconveniences
and allow processing of more fusions, putative hybrids are screened
directly in the 96 well Cluster dishes. When clones cover 10-50% of
the surface of the well, 1 drop of 0.1-0.2% of sterile PC-sheep erythrocytes (5) is added with a Pasteur pipette. During the next 1-24 hrs
wells are scored for microscopic hemagglutination or rosette formation.
Negative wells show an even, near monolayer of RBC which do not show
any hemagglutination or rosettes. Positive wells are readily scored,
and only these have to be processed further (step g). No inhibition of
growth or cessation of antibody production, directly attributable to
PC-SRBC, has been observed.

Table 1. Hybridization frequency in 45.6TG1.7 x mouse spleen fusions

Strain	Fusion no.	Input cells/ml	\% positive wells (day) 8-10	12-16	19-24	Total \% pos.	Clones/ 10^8 cells
BALB/c	F6a	12×10^5	68	18	1	87	610
	F6b	3×10^5	36	11	1	48	1860
AKR	F7a	8×10^5	0	41	21	62	360
	F7b	2×10^5	0	29	12	41	1300
A	F8	8×10^5	13	64	13	90	951

Table 1 shows the results of   typical experiments illustrating hybridization frequencies obtained at different seeding densities. Input concentrations below about 0.2×10^5/ml (not shown) yielded very few hybrids. As can be seen much larger numbers of isolated hybrids can be
obtained when the cell density is diluted to $2-4 \times 10^5$ cells/ml rather
than to $10-12 \times 10^5$ cells/ml. Many of the wells at the higher density
actually contained 1-4 clones/well. Table 1 also shows a reproducible
observation that allogeneic barriers affect hybridization. Fusions
with BALB/c spleens often yield more hybrids than spleens from other
strains and they appear at a faster rate.

Table 2. Anti-PC hybrids; effect of time after immunization[1]

	Days after injection with R36A 3	4	5
PFC/10^6 cells	80	250	810
No. positive hybrids (d20)	21	543	282
Successful hybrids (d60)	12	295	135
Anti-PC clones	4	3	0

[1]Growing hybrids were scored on day 20 (d20) and on day 60 (d60).

Successful hybrids secreting antibody to PC have been obtained with this
procedure, but their frequency is dependent on an additional parameter.
Table 2 shows that the optimal time for obtaining positive hybrids is

prior to maximum PFC production. This suggests that fusion may actually occur at the preplasma cell stage rather than with the mature plasma
cell itself.

C. Concluding Remarks

A modification of existing procedures for generating and screening somatic cell hybrids between mouse myeloma and spleen cells is described.
As many as 2000 hybrids can be obtained in a single fusion with 10^8
spleen cells, and these can be easily screened for specific antibody
production in a single day. The acquisition of sufficient numbers of
desirable hybrids for study thus becomes a practical goal, even when
the frequency of such hybrids is low.

Acknowledgments. I am grateful to M. Scharff and his entire laboratory
personnel for teaching me the science and art of somatic cell fusion.
This work was supported by grants IM-157 (American Cancer Society,
Michigan Division) and AI-12533 (USPHS).

References

1. Claflin, J.L., Rudikoff, S.: Uniformity in a clonal repertoire: A
   case for a germ line basis of antibody diversity. Cold Spring Harbor
   Symp. Quant. Biol. 41, 725-733 (1977)
2. Köhler, G., Milstein, C.: Derivation of specific antibody-producing
   tissue culture and tumor lines by cell fusion. Eur. J. Immunol. 6,
   511-519 (1976)
3. Marguiles, D.H., Kuehl, W.M., Scharff, M.D.: Somatic cell hybridization of mouse myeloma cells. Cell 8, 405-415 (1976)
4. Littlefield, J.W.: Selection of hybrids from matings of fibroblasts
   in vitro and their presumed recombinants. Science 145, 709-710 (1964)
5. Claflin, J.L., Davie, J.M.: Clonal nature of the immune response to
   phosphocholine. III. Species-specific binding characteristics of ro-
   dent anti-PC antibodies. J. Immunol 113, 1678-1684 (1974)
6. Gefter, M.L., Marguiles, D.H., Scharff, M.D.: A simple method for
   polyethylene glycol-promoted hybridization of mouse myeloma cells.
   Somatic Cell Genetics 3, 231-236 (1977)

Murine Anti-α (1→3) Dextran Antibody Production by Hybrid Cells

B. Clevinger, D. Hansburg, J. Davie

Introduction

An understanding of the mechanism of antibody diversification requires, of course, a characterization of the extent of antibody
diversity. Over the past few years, we have attempted to characterize the diversity of the mouse repertoire to $\alpha(1→3)$ dextran, in part
because it was known that the antibodies elicited to this antigen
were of limited heterogeneity and we could generate large amounts
(10-20mg/ml) of serum antibody (1). The bulk of the antibodies are
λ-bearing IgM and IgG$_3$ which, by isoelectric focusing (IEF) analysis, show considerable spectrotypic sharing between animals. In
addition, idiotypic analysis shows extensive sharing of variable
region determinants of serum antibody and one or more of the three
dextran-binding myeloma proteins (2). The combination of IEF and
idiotypic analyses defined at least eight distinct clonotypes and
suggested an even larger repertoire since it was shown that antibodies indistinguishable by IEF could sometimes differ idiotypically
and vice versa (3). Thus, it became apparent that a study of monoclonal antibody was necessary for precise characterization of the
murine anti-$\alpha(1→3)$ dextran repertoire. In this paper, we will
summarize preliminary results on the production of monoclonal anti-
dextran antibodies by somatic cell hybrids.

Generation of Anti-Dextran Hybridomas

Somatic cell hybrids were generated as described by Galfre et al.
(4) using drug resistant MPC-11 ($\gamma_{2b}\kappa$) myeloma cells obtained
through the generosity of Matthew Scharff. Normally, 10^8 spleen
cells from mice immunized according to the procedure of Hansburg
et al. (1) and which were producing several mg/ml of anti-dextran
antibodies, were fused with 10^7 myeloma cells using PEG as the
fusogenic reagent. The cells were then divided into 48 cultures,
each containing approximately 10^6 cells in 2 ml of growth media.
Two to three weeks later, all 48 cultures contained cell growth in
the presence of HAT selection media and, by using a variety of
assays, more than 90% of the cultures contained anti-dextran activity. Antibody activity was detectable by a variety of methods,
including rosette or plaque formation by individual cells, or by
direct binding of radiolabelled dextran by culture supernates. Two
assays in particular were useful in selecting cultures for cloning
and further characterization. Rosette formation between hybrid
cells and dextran-coated erythrocytes provided an estimate of the
frequency of dextran-specific cells in a culture, which therefore
predicted the ease of cloning. The second assay that was found to
be particularly useful as a screening technique was IEF in acrylamide gels (1,3). Since fusion probably resulted in several hybridomas producing the same product, IEF of culture media identified
clones producing different and interesting anti-dextran antibodies.
An example of such a screening gel using an ^{125}I-dextran overlay to
detect dextran binding activity is shown in Fig. 1.

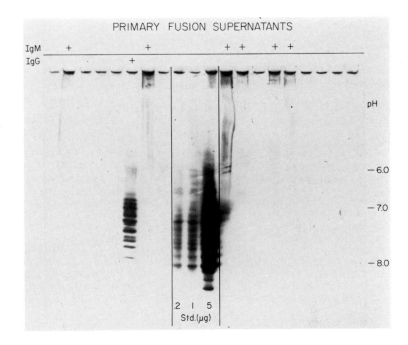

PRIMARY FUSION SUPERNATANTS

IgM   +   +   + +   + +

IgG   .   +

pH

—6.0

—7.0

—8.0

.2 I 5
Std.(µg)

Figure 1:

Screening of
culture media
for ^{125}I-dex-
tran binding
after IEF in
acrylamide
gels

The sensitivity of this technique is demonstrated by the detection
of as little as 0.2 µg of pooled 7S anti-dextran serum antibody
shown in the middle. Subsequent analysis of the products shown in
this gel demonstrated that the smears of labeled material at the top
of six indicated tracts were IgM anti-dextran antibodies while the
focused pattern in the sixth tract from the left was an IgG hybridoma product.

Cells from selected wells were cloned in soft agar over mouse embryo
fibroblast feeder layers essentially as described by Coffino et al.
(5). Clones that demonstrated anti-dextran activity were then grown
until 10^7 cells could be injected i.p. into pristane primed, histocompatible mice to induce ascites. IgM anti-dextran hybridomas
induced a vigorous ascites in 7 to 10 days which contained 7 to
16 mg/ml of dextran-precipitable protein.

Characterization of Hybridoma Protein

Since serum 19S anti-dextran antibody is frequently spectrotypically
(3) and idiotypically similar to MOPC-104 (2), a dextran binding µ,
λ myeloma protein, comparisons were made of IEF patterns and idiotypes of MOPC-104 and seven reduced and neuraminidase treated 19S
fractions of ascites protein derived from three primary cultures
(A, B, C) of a single fusion (Fig. 2). The upper panel shows the
negative image of the protein bands after salt precipitation, while
the lower panel is an autoradiograph of the same gel after an ^{125}I-
dextran overlay. It is clear from the autoradiograph that these
hybridoma proteins are spectrotypically quite different from MOPC-
104, clones derived from the same cultures are similar, and spectrotypes of clones A and B differ from C.

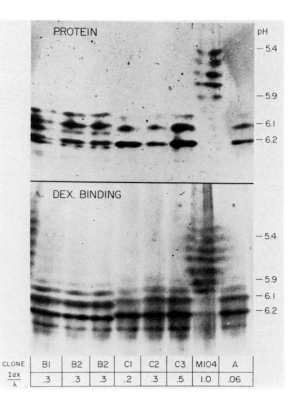

Figure 2:

IEF analysis of
19S hybridomas

CLONE	B1	B2	B2	C1	C2	C3	M104	A
$\dfrac{IdX}{\lambda}$	.3	.3	.3	.2	.3	.5	1.0	.06

The relative amount of cross-reactive idiotypic determinants (IdX)
present on 19S ascites proteins as compared to MOPC-104 was studied
using an idiotype assay as described previously (2). An expression
of IdX/λ is generated from this assay in which a value of 1 for a
protein indicates that it shares all of the cross-reactive determinants with MOPC-104; fractional values indicate the sharing of fewer
determinants. The IdX/λ values for clones B and C are similar although their spectrotypes differ. However, clones A or C which are
spectrotypically similar show significantly different expression of
cross-reactive idiotypic determinants. Thus, as has been shown previously with serum antibody, the comparison of idiotypic analysis
with IEF patterns can detect molecular heterogeneity not evident by
either technique alone.

Nearly all normally occurring anti-α(1→3) dextran antibodies possess
λ light chains (6). Hybrid cells produce the products of the myeloma and the normal cell, which in the case of the IgM anti-dextran
hybrid lines, include μ, $\gamma2b$, κ and λ. The possibility of mixed
molecules with variable binding activity being secreted is very real
and may explain the observation in Fig. 2 that the B clones produce
two IgM molecules only one of which binds dextran.

To examine this problem, SDS-PAGE was performed (Fig. 3) on sucrose
gradient fractions of the same six proteins described above.

112

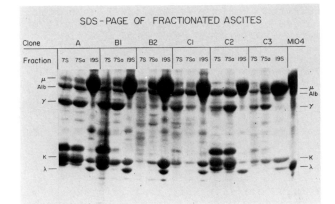

SDS-PAGE OF FRACTIONATED ASCITES

Figure 3:

Chain composition of 7S and
19S serum
fractions of
hybridoma-bearing mice

The positions of μ, γ, κ and λ chains are shown on the edges. The
19S fractions of all six clones clearly contain significant amounts
of both κ and λ chains. That the presence of κ chains is not due to
MPC-11 contamination of the 19S fraction is indicated by the absence
of γ chains. Interestingly, there is an absence of λ chains in both
the ascending (7Sa) and the descending (7S) sides of the 7S peak.
This could be due to either a limitation on this particular γ-λ
pairing, or more likely, to a higher production of κ light chains
allowing relatively few λ chains to pair with MPC-11 γ chains.

Finally, we determined if κμ pairs bind dextran. Since it was considered likely that heavy and light chain pairing was random and
that nearly all IgM pentamers would contain both λ and κ chains,
while a larger fraction of monomers would contain only κ or only λ,
binding activity was measured both at the pentamer and monomer level
as shown in Fig. 4.

Figure 4:

Dextran-binding
activity of
hybridoma
products

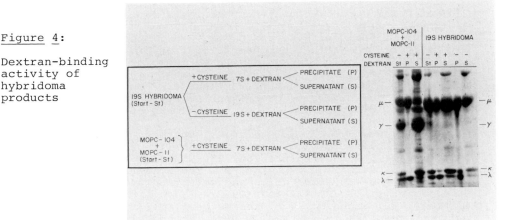

Aliquots of 19S sucrose gradient fractions were reduced to 7S sub-
units with cysteine. These and unreduced aliquots, and control
mixtures of MOPC-104 and MPC-11, were precipitated with dextran
followed by SDS-PAGE analysis of the precipitate and supernate.
Results of the control experiment indicate that the gels can distin-

113

guish the various chains in the precipitates and supernates; either
before or after reduction, the dextran precipitate contains primarily MOPC-104 μ and λ chains while the supernate contains the
majority of κ and γ chains of MPC-11. Dextran precipitable hybridoma pentamers contain significant amounts of both κ and λ chains.
The supernate contains a large amount of μ and κ chains with barely
detectable λ chains. This result is consistent with κ and λ chains
existing in the same dextran precipitable pentamer and shows that
pentamers containing predominantly κ chains bind dextran poorly, if
at all. After reduction, the amount of κ in the dextran precipitate decreases drastically which demonstrates that binding activity
is largely restricted to μλ pairs.

Summary

Hybrid cell lines are readily produced which secrete antibodies
directed to α(1→3) dextran. These antibodies are probably representative idiotypically and isotypically of those produced by the
animal at the time of fusion: 6 of 7 cloned hybrids secrete μλ,
the other secretes γλ. By a combination of IEF and idiotype analyses, the 6 IgM producing hybrids were shown to be derived from at
least 3 separate clones. Each IgM producing hybrid secreted both
myeloma and splenocyte immunoglobulins as well as mixed IgM molecules containing both κ and λ light chains. However, the μκ pairs
bound dextran poorly.

Acknowledgements: We thank Dr. Matthew Scharff for his generous
gift of drug resistant myeloma cell lines. This research was supported by U.S. Public Health Service Grants AI-11635, CA-09118,
and AI-05344 and by National Science Foundation Grant PCM78-15318.

References:

1. Hansburg, D., Briles, D.E. and Davie, J.M.: Analysis of the
   diversity of the murine response to dextran B1355. I. Generation of a large pauci-clonal response by a bacterial vaccine.
   J. Immunol. 117, 569-575 (1976)
2. Hansburg, D., Briles, D.E. and Davie, J.M.: Analysis of the
   diversity of the murine response to dextran B1355. II. Demonstration of multiple idiotypes with variable expression in
   several strains. J. Immunol. 119, 1406-1412 (1977)
3. Hansburg, D., Perlmutter, R.M., Briles, D.E. and Davie, J.M.:
   Analysis of the diversity of the murine response to dextran
   B1355. III. Idiotypic and spectrotypic correlation. Europ.
   J. Immunol. in press
4. Galfre, G., Howe, S.C., Milstein, C., Butcher, G.W. and Howard,
   J.C.: Antibodies to major histocompatibility antigens produced
   by hybrid cell lines. Nature 266, 550-552 (1977)
5. Coffino, P., Baumal, R., Laskov, R., and Scharff, M.D.:
   Cloning of mouse myeloma cells and detection of rare variants.
   J. Cell. Physiol. 79, 429-440 (1972)
6. Blomberg, B., Geckeler, W.R. and Weigert, M.: Genetics of the
   antibody response to dextran in mice. Science 177, 178-180
   (1972)

Properties of Monoclonal Antibodies to Mouse Ig Allotypes, H-2, and Ia Antigens

V.T. Oi, P.P. Jones, J.W. Goding, L.A. Herzenberg, L.A. Herzenberg

Advances in somatic cell hybridization techniques have made it possible to generate hybrid cell lines producing monospecific antibodies directed at desired antigenic determinants (1). In this paper a modification of the cell fusion procedure (2,3) was used to recover stable hybrid cell lines secreting IgG antibodies to: (a) mouse major histocompatibility complex (MHC) alloantigens (H-2K and I-A); and (b) mouse immunoglobulin (Ig) allotypes (Ig-1b, Ig-5a, and Ig-5b).

I. Production of NS-1 Derived Hybrid Cell Lines Secreting Monospecific Antibodies

The parental myeloma cell line used was the NS-1 variant of the P3 (MOPC 21) line (4). Cells of the NS-1 line do not synthesize the MOPC 21 γ_1 heavy chain; although NS-1 synthesizes the MOPC 21 κ chain, it is not secreted. NS-1 derived hybrid cell lines secreting the parental spleen cell Ig's will also secrete hybrid molecules containing the MOPC 21 κ chain.

Immune and hyperimmune mice were compared as sources of spleen cells for generating antibody-producing hybrid cell lines. This was examined with BALB/c mice immunized with C57BL/10 (Ig^b) anti-B. pertussis-pertussis complexes, an immunization that generates anti-$\overline{Ig}^b$ allotype antibodies (5). Spleen cells from hyperimmune animals three, six and eight days after their last antigen boost, and spleen cells from mice primed with a single boost antigen dose, also three, six, and eight days after a single boost were used for hybridization. Table 1 summarizes the recovery frequencies of antibody-producing hybrid cell lines from these experiments. The data suggest that primed and once-boosted spleen cells may be more effective in generating antibody-producing hybrids. The subsequent anti-spleen cell alloantigen immunizations consisted of an initial i.p. injection of $\sim 2 \times 10^7$ spleen cells per animal and a single boost also of $\sim 2 \times 10^7$ cells (Table 1).

Using polyethylene glycol (PEG) 1500 (BDH Chemicals Ltd., Poole, England) as the fusion agent, NS-1 cells from log-phase growth were fused with immune spleen cells at a 1:4 or 1:2 ratio. A total of 3×10^8 cells were treated with 1 ml of 50% PEG in serum-free RPMI-1640 medium. The cell mixture was plated in three 96-well microculture plates (Costar, Cambridge, MA) in 0.1 ml of 15% FCS-RPMI-1640 per well. Progressive HAT selection was carried out for two weeks. HAT medium, 0.1 ml, was added on day 1 and half the medium (0.1 ml) was replaced with fresh HAT medium on days 2, 3, 5, 8, 11 and 14.

During the third and fourth weeks, supernates from those microcultures containing growing hybrid cells were tested for antibody activity. Reactivity against Ig's was measured by a plate-binding assay (5). Ig adsorbed onto wells of flexible plastic microtiter plates (Cooke Lab. Prod., Alexandria, VA) was reacted with 5-20 µl of culture supernate for 1 hr at room temperature. ^{125}I-labeled anti-allotype antibodies (either a mixture of anti-Ig-1b and anti-Ig-4b or anti-Ig-1a

115

TABLE 1

PRODUCTION OF NS-1 (MYELOMA) HYBRIDS

Priming (i.p.)	Days past boost (i.p.)	No. of Expts.	Wells c̄ hybrids[*] Total wells		Antibody Initial pos. wells Total hybrids		Final pos. hybrids Total hybrids	
			No.	(%)	No.	(%)	No.	(%)
αIg[b]	Hyper-immune	3	2	187/271 (69)	2/187	(1)	0	
"	"	6	2	218/346 (63)	4/218	(2)	0	
"	"	8	1	180/180 (100)	0		0	
"	1 wk	3	2	120/288 (42)	18/120	(15)	4/120	(3)
"	"	6	1	93/182 (51)	4/93	(4)	0	
"	"	8	2	75/288 (26)	3/75	(4)	0	
αspleen cell	3 wk	3	2	576/576[+] (100)	42/576	(7)	12/576	(2)

[*] 10^6 cells/well (2×10^5 NS-1, 8×10^5 spleen cells/well)

[+] 10^6 cells/well (3.3×10^5 NS-1, 6.6×10^5 spleen cells/well)

and anti-Ig-4a) were used to detect bound antibody. Labeling with ^{125}I was done with antibodies adsorbed to immunoadsorbants to protect the combining sites, and then eluted with 0.2 M glycine-HCl (pH 2.3) containing carrier protein (5). Antibodies to spleen cell alloantigens were detected by reacting 4×10^5 spleen cells with culture supernates in microtiter wells for 1 hr. Bound antibodies were also detected with ^{125}I-anti-allotype antibodies or with ^{125}I-protein A. Screening for IgM antibodies was not done.

Microcultures producing desired antibodies were transferred into 1 ml cultures (Costar, Cambridge, MA) with 5×10^6 thymocytes as feeder cells. These cultures were expanded into flasks (6-14 days) and assessed for continued antibody production. Cells from 10 ml of high density cultures (containing $2-5 \times 10^6$ cells) were frozen in 0.5 ml of 90% FCS-10% DMSO. The anti-spleen cell hybrids were cloned by limiting dilution from the initial microculture wells. Cultures of 0.1 ml were plated with 10^6 thymocytes/well as feeder cells. These were fed with 0.1 ml of medium on days 7 and 12. At the beginning of the third week, microcultures with growing clones were tested for activity. The anti-Ig[b] hybrid cell lines were cloned from cells grown in flasks either by limiting dilution or in soft agar. It was generally observed that those hybrid cell lines that maintained antibody production in 1 ml and flask cultures had high cloning efficiencies and good recoveries of antibody-producing clones.

At least four antibody-producing clones from each hybrid cell line were expanded and frozen (6 vials, each containing $2-5 \times 10^6$ cells). A total of 24 clones were injected subcutaneously into syngeneic mice (at least five mice/clone, each receiving $2-5 \times 10^6$ cells). Twenty-two (92%) of these clones produced tumors (hybridomas) within 10-30 days. All hybridoma-bearing mice produced myeloma-like proteins and 20 of the 22 hybridomas (91%) produced the desired antibody activity. These tumors were transplantable and continued to produce antibody.

The chain composition of the antibodies produced by these hybrid cell lines was assessed by two-dimensional (2-D) polyacrylamide gel electrophoresis (PAGE) (see below).

II. Antibodies to Alloantigens of the Major Histocompatibility Complex

Two hybridizations were done with spleen cells from mice immunized with allogeneic cells with the goal of obtaining monospecific antibodies reacting with antigens controlled by the H-2 complex or with cell surface Ig's. Antibodies from one hybridization (H10), in which the donor spleen cells were from CWB mice immunized with C3H spleen cells, should react only with H-2^k or Ig^a determinants. Antibodies from the second hybridization (H11), in which the BALB/c spleen cell donor had been immunized with CKB cells, could potentially react with H-2^k or Ig^b determinants or with the products of other genes.

Antibody production by hybrid cells was determined by testing the reactivities of supernates from the initial microcultures in the cell-binding assay, using as targets spleen cells from H-2 and Ig congenic strains of mice. As shown in Table 2, the reactivities of five of the H11 and two of the H10 supernatant antibodies are against antigens linked to the MHC, while one antibody each from H10 and H11 seems to react with cell surface Ig. The 10.1 hybrid supernate reacted with cells from all four congenic strains, including CWB, the spleen cell donor. This hybrid cell line apparently is producing an autoantibody against an undefined cell surface antigen.

TABLE 2

LINKAGE ANALYSIS OF HYBRID CELL ANTIBODY REACTIVITY

Strain	H-2	Ig	BALB anti-CKB (H11) (anti-H-2^k, Ig^b, . . .)						CWB anti-C3H (H10) (anti-H-2^k, Ig^a)			
			11-1*	11-2*	11-3*	11-4†	11-5†	11-6†	10-1*	10-2*	10-3Ψ	10-4Ψ
C3H	k	a	924§	1216	2672	1615	165	74	1761	1349	nd	nd
CWB	b	b	30	64	176	83	13	567	921	71	nd	nd
CKB	k	b	1278	1236	3072	nd	nd	nd	2213	1290	1896	66
CSW	b	a	136	142	244	nd	nd	nd	1790	157	139	1409
Linkage			MHC	MHC	MHC	MHC	MHC	Ig	Auto	MHC	MHC	Ig

nd: not determined

* ^{125}I-protein A

† ^{125}I-anti-Ig-1a + anti-Ig-4a

Ψ ^{125}I-anti-Ig-1b + anti-Ig-4b

§ Counts per minute bound to 4×10^5 spleen cells

The reactivity of the seven antibodies directed against products of the MHC were mapped to regions of the complex using a series of recombinant strains of mice (Table 3). The reactivity patterns of five of these antibodies are consistent with their detecting I-A^k antigens. The remaining pair of antibodies apparently react with H-2K^k antigens. None of these antibodies detect H-2D antigens or the products of other I subregions.

117

TABLE 3

MHC MAPPING OF HYBRID CELL ANTIBODY REACTIVITY

Strain	K	A B J E C	S G D	Medium	H10		H11				
		I			10-3.6	10-2.15	11-4.1	11-5.2	11-2.12	11-1	11-3
CKB	k	k k k k k	k k k	86*	1636	891	1358	736	1390	1634	2586
A.TL	s	k k k k k	k k d	70	656	338	58	196	493	112	1166
B10.A(4R)	k	k b b b b	b b b	64	1624	766	1081	606	1280	1436	2446
B10.A(3R)	b	b b k k d	d d d	41	100	54	84	69	76	69	77
C3H.OH	d	d d d d d	d d k	141	160	154	126	223	188	170	161
Reactivity					I-A	I-A	H-2K	I-A	I-A	H-2K	I-A

*^{125}I-protein A counts per minute bound to 4×10^5 spleen cells

To confirm the specificities of these anti-MHC antibodies, the proteins they react with were analyzed biochemically. Antibodies produced by cloned cell lines were used for this analysis; evidence that the antibodies themselves are clonal products is presented below. Earlier studies have shown that 2-D PAGE of H-2 and Ia antigens produces patterns that are characteristic both of the region or subregion coding for the precipitated antigen and of the haplotype (6,7). As shown in Fig. 1, antibody from the cloned hybrid cell line 11-4.1 precipitates molecules from ^{35}S-methionine-labeled CKB extracts identical to those precipitated by an alloantiserum, (A.TL x C3H.OL)F$_1$ anti-C3H, which is directed against H-2K^k. Similarly, antibody from the cloned line 10-2.16 precipitates the same I-A^k molecules from C3H extracts as does A.TH anti-A.TL (anti-I^k) alloantiserum from B10.A(4R) (Fig. 2). It is interesting to note that all of the I-A^k molecules precipitated by the A.TH anti-A.TL serum, which probably represent three distinct gene products (P.P. Jones, D.B. Murphy and H.O. McDevitt, in preparation), are also precipitated by monospecific anti-I-A antibodies. Unless all three proteins have in common the antigenic determinant recognized by the 10-2.16 antibody and are independently precipitated, the three polypeptide chains probably are precipitated because they exist as a molecular complex on the cell surface. Anti-I-A antibodies produced by clones 10-3.6 and 11-5.2 precipitate the same molecules as 10-2.16; the other two, 11-2.12 and 11-3.25, are currently being examined.

The reactivities of the anti-H-2K and the anti-I-A antibodies were also analyzed by two-color immunofluorescence, using the fluorescence-activated cell sorter (FACS) (9). C3H spleen cells were stained with rhodamine-conjugated rabbit anti-μ antibodies, to label B lymphocytes, and with either anti-H-2K^k (11-4.1) or anti-I-A^k (10-2.16), followed by a fluorescein-conjugated rabbit anti-γ. As shown in Fig. 3-b, the 11-4.1 anti-H-2K stains most of the spleen cells, as would be expected for an anti-H-2K antibody. Fig. 3-c indicates that the μ$^-$ population contains a small population of cells that have little or no H-2K, but the vast majority of μ$^-$ (Fig. 3-c) and μ$^+$ (Fig. 3-d) cells have H-2K determinants recognized by this antibody. The other anti-H-2K antibody (11-1.23) shows a similar reactivity pattern.

The FACS analyses of C3H cells stained with rhodamine-conjugated anti-μ and 10-2.16 anti-I-A^k followed by fluorescein-conjugated anti-γ are presented in Fig. 4. The μ$^+$ peak in Fig. 4-a and the I-A$^+$ peak in

Figure 1   Figure 2

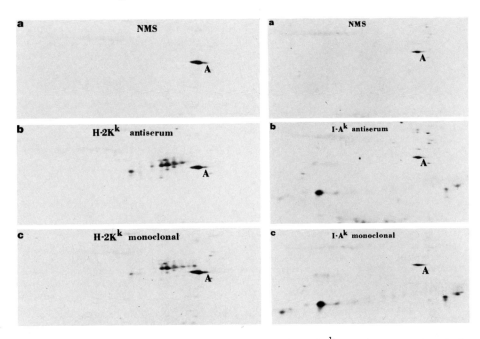

Fig. 1.   Autoradiograms of 2-D gels of H-2K^k antigens.   Proteins were precipitated from NP-40 extracts of ^{35}S-methionine labeled CKB spleen cells (6) by (a) normal mouse serum, (b) (A.TL x C3H.OL) anti-C3H, and (c) antibody from clone 11-4.1.   The first dimension separation was by non-equilibrium pH-gradient electrophoresis (8) (acidic proteins are on the right and basic proteins on the left).   The second dimension separation was by SDS PAGE (from top to bottom).   Both separations were done under reducing conditions.   Only the relevant portions of the autoradiograms are shown.   The position of actin is indicated by the letter A.

Fig. 2.   Autoradiograms of 2-D gels of I-A^k antigens.   Proteins precipitated from (a) C3H extract by normal mouse serum, (b) B10.A(4R) extract by A.TH anti-A.TL, and (c) C3H extract by antibody from clone 10-2.16 were electrophoresed as described in the legend to Fig. 1.

* * *

Fig. 4-b each contain 60-65% of the cells.   Analysis of the I-A profiles of μ^- cells (Fig. 4-c) and μ^+ cells (Fig. 4-d) indicates that nearly all the μ^+ cells also are I-A$^+$, while few of the μ^- cells have detectable I-A determinants.   Two other anti-I-A^k antibodies reacted in a similar fashion; the final two anti-I-A antibodies are currently being examined.

Several unique antigenic determinants have been described for both H-2K^k and I-A^k based on the cross-reactivities of molecules of k haplotype with those of other haplotypes as detected by alloantisera. To test whether the determinants detected with monospecific antibodies correspond to those described previously, the five anti-I-A^k and the two anti-H-2K^k antibodies were tested for reactivity against cells of eight independent H-2 haplotypes.   As seen in Table 4, three of the

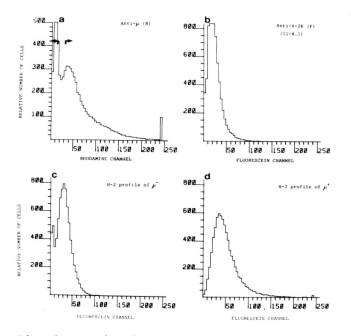

Fig. 3. Anti-H-2K and anti-μ chain immunofluorescence histograms.
C3H spleen cell stained with rhodamine (R) conjugated rabbit anti-μ
and anti-H-2K antibody from clone 11-4.1 followed by fluorescein (F)
conjugated rabbit anti-γ were analyzed with the FACS; (a) R-anti-μ
profile, (b) F-anti-H-2K profile, (c) F-anti-H-2K profile of μ⁻ cells
(cells to the left of channel 15 in panel a, indicated by the left
arrow), and (d) F-anti-H-2K profile of μ⁺ cells (cells to the right
of channel 30 in panel a, indicated by the right arrow).

* * *

anti-I-A^k antibodies react with cells of f, k, r, and s haplotypes,
but not with b, d, p or q haplotypes. This pattern of reactivity corresponds to specificity Ia.17 previously described with alloantisera
(10). The other two anti-I-A^k antibodies react only with cells of k
haplotype, consistent with specificity Ia.2, which is unique to I-A^k
(10). One of the anti-H-2K^k antibodies binds to cells of k, p, q and
r haplotype strains; the other reacts with k and perhaps f haplotype
strains. Neither of these reactivity patterns corresponds to an H-2K
antigenic specificity previously identified by alloantisera.

One additional point can be made from Table 4. These monospecific
antibodies give different levels of binding with different haplotypes.
In some cases, the antibodies seem to react more strongly with cells
of haplotypes other than H-2^k. It is possible that the same determinant may be exposed to different degrees on cells of different haplotypes. Alternatively, the antigenic determinants might be similar
but non-identical, allowing differential binding of the monospecific
antibodies to molecules of different haplotypes.

120

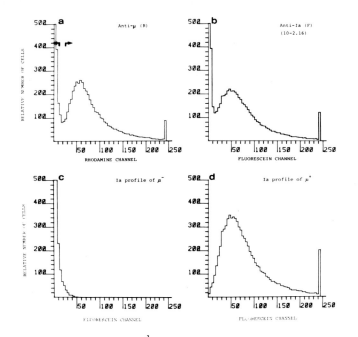

Fig. 4. Anti-I-A^k and anti-μ chain immunofluorescence histograms. C3H spleen cells stained with R-anti-μ and anti-I-A antibody from clone 10-2.16 followed by F-anti-μ were analyzed with the FACS; (a) R-anti-μ profile, (b) F-anti-I-A profile, (c) F-anti-I-A profile of μ⁻ cells (cells to the left of channel 15 in panel a, indicated by the left arrow), and (d) F-anti-I-A profile of μ⁺ cells (cells to the right of channel 25 in panel a, indicated by the right arrow).

TABLE 4

HYBRID CELL ANTIBODY REACTIVITY[*] WITH DIFFERENT H-2 HAPLOTYPES

Strain	H-2K, I-A Haplotype	Anti-I-A					Anti-H-2K	
		10-3.6	10-2.16	11-5.2	11-2.12	11-3.25	11-4.1	11-1.23
B10.D2	d	1.7	1.4	0.9	1.0	1.9	0.9	1.0
B10	b	1.4	1.3	0.8	1.0	2.1	1.0	1.1
B10.A	k	14.7	10.0	5.8	14.2	22.4	16.7	5.9
B10.M	f	22.2	13.6	1.0	1.4	36.5	1.3	3.7
B10.RIII	r	12.9	5.0	1.3	1.3	15.5	7.4	1.8
B10.S	s	13.1	5.4	1.0	1.4	18.8	1.1	1.4
B10.P	p	3.1	2.7	1.2	1.0	1.9	4.7	2.4
B10.T(6R)	q	1.6	1.6	1.1	1.5	2.4	4.4	1.5

[*]Ratio $\dfrac{\text{Specific } ^{125}\text{I-protein A cpm bound}}{\text{Control } ^{125}\text{I-protein A cpm bound}}$

Results are means for three experiments (d, p, q haplotypes) or four experiments (b, k, f, r, s haplotypes)

121

III.   Isolation and Characterization of the Products of Hybrid Cell Lines

The majority of the products of these hybrid cell lines bound staphylococcal protein A in the cell binding assay; therefore affinity chromatography on protein A-Sepharose (11) provided a simple and rapid one-step procedure for their isolation from serum or culture supernates. Yields of IgG from culture supernates were typically 10-20 μg/ml, but were occasionally as high as 50 μg/ml.

Fig. 5 shows the analysis of Ig's by 2-D PAGE. Both heavy and light chains of IgG isolated from normal mouse serum (Fig. 5-a) are heterogenous. In contrast, the pattern obtained for MOPC 21 myeloma protein (Fig. 5-b) is simpler. The presence of 2-3 light and heavy chain spots probably is due to post-translational modifications such as deamidation. These modifications result in horizontal shifts toward the acidic end of the gel. A similar analysis of Ig purified from the culture supernate of clone 10-3.6 is shown in Fig. 5-c. As expected from the contribution of the NS-1 parent, one of the two pairs of light chain spots corresponds to the position of the MOPC 21 κ chain. On the other hand, the single pair of heavy chain spots is distinct from the expected position of the MOPC 21 heavy chain, which is clearly absent.

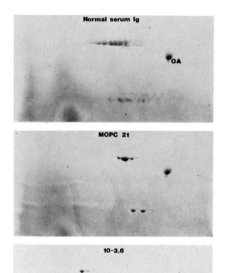

Fig. 5. 2-D PAGE analysis of light and heavy chains from (a) normal BALB/c serum Ig purified on protein A-Sepharose, (b) MOPC 21 myeloma protein, and (c) antibody from clone 10-3.6 (anti-I-A^k) purified from culture supernates on protein A-Sepharose. The gels were run as described in the legend to Fig. 1, and the proteins visualized by Coomassie Blue staining. Ovalbumin (OA), 45,000 daltons, was added as a molecular weight marker.

Schematic representations of the 2-D PAGE analyses of the anti-H-2K and I-A antibodies are shown in Fig. 6. All of these proteins appear to be clonal products. Also shown in Fig. 6 are the chain compositions of the isolated proteins as determined by Ouchterlony gel diffusion using class-specific antisera.

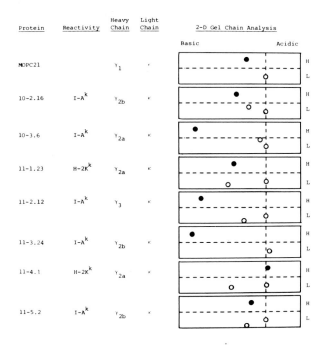

Protein	Reactivity	Heavy Chain	Light Chain	2-D Gel Chain Analysis
MOPC21		γ_1		
10-2.16	I-A^k	γ_{2b}	κ	
10-3.6	I-A^k	γ_{2a}	κ	
11-1.23	H-2K^k	γ_{2a}	κ	
11-2.12	I-A^k	γ_3	κ	
11-3.24	I-A^k	γ_{2b}	κ	
11-4.1	H-2K^k	γ_{2a}	κ	
11-5.2	I-A^k	γ_{2b}	κ	

Fig. 6. Heavy and light chain composition of anti-H-2K^k and anti-I-A^k antibodies from hybrid cell lines. The mobilities of heavy and light chains from MOPC 21 protein and hybrid cell antibodies in 2-D gels (see Fig. 5) are schematized on the right. The spots represent the most basic (unmodified) molecular species of each chain. The vertical dotted line marks the position of the MOPC 21 (NS-1) light chain spot. The heavy and light chain isotypes of each antibody, determined by Ouchterlony analysis, are also given.

IV. Antibodies to Ig Allotypes

A. Ig-5a and Ig-5b δ Heavy Chain Allotypes

In the analysis of hybrids derived from mice immunized with allogeneic spleen cells, two cell lines were found to produce antibody that reacts

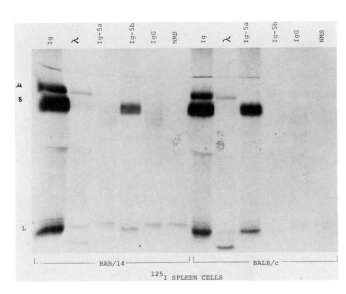

Fig. 7. Autoradiogram of SDS PAGE analysis of immunoprecipitates of NP-40 solubilized BALB/c (Ig^a) and BAB/14 (Ig^b) spleen cells, labeled with ^{125}I by the lactoperoxidase technique. Reactivities of antibodies used for immunoprecipitation are indicated at the top of the gel. Anti-Ig-5a is from clone 10-4.22 and anti-Ig-5b from clone 11-6.3. Fixed S. aureus bacteria were used to bring down the antigen-antibody complexes. Since the 11-6.3 antibody (IgG$_1$) does not bind to S. aureus protein A, a small amount of rabbit anti-mouse γ chain was added to facilitate the binding of the 11-6.3 complexes to the bacteria.

with cell surface antigens controlled by genes linked to the Ig heavy chain gene complex (Table 2). These antigens were likely to be δ or μ allotypic determinants. The proteins precipitated by these antibodies from NP-40 extracts of ^{125}I-labeled BALB/c (Ig[a]) and BAB/14 (Ig[b]) spleen cells were analyzed by SDS PAGE (see Fig. 7). Both antibodies precipitate molecules containing chains with mobilities of light chains and δ heavy chains. However, only molecules of the appropriate heavy chain allotype are precipitated.

The anti-δ antibodies also were analyzed by two-color fluorescence, as described above. Approximately 60% of C3H spleen cells are μ+ (Fig. 8-a) and a similar number are stained by the 10-4.22 anti-δ (Fig. 8-b). The great majority of μ+ cells are also stained by 10-4.22 antibody (Fig. 8-d), while virutally no μ− cells are stained (Fig. 8-c). Analysis of clone 11-6.3 antibody showed identical results when tested on BAB/14 cells. These results are consistent with clones 10-4.22 and 11-6.3 recognizing allotypes of murine δ chain. Clone 10-4.22 recognizes the Ig-5a allele and 11-6.3, the Ig-5b allele.

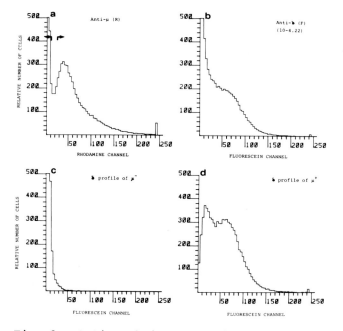

Fig. 8. Anti-δ chain and anti-μ chain immunofluorescence histograms. C3H spleen cells were stained with R-anti-μ and anti-Ig-5a antibody from clone 10-4.22 followed by F-anti-γ was analyzed with the FACS. (a) R-anti-μ profile, (b) F-anti-δ profile, (c) F-anti-δ profile of μ− cells (cells to the left of channel 10 in panel a, indicated by the left arrow), and (d) F-anti-δ profile of μ+ cells (cells to the right of channel 30 in panel a, indicated by the right arrow).

124

TABLE 5

GENETIC ANALYSIS OF ANTI-IgD ANTIBODIES

	Ig Haplotype					
	BALB/c	B10	DBA/2	AKR	A/J	CE
	a	b	c	d	e	f
NMS	62	42	50	39	66	53
11-6.3[*]	102	284	44	44	97	92
NMS	48	23	56	52	32	43
10-4.22[+]	272	57	349	186	27	327

Data presented as counts/min. Mean of 4 replicates. Standard error was typically 10% of mean.

[*]Second step was ^{125}I-anti-Ig-4a   [+]Second step was ^{125}I-protein A

Table 5 shows the results of a more extensive genetic analysis of the reactivities of antibodies 10-4.22 and 11-6.3. On the basis of these results and previous findings using alloantisera, four IgD specificities are defined (Table 6). Specificities 1 and 2 are defined by allo-antisera (12). Specificity 2 is also defined by the 11-6.3 antibody. Specificity 3 is defined by the H6/31 antibody of Pearson et al. (13, and T. Pearson and L. A. Herzenberg, unpublished). Specificity 4 is defined by antibody from clone 10-4.22.

TABLE 6

IgD SPECIFICITIES

	1	2	3	4
a	1	-	-	4
b	-	2	3	-
c	n.d.	-	-	4
d	n.d.	-	-	4
e	1	-	3	-
f	n.d.	-	n.d.	4

B. Ig-1b (γ_{2a}) Heavy Chain Allotype

Only one allotype specificity has been described with conventional alloantisera that distinguishes the γ_{2a} heavy chain of the b (Ig-1b) from the a (Ig-1a) allelic products. The five clones producing anti-Ig-1b antibodies (Ig(1b)1.2, 1.7, 2.4, 2.9, and 3.1) that were recovered from hybridizations with BALB/c spleen cells immunized with Ig^b immunoglobulins define three specificities on C.BPC101 (Ig-1b) myeloma protein. This was determined by testing whether different antibodies could block the subsequent binding of a second antibody to the Ig-1b molecule. Various concentrations of unlabeled antibodies were reacted with C.BPC101 myeloma protein that had been fixed with glutaraldehyde onto wells of flexible plastic microtiter plates. The reaction was allowed to proceed overnight at 4°C. The plates were then washed and a constant amount of ^{125}I-labeled antibody was added for 1 hr at room temperature to react with the remaining available (i.e., unblocked) determinants on the myeloma protein. A typical blocking curve using a constant amount of ^{125}I-labeled Ig(1b)2.9 protein, known to be in excess of the amount of antigen available on the plate, is presented in Fig. 9.

Clearly, unlabeled Ig(1b)2.9 is able to block itself (i.e., the ^{125}I-Ig(1b)2.9), while the same amounts of unlabeled Ig(1b)2.4 and Ig(1b)3.1 proteins did not block the subsequent binding of ^{125}I-Ig(1b)2.9. The same amounts of unlabeled Ig(1b)2.4 and Ig(1b)3.1 antibodies were also capable of blocking themselves (curves not shown), but not each other. Another antibody, Ig(1b)1.2, duplicates the Ig(1b)2.9 specificity profile. The enhancing effect of unlabeled Ig(1b)2.4 on the subsequent binding of ^{125}I-Ig(1b)2.9 is a reproducible observation and also occurs with other antibody combinations. The nature of this interaction is unknown at this point. It may involve an induced conformational change of the C.BPC101 myeloma protein by one antibody, making other allotypic determinants available for other antibodies to bind. This problem is presently being investigated.

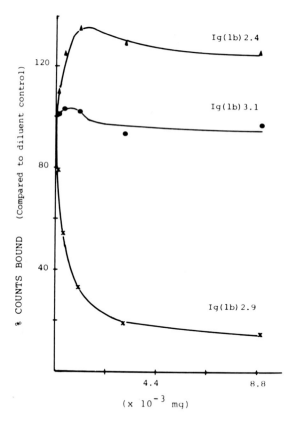

Ig(1b)2.4

Ig(1b)3.1

Ig(1b)2.9

120

80

40

% COUNTS BOUND (Compared to diluent control)

4.4   8.8

$(\times 10^{-3}$ mg$)$

Fig. 9. Anti-Ig-1b blocking curve. Various concentrations of unlabeled Ig(1b)2.4, 3.1, and 2.9 were used (abscissa) to block the subsequent binding of ^{125}I-Ig(1b)2.9. One hundred percent of counts bound represents the number of ^{125}I-Ig(1b)2.9 bound when medium (1% BSA-PBS, pH 7.5) was used in the blocking step of the assay (as described in the text).

126

These new specificities, defined by the inability of the antibodies to block each other, represent distinct antigenic determinants. The members of pairs of determinants are far enough apart not to cause steric hindrance of antibodies binding to both determinants.

The chain compositions of the anti-Ig-1b and the anti-Ig-5a antibodies were examined by 2-D PAGE analysis (Fig. 10). The pattern obtained with (Ig(1b)1.7 protein shows that it actually contains two different heavy chains and three different light chains, despite the fact that it was "cloned" in soft agar. The other antibodies appear to be clonal products. The Ig(1b)2.9 protein and the Ig(1b)1.2 protein (not shown) do not have the MOPC 21 κ chain. Closer examination of the mobility patterns reveal that Ig(1b)2.9 and one of the proteins of Ig(1b)1.7 are identical. These products have similar specificity profiles as determined in the blocking assay. The other protein in Ig(1b)1.7 appears to be identical to Ig(1b)2.4. These could have been derived from independent fusions of cells derived in vivo from the same clone. The anti-Ig-5b antibody, 11-6.3, has not been purified but has been shown by Ouchterlony analysis to be an IgG_1 immunoglobulin.

V. Conclusion

This paper describes the methods that were used to generate stable, cloned hybrid cell lines secreting antibodies to $H-2K^k$, $I-A^k$, Ig-5a, Ig-5b, and Ig-1b antigens. The importance of the results presented here is that they illustrate the used of these monospecific reagents as new serological probes. The monoclonal anti-I-A antibodies provide a means for examining the polypeptide chain structure of Ia antigens.

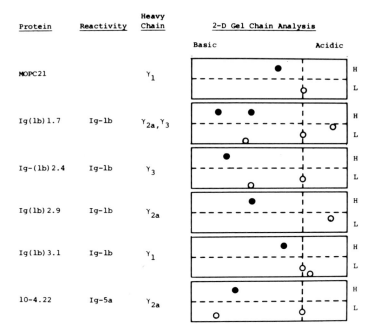

Fig. 10. Heavy and light chain compositions of anti-Ig-1b and Ig-5a antibodies from hybrid cell lines (see legend to Fig. 6).

With the monospecific anti-Ig-1b antibodies, the genetic polymorphisms and antigenic complexities of the γ_{2a} heavy chains can be dissected. Finally, the phenomenon of the enhanced binding of sets of anti-Ig-1b antibodies to Ig-1b protein may provide new information on antigen-antibody interactions. The availability of monospecific antibody reagents in general seems certain to introduce additional new uses of antibodies as molecular probes.

Acknowledgments: The authors wish to thank Dr. Samuel J. Black for the biochemical analyses of the reactivities of the anti-5a and anti-5b monoclonal antibodies and Dr. Hugh McDevitt for generously providing some of the mice and alloantisera used in these studies. Thanks are also extended to Mr. F. T. Gadus, Mr. T. T. Tsu and Mrs. Jennifer Scott for superb technical assistance. We also gratefully acknowledge the skilled assistance of Ms. Jean Anderson in preparation of this manuscript.

This research supported, in part, by grants from the National Institutes of Health, GM-17367, CA-04681, AI-08917 and HD-01287. Dr. Goding is a C.J. Martin Fellow of the National Health and Medical Research Council of Australia; Dr. Jones is a Fellow of the National Science Foundation.

References

1. Köhler, G., Milstein,C.: Continuous cultures of fused cells secreting antibody of predefined specificity. Nature 256, 495-497 (1975)
2. Galfrè, G., Howe, S.C., Milstein, C., Butcher, G.W., Howard, J.C.: Antibodies to major histocompatibility antigens produced by hybrid cell lines. Nature 266, 550-552 (1977)
3. Herzenberg, L.A., Herzenberg, L.A., Milstein,C.: Cell hybrids of myelomas with antibody forming cells and T-lymphomas with T cells In: Handbook of Experimental Immunology, 3rd Edition. Weir, D.M. (ed.). Blackwell Sci. Pub., Oxford, England, Chapter 25 (1978)
4. Köhler, G., Howe, S.C., Milstein,C.: Fusion between immunoglobulin-secreting and non-secreting myeloma cell lines. Eur. J. Immunol. 6, 292-295 (1976)
5. Herzenberg, LA., Herzenberg, L.A.: Mouse immunoglobulin allotypes: Description and specific methodology. In: Handbook of Experimental Immunology, 3rd Edition. Weir, D.M. (ed.). Blackwell Sci. Pub., Oxford, England, Chapter 12 (1978)
6. Jones, P.: Analysis of H-2 and Ia molecules by two-dimensional gel electrophoresis. J. Exp. Med. 146, 1261-1279 (1977)
7. Jones, P., Murphy, D.B., McDevitt, H.O.: Identification of separate I region products by two-dimensional electrophoresis. In: Ir Genes and Ia Antigens. McDevitt, H.O. (ed.). Academic Press, New York, 1978, pp. 203-213
8. O'Farrell, P.Z., Goodman, H.M., O'Farrell, P.H.: High resolution two-dimensional electrophoresis of basic as well as acidic proteins. Cell 12, 1133-1142 (1977)
9. Loken, M.R., Parks, D.R., Herzenberg, L.A.: Two-color immunofluorescence using a fluorescence-activated cell sorter. J. Histochem. Cytochem. 24, 899-907 (1977)
10. David, C.S.: Serologic and genetic aspects of murine Ia antigens. Transplant. Rev. 30, 299-322 (1976)
11. Goding, J.W.: Use of staphylococcal protein A as an immunological reagent. J. Immunol. Methods (in press)

12. Goding, J.W., Scott, D.W., Layton, J.E.: Genetics, cellular expression and function of IgD and IgM receptors. Immunol Rev. 37, 152-186 (1977)
13. Pearson, T., Galfrè, G., Ziegler, A., Milstein, C.: A myeloma hybrid producing antibody specific for an allotypic determinant on "IgD-like" molecules of the mouse. Eur. J. Immunol. 7, 684-690 (1977)

The Antibody Repertoire of Hybrid Cell Lines Obtained by Fusion of X63-AG8 Myeloma Cells with Mitogen-Activated B-Cell Blasts

J. Andersson, F. Melchers

Introduction

Fusion of myeloma cells with antibody-secreting B-cells has been achieved with in vivo antigen-activated spleen cells (1). Such fusions appeared to occur with spleen cells at a stage of differentiation similar to that of the myeloma line, i.e. with antigen-activated, antibody-secreting B-cell blasts. This was concluded from the high incidence of antibody-secreting hybrid cell lines and by the absence of T-cell-B-cell hybrids in these fusions (2).

Immunoglobulin (Ig) secreting B-cell blasts can also be generated in vitro by the activation with mitogens (3). This activation is poly-clonal and, therefore, induces the growth and maturation of resting, small B-cells into clones of Ig-secreting cells synthesizing the entire repertoire of v-regions of the mitogen-reactive B-cell subset. One of three splenic B-cells is activated either by lipopolysaccharide (LPS) (4) or by lipoprotein (LPP) (5) to growth and to IgM-secretion (6,) while one of thirty, i.e. one tenth of them, will also switch to IgG-secretion (8).

This paper reports fusions of mitogen-activated B-cell blasts with the azaguanine-resistant myeloma cell line X63-AG8 (1). If fusions are random events between activated cells and myeloma cells, then the resulting hybrid lines should display a repertoire of Ig-heavy and light chain classes and of V-region specificities characteristic of the previously resting mitogen-activated normal B-cells. The secretion of Ig-molecules of a given class or type can be monitored by the protein A-sheep red cell plaque assay (9), the secretion of Ig-mole-cules with given specificities by the corresponding specific hemolytic plaque assays. In this paper hybrid cell lines have been monitored with plaque assays specific for sheep red cells (SRC), horse red cells (HRC) and heavily substituted, trinitrophenylated sheep red cells (TNP-SRC) (7).

Our results suggest that fusions between mitogen-activated B-cell blasts and the X63-AG8-myeloma cell line are random events. Optimal fusion is observed with B-cell blasts activated for four days by mitogen. Small, resting spleen cells fuse with frequencies which are at least 100 fold lower - and in our experiments undetectable -than large cells in normal spleen which contain "background" plaque forming cells (10).

Materials and Methods

Animals. C_3H/Tif/BOM mice, 6-8 weeks of age, were obtained from Gr. Bomholtgaard, Ry, Denmark. C57Bl/6J mice, 6-8 weeks of age, and Lewis

strain rats, 4 weeks of age, were from the Institut für Biologisch-
medizinische Forschung AG., Füllinsdorf, Switzerland.

Cells, mitogens and culture conditions. The azaguanine-resistant
myeloma cell line X63-AG8 was kindly given to us by Dr Cesar Milstein,
MRC Postgraduate School for Molecular Biology, Cambridge, England. It
was grown in RPMI 1640 medium supplied with 2-mercaptoethanol (5×10^{-5}
M), fetal calf serum (10%, batches K255701D and K653501, Gibco),
antibiotics (Kanamycin (Gibco) or gentamycin (Schering AG., Berlin)),
glutamine (2mM) and azaguanine (20 µg/ml, Sigma). For some experiments
the line was kept in a Dulbecco's modified Eagle's medium, further
modified by Iscove (11), omitting fetal calf serum but adding bovine
serum albumin, transferrin and soybean lipid (11). The tissue culture
line was readily adaptable to growth in either medium. For fusion
experiments X63-AG8 cells were taken at their exponential phase of
growth, i.e. between 2 and 4×10^5 cells/ml.

Spleen cells were grown in either of the two media described above
from an initial concentration of 3×10^5 cells/ml in the presence of
either 50 µg/ml LPS-S (EDTEN 18735, a gift of Drs C. Galanos and O.
Lüderitz, Max Planck Institut für Immunobiologie, Freiburg i.Br. West
Germany) or 5 µg/ml E.coli lipoprotein (purified from E.coli strain
3300 by the methods described by Braun et al. (12,13). Actual cell
concentrations in the mitogen-activated spleen cell cultures were
between 5 and 7×10^5/ml at day 3, 1 and 2×10^6/ml at day 4, 2 and 4
$\times 10^6$/ml at day 5 of culture for both mitogens used. Wherever rat
thymus cells were added to fused cell cultures they were present at 3-
5×10^6 cells/ml (14,15). Small, resting lymphocytes were purified
from larger cells containing dividing and plaque-forming cells (10) by
velocity sedimentation at unit gravity (16).

Fusions. The fusing agent used in all studies was poly(ethyleneglycol)
(PEG, Polysciences, Worthington, Pa., lot No. 2871-125, M.W. 4000).
It was dissolved in 0.9% NaCl. The final solution was 45% PEG, 15%
dimethylsulfoxide (DMSO) (Merck, Darmstadt, West Germany). Two procedures were practiced for fusions. In both procedures equal numbers
of X63-AG8 myeloma cells and of spleen cells were mixed and washed
twice in serum-free pH 8-medium (RPMI 1640, no additions, pH adjusted
to 8 with NaOH) at room temperature.

In procedure I (17) cells were pelleted in a Falcon tube (either 50
ml, or 15 ml). The medium was removed by suction. Then the cells
were brought up to 37°C where upon 1 to 2 ml of the 45% PEG, 15% DMSO
solution (18) at 37°C was slowly (3 min) added to cover the cell
pellet. Thereafter 10 ml pH 8-medium (see above) were layered slowly
(10 min) above the PEG/DMSO-solution, again at 37°C. This mixture was
spun at room temperature for 5 min at 200xg, the supernatant removed
by suction and the cells washed once with serum-free medium at pH 8.
The cells were finally resuspended in full culture medium (see above)
and plated in either Linbro dispo trays (model FB 16-27, TC, Linbro
Scientific, New Haven, Connecticut, USA) at 1 ml per culture or in
microtiter plates (Falcon) at 0.2 ml per culture. Usually 10^7 myeloma
cells and 10^7 spleen cells were fused in an experiment. Below 10^6 of
each cell type per fusion did not yield successful fusion events, when
mitogen-activated spleen cells were employed for fusion.

In procedure II myeloma cells and spleen cells were washed twice in pH
8-medium. Equal numbers of both cell types were then spun together

onto a glass-fiber filter (Whatman GF/A; ∅ 0.8 cm) in a screw-cap glass tube (19). Usually 10^6 of both cell types were mixed. Below 5 x 10^4 cells no successful fusion events were observed when mitogen-activated spleen cells were employed for fusion. The filter was picked up from the glass tube with forceps, transferred into 45% PEG, 15% DMSO (see above) at $37^{\circ}C$ for one minute, then into serum-free medium for one minute, and thereafter cultured in full medium (all done in Linbro dispo trays).

In procedures I and II hypoxanthine (H, 1 x 10^{-4}M), aminopterine (A, 4 x 10^{-7}M) and thymidine (T, 1.6 x 10^{-5}M) were added to the cultures 24 hours after fusions (20). Thereafter cultures were fed every other day by removing half of the supernatant medium, replacing it with full medium (see above) containing HAT. After 10 days to two weeks cells could be seen growing whenever fusion experiments were successful. After 3 weeks cells were subcultured in full H,T-containing medium. Finally, plaque assays for secretion of Ig-classes and specificities by the hybrid cells were performed between 4 and 8 weeks after fusion, usually with 10^3-5 x 10^4 hybrid cells per assay.

Hybrid cell lines were grown in either medium described above. In RPMI 1640-medium single clones of mitogen-stimulated B-cells have been shown to secrete up to 30 ng antibody (25), while in Iscove's modified Dulbecco's MEM-medium up to 150 ng antibody has been measured per clone. In mass cultures of hybrid lines grown to exhaustion of the medium 500 µg-1 mg/ml of Ig was obtained in either medium. These quantities are so high that adaptation of the hybrid lines to growth as a tumor in the mouse was considered unnecessary.

Plaque assays for Ig-secreting cells. For the detection of all cells secreting Ig of a given class or subclass a modified hemolytic plaque assay was used (9). Protein A-coupled sheep red cells and rabbit (anti-mouse Ig) antisera were used in the assay together with properly diluted complement (Gibco). Protein A was obtained from Pharmacia, Uppsala, Sweden or from Dr H. Wigzell, Biomedicum, University of Uppsala, Sweden. Rabbit (anti-mouse IgM) antibodies were raised against purified myeloma MOPC 104E 19S IgM (λ,μ), rabbit (anti-mouse IgG_1) antibodies against purified myeloma MOPC 21 IgG_1 (produced by the X63-AG8 myeloma cell line) and rabbit (anti-mouse IgG_2) against purified myeloma Adj-PC-5 IgG_2 (κ, γ_2). Each antibody preparation was titrated for optimal plaque-developing capacity in the plaque assay (9), using myeloma lines or polyclonally mitogen-activated B-cell blasts producing the appropriate Ig-classes and subclasses.

Direct plaque-forming cells (PFC) with SRC, HRC and TNP-SRC were measured as described previously (21,22). TNP-groups were coupled to SRC by the method of Rittenberg and Pratt (23) with 30 mg trinitro-benzene sulfonic acid (Sigma) per ml of packed, washed SRC.

Results
___

Fusion with mitogen-activated B-cell blasts
___

C57Bl/6J spleen cells were activated with either LPS or LPP for 2,3, 4,5 or 6 days in culture. Approximately 1 x 10^7 activated spleen cells were fused with an equal number of X63-AG8 myeloma cells at

these different times of activation and plated after the fusion
process at 2×10^6, 6×10^5 and 2×10^5 cells per milliliter in microtiter plates at 0.2 ml per culture. Four to six weeks later success-
ful fusion events were visible under the microscope by growth of
cells. The presence of IgM-, IgG- and IgG_2-secreting cells was
analyzed by the protein A-SRC-plaque assay.

The highest incidence of successful fusions yielding growing hybrid
lines was observed with B-cell blasts activated with mitogen for 4
days. All cultures in which growing cells were observed, also contained secreting cells. Table I summarizes the results.

Almost every culture (184 of 186 total positive cultures containing
growing, secreting cells) had IgM-secreting cells. This was expected
since mitogenic activation of B-cells leads, particularily at early
times of activation, to the maturation of IgM-secreting cells. The
two exceptions, where no IgM-secreting cells could be observed,
occurred with B-cells blast activated for 6 days. At this time of
activation a part of the B-cell blasts may have stopped to secrete IgM
and/or may have switched to IgG-secretion (8). The high incidence of
IgG_1-secreting cells (180 of 186 positive cultures) was also expected
since the myeloma line X63-AG8 used in the fusions secretes IgG_1.
With the protein A-SRC-plaque assay used for detection of IgG_1-
secreting cells we, therefore, cannot distinguish between the secretion
of the myeloma IgG_1 and any other IgG_1 secreted as a distinct molecule
by a hybrid line. Eleven of 186 positive cultures secreted IgG_2
molecules. This frequency is similar to frequencies of IgG_2-secreting
cells found in mitogen-activated B-cell blast populations, i.e. 3% of
the number of IgM-secreting B-cell blasts (unpublished observations,
see also ref. 8). LPS and LPP yielded B-cell blasts with similar
fusion efficiencies. When cells were plated after the fusion procedure
I at densities below 2×10^5 cells/ml successful fusion events were
not obtained.

Growth of fused cell lines in the presence of thymus "feeder" cells

Syngeneic, allogeneic or xenogeneic thymus cells support the growth
of B-cells and thereby increase their cloning efficiences in suspension cultures (14,15). We, therefore, attempted to increase the
recovery of successful fusion events by plating the cells after the
fusion procedure in medium containing 3×10^6 rat thymus cells/ml.

The results in Table 2 show a 10 to 30 fold increase in the frequencies
of recovered successful fusions in the presence of "feeder" thymus
cells. Below 2×10^4 cells/ml plated in the presence of thymus cells
successful fusion events were recovered very rarely (in parallel
experiments 2 positive of 192 plated cultures).

Fusion on glass-fiber discs

Fusion procedure I (see Materials and Methods) did not yield successful fusion events when the number of cells in the fusion mixture was
lower than 10^6. In many cases, however, the number of normal, activated B-cells secreting specific antibody which can be obtained experi-
mentally is much lower. As an example, single clones of mitogen-
activated B-cells grow to $1-2 \times 10^3$ cells (6,7,14). If one wants to
make such clones immortal through fusion to the myeloma line procedures should be available which allow such low numbers of cells to

Table 1. Frequencies of fusion events with B-cell blasts activated by mitogen for different lengths of time

Number of positive cultures[a] of total cultures plated after fusion at

B-cell blasts activated in culture for	Mitogen used for activation	2×10^6 cells/ml				6×10^5 cells/ml				2×10^5 cells/ml			
		Total	Secreting			Total	Secreting			Total	Secreting		
			IgM	IgG_1	IgG_2		IgM	IgG_1	IgG_2		IgM	IgG_1	IgG_2
2 days	LPS	1/12	1	1	0	0/12	-	-	-	0/12	-	-	-
3 days	LPS	18/24	18	17	0	3/24	3	3	0	1/24	1	1	0
4 days	LPS	24/24	24	22	1	10/24	10	10	1	3/24	3	3	0
4 days	LPP	34/36	34	33	1	12/36	12	12	2	5/36	5	5	1
5 days	LPS	21/24	21	21	1	9/24	9	9	0	1/24	1	1	0
5 days	LPP	26/36	25	24	2	11/36	11	11	1	1/36	1	1	0
6 days	LPS	5/24	3	5	1	1/24	1	1	0	0/24	-	-	-

[a] Grown in RPMI 1640 + 10% FCS, selected in medium (20).

Table 2. Frequencies of fusion events with B-cell blasts[a] initiated in culture in the presence or absence of feeder rat thymus cells[b]

Number of cells plated after the fusion procedure I per ml	Number of positive cultures[c] of total cultures plated after fusion	
	in the absence of rat thymus cells	in the presence of rat thymus cells
2×10^6	24/24	24/24
6×10^5	10/24	24/24
2×10^5	3/24	18/24
6×10^4	0/24	11/24
2×10^4	0/24	3/24
6×10^3	0/24	1/24

[a] activated for 4 days with LPS

[b] in Iscove's modified, Dulbecco's modified Eagle's medium (ii); 3×10^6 rat thymus cells/ml added once at the initiation of the culture but not when cultures were fed thereafter.

[c] 0.2 ml in microtiter plates; assay after 4-5 weeks by observation under the microscope and positive in the protein A-SRC-plaque assay for IgM-secreting cells.

yield successful fusion events. In principle the method of fusing cells on a filter, developed by Davidson (19) and Buttin (this volume) was employed to lower the number of cells in the fusion reaction. Results in Table 3 III indicate that as few as 2×10^4 B-cell blasts will yield hybrid lines secreting IgM when they are fused by procedure II (for details see the Materials and Methods section).

Table 3. Fusion of limiting numbers of B-cell blasts[a] on glass-fiber filter discs

Number of cells in fusion procedure II[b]	Number of positive cultures[c] of total cultures with filters established after fusion
2×10^6	5/5
6×10^5	10/10
2×10^5	10/10
6×10^4	4/10
2×10^4	2/10
6×10^3	0/10

[a] Activated for 4 days with LPS, fused with an equal number of X63-AG8 cells

[b] See Materials and Methods

135

^{c)}Containing the filter used in the fusion reaction. After 4 to 5 weeks some of the growing cells could be removed from the filter by agitation and gentle disruption of the filter. Cells could then be observed under the microscope and assayed for IgM-secreting cells with the protein A-SRC-plaque assay.

Fusion of small, resting cells and of large, growing, plaque forming cells of spleen

Normal spleen cells of $C_3H/Tif/BOM$ mice were separated according to their size by velocity sedimentation at unit gravity (16). A separation of small, resting cells which do not secrete Ig and do not form plaques, from large cells, which synthesize DNA, divide, secrete IgM and, therefore, form so-called "background" plaques (as described in ref. 10, Figure 1; or in ref. 24, Figure 1) was achieved. A pool of small and one of large cells were subsequently fused with the myeloma cell line. Table 4 summarizes the results.

Table 4. Fusion of small and of large splenic lymphocytes with the myeloma cell line X63-AG8

Number of cells plated after the fusion procedure I[a] per ml	Number of positive cultures of total cultures plates after fusion	
	Spleen cells used in fusion	
	Small cell pool	Large cell pool
6×10^6	0/5	3/3
2×10^6	0/5	9/10
6×10^5	0/5	2/10

[a] Using 3×10^7 small cells and large cells in the fusion mixture

Successful fusion events yielding growing hybrid cells were obtained only with the pool of large cells. The growing hybrid cells secreted IgM. These results support the notion that B-cells have to be activated to growth and maturation to Ig secretion before they can fuse with the myeloma cell line X63-AG8.

The repertoire of antibody specificities of hybrid cells from B-cell blasts

A total of 1880 successful fusion cultures containing IgM-secreting hybrid cells were analyzed for their antibody specificity with plaque assays detecting SRC-, HRC- or TNP-SRC-specific IgM-secreting (i.e. "direct"), plaque-forming cells. Most of these fusion cultures were obtained by fusion of either C57Bl/6J or $C_3H/Tif/BOM$ splenic B-cell blasts, activated for 4 days with mitogen and plated after fusion at cell concentrations between 2 and 8×10^5 cells/ml without feeder thymus cells, i.e. at concentrations limiting the successful fusion events to near one per culture (see Table 1). The results of these analyses are summarized in Table 5.

136

Table 5. Frequencies of hybrid cells with mitogen-activated B-cell blasts and of the corresponding clones of normal, mitogen-activated B-cells producing specific antibodies.

	SRC-specific	HRC-specific	TNP-SRC-specific
Fusion cultures with hybrid cells[a]	1 in 940	1 in 376	1 in 94
Normal, mitogen-reactive B-cells[b]	1 in 1000	1 in 500	1 in 50

[a] 1880 cultures with hybrid cells secreting IgM assayed 4 to 6 weeks after fusion

[b] Numbers taken from Ref. 7

The frequencies of specific antibody-producing hybrid cells amongst all fusion cultures are remarkably similar to those found for normal, mitogen-activated B-cell clones (7). This suggests that fusion is a random event which depends on the frequency of a specific B-cell in its activated blast form in the fusion mixture.

Concluding remarks

Mitogen-activated B-cell blast can be fused with the myeloma line X63-AG8 to yield Ig-secreting hybrid cell lines. Over 90% of them secrete IgM, less than 10% IgG_2. Fusion appears to require B-cells in their activated growing state since small, resting B-cells could not be fused to yield growing hybrid lines. The antibody repertoire of the fused, hybrid lines closely resembles that found for normal, mitogen-activated B-cells, suggesting that fusion is a random event between activated B-cells and the myeloma cells. The frequency of successful fusions can be increased by a fusion procedure which brings together the cell partners in the fusion on a glass filter disc, and by plating the fused cells in the presence of feeder thymus cells. So far these frequencies have, however, not reached a level which would permit to produce a hybrid cell line from the cells of a single mitogen-activated B-cell clone of which the antigen specificity could be determined prior to fusion. Hybrid B-cell lines producing antibodies with given wanted specificities will, therefore, at present be obtained more easily by in vivo activation of B-cells with the given corresponding antigen and subsequent fusion of the activated cells. Even spleen cells of mice with memory for a given antigen are not more suitable in this search for specific hybrid cell lines when their corresponding in vitro mitogen-activated B-cell blasts are fused, since memory results only in a 10-20 fold increase in the absolute frequency of specific B-cell precursors (25). The frequencies of specific B-cell blasts and, thereby, the chances of obtaining specific hybrid B-cell lines can, however, be increased if the corresponding specific mitogen-reactive precursor B-cells can either be purified hapten-specifically (26), or if these precursors can be activated selectively in vitro by low concentrations of hapten-mitogen-conjugates (27). In vitro activation of B-cells by mitogen offers advantages in all cases where the antigen is either toxic, or contained in or crossreactive with the universe of self antigens. The polyclonal nature of mitogenic activation allows the expression of B-cells secreting the entire repertoire of v-regions.

Hybrid cell lines emerging from fusions with mitogen-activated B-cell blasts should, therefore, be ideal tools to study V-region interactions in a postulated network of lymphocyte interactions (28).

Acknowledgements. The able technical assistance of Ms Helen Campbell, Joy Monckton and Debbie Norman is gratefully acknowledged.

References

1. Köhler, G. and Milstein, C. Continuous cultures of fused cells secreting antibody of predetermined specificity. Nature 256, 495 (1975).
2. Köhler, G. and Milstein, C. Derivation of specific antibody-producing tissue culture and tumor lines by cell fusion. Eur. J. Immunol. 6,511 (1976).
3. Möller, G. (ed.): Lymphocyte Activation by Mitogens. Transplant. Rev. 11, (1972)
4. Andersson, J., Sjöberg, O. and Möller, G.: Induction of immunoglobulin and antibody synthesis in vitro by lipopolysaccharides. Eur. J. Immunol. 2, 349 (1972)
5. Melchers, F., Braun, V. and Galanos, C.: The lipoprotein of the outer membrane of Escherichia coli: a B-lymphocyte mitogen. J. Exp. Med. 142, 473 (1975)
6. Andersson, J., Coutinho, A. and Melchers, F.: Frequencies of mitogen-reactive B-cells in the mouse. I. Distribution in different lymphoid organs from different inbred strains of mice at different ages. J. Exp. Med. 145, 1511 (1977)
7. Andersson, J., Coutinho, A. and Melchers, F.: Frequencies of mitogen-reactive B-cells in the mouse. II. Frequencies of B-cells producing antibodies which lyse sheep or horse erythrocytes, and trinitrophenylated or nitroiodophenylated sheep erythrocytes. J. Exp. Med. 145, 1520 (1977)
8. Andersson, J., Coutinho, A. and Melchers, F.: The switch from IgM- to IgG-secretion in single mitogen-stimulated B-cell clones J. Exp. Med, in press (1978)
9. Gronowicz, E., Coutinho, A. and Melchers, F.: A plaque assay for all cells secreting Ig of a given type or class. Eur. J. Immunol. 6, 588 (1976)
10. Andersson, J., Lafleur, L. and Melchers, F.: Immunoglobulin M in bone marrow-derived lymphocytes. Synthesis, turnover and carbohydrate composition in unstimulated mouse B-cells. Eur. J. Immunol. 4, 170 (1974)
11. Iscove, N.N. and Melchers, F.: Complete replacement of serum by albumin, transferrin and soybean lipid in cultures of lipopolysaccharide-reactive B-lymphocytes. J. Exp. Med. 147, 923 (1978)
12. Braun, V. and Rehn, K. Chemical characterization, spatial distribution and function of lipoprotein (Murein-lipoprotein) from E. coli cell wall. The specific effect of trypsin on the membrane structure. Eur. J. Biochem. 10, 426 (1967).
13. Braun, V. and Sieglin, U. The covalent murein-lipoprotein structure of the Escherichia coli cell wall. The attachment of the lipoprotein on the murein. Eur. J. Biochem. 13, 336 (1970).
14. Andersson, J., Coutinho, A., Lernhardt, W. and Melchers, F.: Clonal growth and maturation to immunoglobulin secretion in vitro of every growth-inducible B-lymphocyte. Cell 10, 27 (1977)

15. Lernhardt, W., Andersson, J., Coutinho, A. and Melchers, F. Cloning of murine transformed cell lines in suspension culture with efficiencies near 100%. Exp. Cell Res. 111, 309 (1978).
16. Miller, R.G. and Phillips, R.A.: Separation of cells by velocity sedimentation. J. Cell. Physiol. 73, 191 (1969)
17. Galfré, G., Howe, S.C., Milstein, C., Butcher, G.W. and Howard, J.C. Antibodies to major histocompatibility antigens produced by hybrid cell lines. Nature 266, 550 (1977).
18. Norwood, T.H., Ziegler, C.J. and Martin, G.M. Dimethyl Sulfoxide enhances polyethylene glycol-mediated somatic cell fusion. Somatic Cell Genetics 2, 263 (1976).
19. O'Malley, K.A. and Davidson, R.L. A new dimension in suspension fusion techniques with polyethylene glycol. Somatic Cell Genetics 3, 441 (1977).
20. Littlefield, J.W. Selection of hybrids from matings of fibroblasts in vitro and their presumed recombinants. Science 145, 709 (1964).
21. Andersson, J., Bullock, W.W. and Melchers, F.: Inhibition of mitogenic stimulation of mouse lymphocytes by anti-mouse immunoglobulin antibodies. I. Mode of action. Eur. J. Immunol. 4, 715 (1974).
22. Andersson, J., Melchers, F., Galanos, C. and Lüderitz, O.: The mitogenic effect of lipopolysaccharide on bone marrow-derived mouse lymphocytes. Lipid A as the mitogenic part of the molecule. J. Exp. Med. 137, 943 (1973)
23. Rittenberg, M.B. and Pratt, K.L.: Anti-trinitrophenyl (TNP) plaque assay. Primary response of Balb/c mice to soluble and particulate immunogens. Proc. Soc. Exp. Biol. Med. 132, 575 (1969)
24. Melchers, F. and Cone, R.E.: Turnover of radioiodinated and of leucine-labeled immunoglobulin M in murine splenic lymphocytes. Eur. J. Immunol. 5, 234 (1975)
25. Eichmann, K., Coutinho, A. and Melchers, F. Absolute frequencies of lipopolysaccharide-reactive B cells producing A5A idiotype in unprimed, streptococcal A carbohydrate-primed, anti-A5A idiotype-sensitized and anti-A5A idiotype-suppressed A/J mice. J. Exp. Med. 146, 1436 (1977).
26. Haas, W. and Layton, J.E. Separation of antigen specific lymphocytes. Enrichment of antigen-binding cells. J. Exp. Med. 141, 1004 (1975).
27. Coutinho, A. and Möller, G.: Immune activation of B-cells: Evidence for "one nonspecific triggering signal" not delivered by the Ig receptors. Scand. J. Immunol. 3, 133 (1974)
28. Jerne, N.K. The somatic generation of immune recognition. Eur. J. Immunol. 1, 1 (1971).

Mouse λ_1 Hybridomas

W.R. Geckeler, W.C. Raschke, R. DiPauli, M. Cohn

A. Introduction

Hybridomas which secrete immunoglobulin of the λ_1 light chain class
are described. The λ_1 light chains produced by individual clones
are expected to fall into two classes on the basis of their variable
region (v_λ) sequence; those which are the products on the single
germ-line λ_1 gene encoding v_λ (1) and those λ_1 variants which contain
amino acid substitutions in their complementarity-determining resi-
dues (CDR's) (2). The λ_1 variants result from the combined effects
of mutation and antigenic selection processes which act upon cells
in the lineage of λ_1 antigen-sensitive B cells (λ_1ASC's). The rules
which govern the amino acid replacements seen in CDR's remain ob-
scure. Two examples illustrate the problem. First, on the basis of
X-ray crystallography λ_1 CDR's may be divided into contact residues
in the combining site and adjacent residues which alter the position
of the contact residues within the combining site (3). An analysis
of the CDR replacements in the small collection of known λ_1 variant
sequences (4) suggests that all amino acid replacements may occur in
adjustor residues. Second, there is evidence (5,6) suggesting that
mutations in the third hypervariable region of v_{λ_1} may be generated
by the translocation of v_λ to a transcription unit containing an
intron and c_λ. Mutations in the first and second hypervariable
regions must be generated by a different mechanism, and might be
expected to show a different pattern of codon alteration. A large
collection of λ_1 light chain variants would allow a detailed examination of the pattern of amino acid replacements in CDR's. By comparing such a pattern with X-ray crystallographic data and the DNA
sequence of the germ-line v_{λ_1} gene, one may be able to infer both
the mutational mechanism and also the antigenic selection rules
which operate in the generation of antibody diversity.

B. Materials and Methods

In order to produce a large collection of hybridoma clones expressing
the λ_1 light chain, a fusion employing LPS blasts from the splenic
lymphocytes of κ-suppressed mice and MPC 11 was performed (see
W.C. Raschke, this volume). κ-suppressed mice were produced as
follows: Female BALB/c mice were injected intraperitoneally daily
from birth with 100 µg of immunoabsorbed rabbit anti-mouse µ chain
antibody. These females were mated and their progeny were likewise
injected with 100 µg immunoabsorbed rabbit anti-mouse κ chain antibody.

C. Results

In the κ-suppressed mice used for this fusion the absolute level of
λ_1 associated with normal immunoglobulin is increased ten-fold,
while the amount of κ is reduced by one-half. The estimated ratio

of splenic ASC's expressing λ_1 to those expressing κ is 2:5. At the
lowest dilution of cells plated after the fusion, approximately 7
clones were recovered in each microtiter well. Of approximately 200
wells screened by a radioimmunoassay for λ_1 light chain secreting
clones, 89% were positive. From the fraction of non-responding wells
it is estimated by the Poisson distribution that, on the average, 2
of each 7 clones per well were λ_1 producers. If the majority of the
other clones represent fusions of κ LPS blasts, the ratio of λ_1 to κ
hybridomas recovered in the fusion is 2:5, the same as the estimated
ratio of $\lambda_1:\kappa$ splenic ASC's in the κ-suppressed donor spleen used in
the fusion.

Microtiter wells with λ_1 hybridomas were recloned. Of the first 7
recloned λ_1 hybridomas, 5 are associated with the μ heavy chain and
two have the $\gamma 3$ heavy chain. Of the estimated 400 λ_1-secreting
clones recovered in the initial cloning, 4 clones bind ^{125}I-labelled
$\alpha-(1,3)$ dextran and are presumably expressing λ_1 associated with a
V_H^{DEX} gene product (7).

Acknowledgements. This work was supported by Grant Number CA21531
awarded by the National Cancer Institute to W.C.R., Grant Number
RO1 AI05875 awarded by the Allergy and Immunology Branch of the
National Institutes of Health to M.C., and a Fellowship from the
Deutsche Forschungsgemeinschaft to R.DP. Additional support was
provided by Mrs. Jean James.

References

1. Tonegawa, S., Hozumi, N., Matthyssens, G., Schuller, R.: Somatic
    changes in the content and context of immunoglobulin genes. In:
    Cold Spring Harbor Symposium Quant. Biol. 41, 877-889 (1976).
2. Weigert, M., Cesari, I.M., Yonkovitch, S., Cohn, M.: Variability
    in the lambda light chain sequences of mouse antibody. Nature
    228, 1045-1047 (1970).
3. Poljak, R.J., Amzel, L.M., Chen B.L., Chiu, Y.Y., Phizackerley,
    R.P., Saul, F., Ysern, X.: Three-dimensional structure and diver-
    sity of immunoglobulins. Cold Spring Harbor Symposium Quant.
    Biol. 41, 639-645 (1976).
4. Geckeler, W.R.: On the regulation of expression of the murine
    λ_1 light chain. University Microfilms International, Ann Arbor,
    Michigan (1977).
5. Tonegawa, S., Maxam, A.M., Tizard, R., Bernard, O., Gilbert, W.:
    The sequence of a mouse germ-line gene for a variable region of
    an immunoglobulin light chain. Proc. Natl. Acad. Sci., in press.
6. Brack, C., Tonegawa, S.: The V- and C-parts of the light chain
    gene of a mouse myeloma cell are 1250 non-translated bases apart.
    Proc. Natl. Acad. Sci. 74, 5652-5656 (1977).
7. Blomberg, B., Geckeler, W.R., Weigert, M.: Genetics of the anti-
    body response to dextran in mice. Science 177, 178-190 (1972).

Rapid Binding Test for Detection of Alloantibodies to Lymphocyte Surface Antigens

R.M.E. Parkhouse, G. Guarnotta

A major problem in the recently developed hybridoma technology (1) is screening very large numbers of culture supernatants for antibodies to cell surfaces. Whilst the radioactive anti-immunoglobulin (Ig) binding method has the advantage of detecting all Ig isotopes, it is nonetheless more time consuming than cytotoxic procedures. In addition it is inconvenient for detecting alloantibodies to B lymphocytes since of necessity the radioactive anti-Ig probe will bind to pre-existing B cell sIg. The modification to pre-existing procedures reported here were designed to meet these problems.

Increased speed in the binding assay was achieved by conducting the tests on microtitre plates and revealing positives and negatives by autoradiography rather than counting. Alloantibodies to B cells were screened for by this procedure using cells made sIg-negative by modulation with anti-Ig prior to testing. The binding assay was conducted by standard procedures (described elsewhere in this book) using 96-well microtitre plates with conical shaped wells. The probe was polyspecific ^{125}I-labelled goat anti-mouse Ig. It was that fraction of an anti-TEPC 183 (μK) serum which bound to a column of Adj PC5 (γ2aK), and was recovered by acid elution. The essential point to be made here is that the washed microtitre plates were simply taped to the outside of a black plastic casette containing a pre-exposed X-ray film and an intensifying screen (2). Exposure was at -70º and for 3-16 hr., and the film was processed according to the manufacturer's instructions. The test was easy to score visually, positives and negatives being obvious. When the results were checked by counting there was never any confusion, but one interesting point did emerge. It became clear that non-specific binding of radioactivity to the microtitre plate was less of a problem for the autoradiography than for the more time consuming procedure of counting. This was simply because the cells were confined to the tip of the well (i.e. adjacent to the X-ray film), whereas the non-specifically bound radioactivity was distributed over the total well surface.

Finally the background uptake of ^{125}I-anti-mouse Ig to murine B cells could be totally eliminated by pretreating the cells with polyspecific anti-Ig (~1 mg/ml., 30 min., room temperature). Cells were then washed, incubated at 37º in RPMI 1640-10% FCS for 1½ hr. and washed again. Such cells were clearly positive when subsequently tested with anti-H-2 sera and the labelled anti-Ig, but uptake of the ^{125}I-anti-mouse Ig, with or without prior exposure of the cells to normal mouse serum, was negligible. This approach is also useful for selectively screening alloantibodies to sIg, and could be extended to any situation where a given surface protein can be selectively modulated by antibody.

Acknowledgments: Gabriella Guarnotta thanks the Wellcome Foundation for a research fellowship.

References

1. Köhler, G., Milstein, C.: Continuous cultures of fused cells secreting antibody of defined specificity. Nature. 256, 495-497 (1975).
2. Laskey, R. A., Mills, A. D.: Enhanced autoradiographic detection of ^{32}P and ^{125}I using intensifying screens and hypersensitised film. FEBS Lett. 82, 314-316 (1977).

Cellular and Molecular Restrictions of the Lymphocyte Fusion

G. Köhler, M.J. Shulman

I. Introduction

Since 1975 when the first hybridomas secreting antibodies of predetermined specificity were generated (1) the fusion technique has been
applied in many laboratories for many different purposes. In the
same time several restrictions on cell fusion have become apparent.
Some of these restrictions work to the advantage of the experimenter
and some to his disadvantage. We wish here to describe how it was
possible to exploit some of the benefits and to overcome some of the
hindrances that these restrictions imposed on the fusion process.

II. Cellular Restrictions

1. Selective Fusion to Ig-secreting Spleen Cells

The highest number reported for sheep red cell (SRC) precursor
spleen B cells in unimmunized mice is 0.1% as tested by *in vitro*
lipopolysaccharide stimulation (2). SRC hyperimmunized mice show
about a 5-fold increase of LPS-stimulatable SRC precursor cells (A.
Coutinho, personal communication). SRC specific plaque forming
cells of twice SRC immunized mice comprised 0.05% of the total
spleen cell population at the day of fusion (Table 1). If we assume
random fusion of the myeloma cells with the spleen B cells we would
expect at most 0.5% of the hybrids to be SRC specific. The observed
frequency however was in the range of 10%. Fusion or survival of
hybrid cells therefore must be a nonrandom event. The fact that
antigen specific hybrid cells were observed only when the number of
specific plaque forming cells was at its peak (4) prompted us to
think that the immunoglobulin secreting cell was preferentially
giving rise to hybrid lines. Table 1 summarizes the data of many
fusions, comparing the frequency of specific and nonspecific hybrids
with the frequency of specific and nonspecific plaque forming spleen
cells. There are about 5 background SRC plaque forming cells per 10^6
spleen cells in unimmunized mice. In the spleen, plaque forming
cells comprise 1% of the total cells (2) so that if there is preferential fusion with Ig-secreting cells, one would expect 0.05% of the
hybrids to be SRC specific. In fact we found no SRC specific hybrid
among the more than 500 hybrids examined (<0.2%). After immunization
one would expect 5% of the hybrids to be specific and we found that
20% were specific, reflecting perhaps a slight preference of recently
triggered plasma cells to give rise to viable hybrid cells. The
IgG/IgM class distribution of specific and nonspecific hybrids corresponds to that seen for plaque forming spleen cells. From these
results we conclude that the Ig secreting spleen cells are the major
fusion partners of the myeloma cell. Therefore the successful gener-

ation of hybrid cells secreting antigen specific antibodies is
mainly determined by the frequency of antigen specific plaque forming cells.

Table 1. Comparison of Ig-secreting spleen cells to hybridoma cells
obtained in myeloma X spleen cell fusions.

	Ig secretors			4 days after 2nd stim. SRC spec. secretors			after Imm.	before Imm.
	Total	IgM	IgG	Total	IgM	IgG		
in spleen	1%	90%	10%	0.05%	50%	50%	5%	0.05%
in hybrids	50%* (49/95)	93% (91/98)	7% (7/98)	10% (5/54)	40% (2/54)	60% (3/54)	20%	<0.2%

---
* In these hybrids the myeloma IgG and additional new immunoglobulin could be detected.

The class distribution and frequency of Ig secreting spleen cells
was taken from ref. 2. The numbers obtained for Ig secreting hybridomas were not corrected for multiple hybrids growing in one culture,
but only those data were pooled where less than 50% positive, independent cultures were obtained per fusion so that most cultures
probably included only one hybrid. The class distribution of hybrid
lines was examined by analysing ^{14}C-leucine labelled culture supernatants on SDS polyacrylamide gels under reducing conditions. The
fusions were done with X63-Ag8 and Sp2/HL-Ag which secrete γ chains
(of different mobility in SDS-gels). We might therefore, have
missed some spleen cell derived γ chains comigrating with the respective myeloma γ chains, resulting in an underrepresentation of IgG
secreting hybridomas.

2. Phylogenetic Restrictions

The mouse myeloma line X63-Ag8 (derived from MOPC 21 (3)) has been
fused to syngeneic and allogeneic mouse spleen cells (4,5) and with
similar high fusion and success rates to rat spleen cells (6). In
collaboration with A. Kelus we have fused X63-Ag8 to rabbit splenocytes. The derivation of hybrids was about 10 fold lower and only
occasionally a hybrid secreted a rabbit light chain, but never a
complete rabbit Ig. Fusions to human and frog lymphocytes resulted
in even fewer hybrids, and they exhibited no stable Ig secretion (H.
Hengartner, this volume). The lack of rabbit, human, or frog Ig
expression seems to be due to the preferential loss of non-mouse
chromosomes.

3. Ontogenetic Restrictions

Tumor cells of various kinds have been fused with Ig producing
cells, and the resulting hybrids have been examined for Ig production
and other markers. Fusion of myeloma cells to spleen cells rescues
spleen cell Ig production but not the T-cell marker Thy 1 (5).

Fusion of thymoma cells to spleen cells rescues the spleen cell Thy 1 marker but not Ig production (7,8). Fibroblasts, ontogenetically less related to the lymphoid cell lineage have been fused to thymoma cells or myeloma cells. The hybrids showed extinction of Thy 1 and Ig production respectively (9,10). These results suggest a simple rule for function rescue fusions. Only tumor lines ontogenetically closely related to the normal cell fusion partner are able to express the specialized function of normal cells. This requirement seems to be quite stringent since the myeloma cells apparently fuse much better with Ig secreting cells than with B cells in other states of differentiation. The following rules therefore should be observed to optimize function rescue fusions: 1) The normal cell fusion partner should be enriched, (for example by antigen stimulation), or by other suitable techniques. 2) The tumor fusion partner should be chosen to obtain maximum ontogenetic and phylogenetic matching with the normal specialized cell.

III. Molecular Restrictions

1. Hybrid Antibody Molecules

The first hybrids were obtained by fusing the BALB/c myeloma cell X63-Ag8, derived from MOPC 21 (3) which secreted an IgG_1 (κ) with unknown specificity. The resulting hybrid cells expressed not only the specific heavy and light chains of the spleen cell fusion partner but also the nonspecific myeloma Ig-chains. We have observed that some such hybrid cells synthesize hybrid antibody molecules. The test for hybrid heavy chain molecules, took advantage of the fact that γ_{2b} chains bind strongly to the protein A of $Staphylococcus$ $aureus$ whereas μ and γ_1 do not. When a mixture of radiolabelled IgG_1 and IgG_{2b} molecules was mixed with $Staphylococcus$ $aureus$ we found that only IgG_{2b} could be eluted from the bacteria whereas IgG_1 remained in the supernatant (Fig. la,B: e versus s). From radiolabelled supernatants of a hybrid line secreting both IgG_1 and IgG_{2b} molecules about half of the γ_1 chains could be eluted together with IgG_{2b} (Fig. la,A: e versus s). This indicates that γ_1-γ_{2b} hybrid molecules were secreted and that the association of these chains is about random. There was, however, no hybrid molecule formation between μ and γ_{2b} (Fig. lb). The light chains L_1 (μ origin) and L_2 (γ_{2b} origin) were evenly distributed in the IgM molecules but there was a preferential association of L_2 with γ_{2b} (Fig. lb). As a first approximation it seems reasonable to assume that a hybrid cell randomly associates the two heavy chains of the γ class and the two light chains into complete IgG molecules. In this case ten different IgG molecules are made, one of which is the antigen binding monospecific IgG which represents only 1/16 of the total amount of IgG secreted. Even after purification with antigen columns, (assuming that one antigen binding site per IgG molecule is sufficient for binding), there are 4 different antibodies, only one of which is monospecific and represents 1/7 of the total amount.

For some applications the heterogeneous antibody production poses no problem. However, if the hybridoma antibodies are to be used, for example, to derive anti-idiotype antibodies, truly monospecific preparations are required. There are two ways to overcome this problem. Loss of the nonspecific chains occurs frequently (in about every 50th reclone analysed). In this way clones have been selected which expressed only the specific chains (4). Another possibility

145

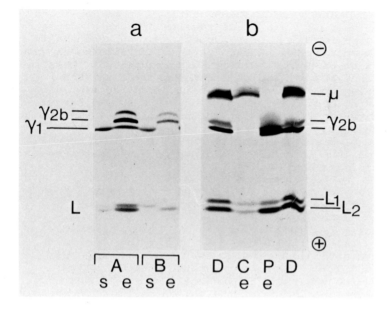

Fig. 1. Hybrid lines secrete hybrid immunoglobulins. (a) ^{14}C-leucine labelled supernatants were adsorbed to *Staphylococcus aureus*. After adsorption, supernatant (s) and eluted material (e) were analysed by SDS-polyacrylamide gel electrophoresis under reducing conditions. A mixture of X63-Ag8 (γ_1,κ) and Sp2/HL-Ag (γ_{2b},κ) (4) supernatants was analysed in slot B whereas the hybrid (X63-Ag8 x Sp2/HL) supernatant is shown in A. (b) ^{14}C-leucine labelled supernatant of a hybrid (SpHL$_2$ x ML$_1$) between Sp2/HL-Ag and a BALB/c spleen cell secreting an IgM in addition to the Sp2 IgG$_{2b}$ is shown in slot D. This supernatant was adsorbed and reeluted from a protein A Sepharose column which binds IgG$_{2b}$ but not IgM (Pe). To remove free light chains the material passing through the protein A column was bound to and reeluted from a concanavalin A sepharose column. (Ce).

is to use a myeloma line which does not synthesize Ig-chains but supports synthesis of the spleen cell derived Ig chains. A subclone (NSI-Ag4/1) of the MOPC 21 myeloma which synthesizes only the myeloma κ but not the γ_1 chain has been successfully used (3).

We recently used the line Sp2/0-Ag for hybridization. It is derived from Sp2/HLGK (3), a hybrid between a BALB/c spleen cell contributing a γ_{2b}(H) and κ (L) chain with anti-SRC activity and X63-Ag8 (γ_1 (G) and κ (K)). The line Sp2/0-Ag is 8-azaguanine resistant, dies in HAT supplemented medium and synthesizes no Ig chains. It was isolated by D. Wilde (MRC, Cambridge) as a reclone from Sp2/HL-Ag. It has about 72 chromosomes which is only 7 more than the chromosome number of X63-Ag8.

146

Although Sp2/HL-Ag is as efficient as X63-Ag8 in yielding viable hybrids when fused with mouse spleen cells, the Sp2/0-Ag line consistently generated 10-20 fold fewer hybrids using either the Sendai or the polyethylene glycol fusion protocol.

We fused 1.5×10^7 Sp2/0-Ag cells with 2.5×10^7 spleen cells of BALB/c immunized against TNP-KLH, using PEG 1500 for the fusion. We derived 8 hybrid lines each synthesizing a monoclonal antibody as judged by SDS-PAGE of radiolabelled supernatants. Four of them synthesizing IgG were found to be specific for the TNP hapten. Since no reexpression of the Sp2/0-Ag parental Ig chains was observed this line is indeed suitable for producing hybrid lines secreting monospecific antibodies.

3. No V-C Scrambling

Different models have been suggested to explain how the expression of Ig class is regulated. On the one hand are models in which the DNA is different in IgG and IgM producing cells. In this model, for example, in an IgM producing cell the DNA for the variable (V) region would be brought in tandem to the DNA for the μ constant (C_μ) region, whereas in an IgG producing cell V would be adjacent to C_γ. On the other hand, the phenomenon of "RNA splicing" suggested a model for switching in which the DNA and primary RNA transcript are the same in both IgM and IgG producing cells. For an IgM producer the transcript would be processed to yield an RNA in which the V region would be in tandem with C_μ whereas for an IgG producer the RNA would be processed so that V^μ and C_γ would be in tandem (11). The fusion of an IgG producing cell (making $V_1 C_\gamma$) with an IgM producing cell (making $V_2 C_\mu$) yields a hybrid producing both IgG and IgM. The RNA processing model predicts that such a hybrid should have both γ and μ processing enzymes and that these enzymes should act on the primary transcripts of both IgG and IgM encoding chromosomes, thus predicting that the hybrid should make the four heavy chains $V_1 C_\gamma$, $V_2 C_\gamma$, $V_1 C_\mu$, and $V_2 C_\mu$. These models can be applied as well to explain regulation of the different IgG subclasses. It was shown by isoelectric focussing analysis that a hybrid between two myeloma lines (Adj PC5 (γ_{2a}, κ) and MOPC 21 (γ_1, κ) does secrete hybrid IgG of both classes but does not secrete molecules having the variable region of one parent cell line coupled to the constant region of the other (1). However, if such "scrambled molecules" represented less than 10% of the total IgG synthesized they would not have been detected. Evidence of scrambling was also sought using a plaque assay on the hybridoma Spl/HLGK which secretes an anti-SRC IgM together with the X63-Ag8 IgG_1. Analyzing 5×10^5 cells for anti-SRC IgG plaque forming cells no such switch could be detected (12). A criticism of this experiment is that we could not be sure of detecting indirect plaques when the variable region of the pentameric IgM molecules is used in the less avid dimeric IgG_1. We have done similar experiments using the cell line (Sp2HL$_2$ x ML$_1$, Fig. 1b) in which the anti-SRC specificity was expressed in a IgG_{2b} molecules which was secreted together with an IgM of unknown specificity. We analysed more than 6×10^6 cells and could not reveal direct IgM anti-SRC plaques. Furthermore, we analysed the supernatant of this line for IgM specific SRC agglutination using an anti-μ specific enhancing serum. We were unable to detect anti-SRC binding IgM molecules although we would have detected anti-SRC IgM activity if it represented 0.1% of the total IgM.

We conclude from these experiments that (a) V-C expression is stable, (b) the RNA processing model in its simplest form is incorrect, and (c) the regulation of class expression might be determined by some mechanism which acts prior to RNA transcription.

References

1. Köhler, G. and Milstein, C.: Continuous cultures of fused cells secreting antibody of predefined specificity. Nature 256: 495-497 (1975).
2. Andersson, J., Coutinho, A. and Melchers, F.: Frequencies of mitogen-reactive B cells in the mouse. J. exp. Med. 145: 1520-1530 (1977).
3. Potter, M.: Immunoglobulin-producing tumors and myeloma proteins of mice. Physiol. Rev. 52: 631-719 (1972).
4. Köhler, G. and Milstein, C.: Derivation of specific antibody-producing tissue culture and tumor lines by cell fusion. Eur. J. Immunol. 6: 511-519 (1976).
5. Köhler, G., Pearson, T. and Milstein, C.: Fusion of T and B cells. Som. Cell. Genet. 3: 303-312 (1977).
6. Galfre, G., Howe, S.C., Milstein, C., Butcher, G.W. and Howard, J.C.: Antibodies to major histocompatibility antigens by hybrid cell lines. Nature 266: 550-552 (1977).
7. Goldsby, R.A., Osborne, B.A. and Herzenberg, L.A.: Hybrid cell lines with T-cell characteristics. Nature 267 707-708 (1977).
8. Hämmerling, G.J.: T lymphocyte tissue culture lines produced by cell hybridization. Eur. J. Immunol. 7: 743-746 (1977).
9. Hyman, R. and Kelleher, T.: Absence of Thy 1 antigen in L-cell x mouse lymphoma hybrids. Som. Cell. Genet. 1: 335-343 (1975).
10. Coffino, P., Knowles, B., Nathenson, S. and Scharff, M.D.: Supression of immunoglobulin synthesis by cellular hybridization. Nature New Biol. 231: 87-88 (1971).
11. Gilbert, W.: Why genes in pieces. Nature 271: 501 (1978).
12. Milstein, C., Adetugbo, K., Cowan, N.J., Köhler, G., Secher, D.S. and Wilde, C.D.: Somatic cell genetics of antibody-secreting cells: studies of clonal diversification and analysis by cell fusion. Cold Spring Harbor Symp. Quant. Biol. 41: 793-803 (1977).

Production of Specific Antibody Without Specific Immunization

R.A. Goldsby, B.A. Osborne, D. Suri, A. Mandel, J.Williams, E. Gronowicz,
L.A. Herzenberg

Under appropriate conditions, the fraction of B lymphocytes stimulated in a population of mouse spleen cells may be large enough to constitute a representative
sample of the total repertoire of antibody specificities which reside in the entire
set of B lymphocytes (1). However the transitory nature of the polyclonal response
(depending on culture condition, it declines and eventually ceases after 4-8 days)
has prevented its exploitation as a strategy for the production of useful amounts
of antibodies to particular antigens. The demonstration (2) that continuous cultures
secreting antibody could be obtained by the hybridization of mouse myeloma and mouse
spleen cells suggested a means of preserving the polyclonal response for an indefinite period of time. It seemed reasonable to assume that the hybridization of
spleen cells in which a polyclonal response had been elicited by LPS with an appropriate myeloma cell line would result in continuous cultures displaying polyclonal
antibody production. Using this approach, it was possible to isolate from the fusion
of unimmunized (Balb/c x SJL)F_1 spleen cells with NS1, a nonsecreting, HAT sensitive variant of MOPC 21, hybrid populations in which the polyclonal response as
evidenced by the elaboration of antibodies to human hemoglobin A, KLH, DNP and human
RBC's was preserved (see Figure 1). Furthermore monospecific production of anti-
DNP antibody was successfully factored out of the polyspecific production of antibodies by a hybrid population through the use of cloning. The expansion and subsequent injection of an active anti-DNP producing clone into a (Balb/c x SJL)F_1 mouse
resulted in the formation of an antibody producing tumor. Collection of serum from
an animal bearing such a tumor allowed us to obtain a useful quantity of an anti-
DNP antiserum of high titer without resort to any program of immunization whatso-
ever (see Figures 2 and 3).

Hybrids were derived as follows: 2×10^8 LPS blasts obtained by the stimulation of
spleen cells from 8-12 week old (Balb/c x SJL)F_1 female mice were mixed with $1 \times$
10^8 NS1 cells in 50 mls of serum-free Dulbecco's Modified Eagle's Medium (DME) and
copelleted by centrifugation for 15 minutes at 400 g. The medium was removed and
the pellet gently resuspended in 2 mls of PEG 1540 and incubated at 37° C for 1.5
minutes. During the 10 minutes following fusion, the suspension was gradually diluted up to 50 mls with serum-free DME and the cells harvested by centrifugation.
The pellet was resuspended in 100 mls of HAT medium which had the following composition: 100 μM hypoxanthine, 10 μM aminopterin and 30 μM thymidine in DME containing
10% fetal calf serum, 2mM glutamine, 5mg/ml glucose, 100 units/ml penicillin and
100 units/ml streptomycin. After 48 hours of incubation at 37° C under an atmosphere
of 90% air/ 10% CO_2 in mass culture, the cells were harvested by centrifugation and
resuspended in 50 mls of HAT-DME. With the aid of a mechanical device for seeding,
sampling and feeding cultures (3), 0.1 ml of the cell suspension was delivered into
the wells of three 96 well microtiter plates onto a feeder layer of 2×10^5 Balb/c
thymocytes. The hybrid populations which appeared after 9-11 days were transferred
by the mechanical replicator on day 14. Fluids from a replica were harvested, diluted three fold and tested for the presence of antibody to the antigens indicated
in Figure 1.

Anti-DNP secreting clones were isolated from hybrid populations of a similar LPS
x NS1 hybridization by the method of limiting dilution in microtiter plates, expanded
to 10^6 cells and injected subcutaneously into 3-5 month old male (Balb/c x SJL) Fl
mice. A tumor bearing mouse was exsanguinated and a portion of the excised tumor
was returned to culture and another portion injected into a (Balb/c x SJL)F_1
recipient.

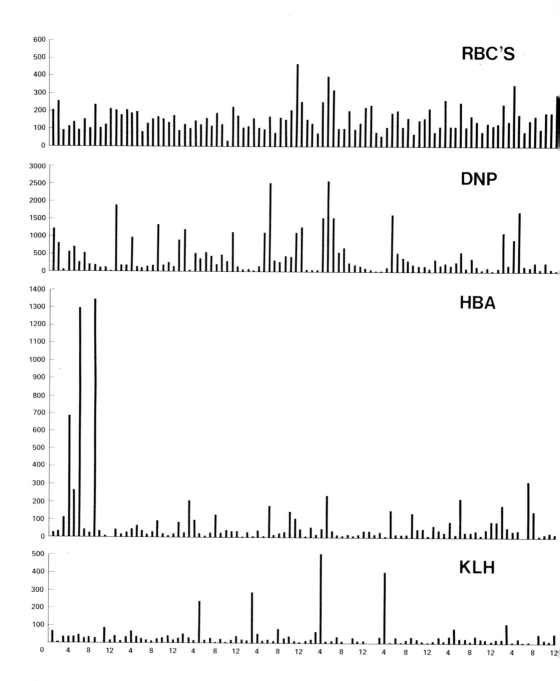

Figure 1. Preservation of the polyclonal response by hybridization. The fluids from 96 hybrid populations were assayed after fusion of 2×10^8 LPS blasts with 1×10^8 NS1 using 52% PEG 1540. The amount of antibody to the indicated antigen was determined by plate or cell binding radioimmunoassay using ^{125}I - labelled goat anti-mouse kappa chain antibody. It is clear that there is a spectrum of reactivity toward each of the antigens.

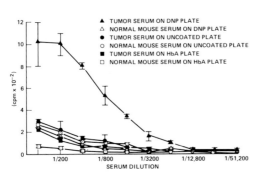

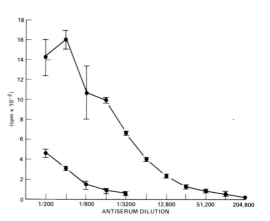

Figure 2. Antigen specificity of serum from an animal bearing an anti-DNP secreting tumor. Serum from an animal bearing a tumor raised by injection of an LPS x NSI anti-DNP producing hybridoma was examined for antibody to DNP or human hemoglobin A in a plate binding radioimmunassay using an ^{125}I-labelled goat anti-mouse kappa chain antibody as a second step. Only plates coated with DNP conjugated BSA and treated with dilutions of serum from tumor bearing animals bound significant amounts of mouse Ig.

Figure 3. Reactivity of the anti-DNP heavy chain with an anti - μ. The hybridoma anti-DNP binds an ^{125}I-labelled rabbit anti-mouse u chain antibody, thus permitting the serological assignment of this antibody to the μ class. The upper curve represents the titration of the tumor serum on a DNP coated plate while the lower curve represents the normal mouse serum control.

References

1. Andersson, J., Coutinho, A., Melchers,F.: Frequencies of mitogen-reactive B cells in the mouse. J. Exp. Med. 145, 1520-1530 (1977).

2. Kohler, G. and Milstein, C.: Continuous cultures of fused cells secreting antibody of predefined specificity. Nature 256, 495-497 (1975).

3. Goldsby, R. A. and Mandel,N.: The isolation and replication of cell clones. In: Methods in Cell Biology. Prescott, D. (ed.) Academic Press 1973, p 261.

Clones of Human Lymphoblastoid Cell Lines Producing Antibody to Tetanus Toxoid

V.R. Zurawski, Jr., S.E. Spedden, P.H. Black, E. Haber

Introduction

We have previously reported (1,2) the establishment of a continuous rabbit cell line synthesizing and secreting antibody to type III pneumococcal polysaccharide. This line was established from a simian virus 40 (SV40) transformed spleen cell suspension obtained from a rabbit hyperimmunized with type III pneumococcal vaccine. This approach has not been generally applied because the quantity of antibody produced by the cell line, TRSC-1, was small (3) and resulted from an infrequently occurring transformation event. Furthermore, as reported in this volume by several laboratories, cell fusion methods (4) have regularly resulted in the establishment of murine hybridomas producing specific antibody.

Synthesis and secretion of immunoglobulin (Ig) by human lymphoblastoid cell lines is well documented (5). Therefore, to obtain cell lines synthesizing specific human antibodies, we have extended the viral transformation approach to human cells. We have effected viral transformation of human peripheral blood lymphocytes with Epstein-Barr virus (EBV), and demonstrated (6) that such EBV transformed human cell lines can be established that synthesize and secrete specific antibody to tetanus toxoid. Moreover, Steinitz et al (7) have independently reported establishing EBV transformed lines that produce antibody to another antigen, 4-hydroxy-3,5-dinitrophenacetic acid (NNP). This suggests that with appropriate selection of immunocompetent lymphocytes, the production of human antibodies to any antigen of interest might be accomplished utilizing EBV "immortalized" B cells.

Central to the problem of long-term antibody production by such cells is retention of their differentiated function despite the fact that they have been transformed to continuously proliferating lines. Our parent lines secreting antibody to tetanus toxoid did not retain their ability to secrete specific antibody after extended time in tissue culture (6). We report here the establishment of limiting dilution clones of these parent lines, which appear to be stable producers of antibody to tetanus toxoid.

Results and Discussion

Culturing cells at limiting dilution on feeder layers has proven to be the best method for obtaining clones of our lyphoblastoid cell lines producing antibody to tetanus toxoid. Critical to successful establishment of such clones has been the type of cells in the feeder layer. We found that human foreskin fibroblasts (HFF) provided an excellent monolayer that was stable for several weeks and evidently provided appropriate nutrients and/or growth factors to colony forming lymphoblasts. Human embryonic kidney, amnion, or embryonic lung cells were not adequate for this purpose. Placental cells (8) have not yet been evaluated.

Cells of the parent lines in log phase growth were suspended in RPMI 1640 medium, supplemented with 20% fetal calf serum and 2 mM glutamine. No doublets were detected in at least 5000 cells counted. Serial five- or tenfold dilutions of parent cell lines were made in the supplemented RPMI 1640 medium. Addition of conditioned medium did not seem to affect the efficiency of cloning. Aliquots of 100 μl of each dilution at 10^4 through 1 cell per well were placed in 96-well Costar microtiter plates containing confluent HFF monolayers. HFF cells at lower passage number (<20) were most effective as feeder layers. The peripheral wells of the microtiter plates were filled with phosphate buffered saline

(PBS), and were not used for culture, leaving 60 wells per plate containing cells. At least 10 plates were seeded for each cloning experiment. Inspection of microtiter plates indicated that lymphoblastoid colonies began appearing in those wells where only 1 cell had been seeded as early as two weeks after initiation of the cultures of most parent lines. After eight weeks, no new colonies were observed in any wells. Therefore, wells with viable cells were counted several times between two and eight weeks to determine approximate cloning efficiencies.

Table 1 presents cloning data for some parent lines at various passage levels. Cloning efficiencies for the lines were variable. Antibody to tetanus toxoid was detected in only a small percentage of the wells seeded at 1 cell per well. For some parent lines no antibody was found in wells so seeded, although at 10^5 or 10^4 cells per well, antibody was detectable (data not shown).

Table 1: Cloning efficiencies of various cell lines on HFF monolayers

Parent line	Passage	Cells seeded per well	Wells with lymphoblastoid colonies (%)	Wells with colonies producing antitetanus antibody (%)
3GC-C2(2H8)	5	100	100	N.D.
		10	82	5
		1	36	0.8
3GC-C2(F3)	1	10	100	8
		1	1	0
3GC-C2	6	1	3	2
3GC-C2	7	10	35	14
		1	5	0
3GC-C2	8	1	7	0
3GC-C5	5	100	6	0
		10	0	0
		1	0	0
4LP-C3	6	1000	92	0
		100	72	0
		10	17	0
		1	1	0
4LP-B4	8	10	100	N.D.
		1	13	N.D.
4LP-B4	9	10	100	N.D.
		1	83	2

When 4LP-B4 at passage 9 was seeded at 1 cell per well, 450 out of 540 wells (83%) contained lymphoblastoid colonies. Of these, probably no more than 38% were true clones, as computed from Poisson distribution curves. Only 7 of the 450 clones (2%) showed evidence of antitetanus antibody production. Six of these 7 were successfully established in continuous culture for more than five months, and were routinely passaged for more than 10 weeks. For this entire time, they retained their ability to synthesize and secrete antibody to tetanus toxoid, achieving concentrations of 10 ng/ml in the culture supernatants, as determined by radioimmunometric assay (6). These antibodies also bound a highly purified sample of tetanus toxin (9). The stability of antitetanus antibody production by each of the six clones far exceeded that for the parent line 4LP-B4.

Cytoplasmic and surface staining for Ig light and heavy chains with specific fluoresceinated antibodies revealed that only λ light chains and γ heavy chains were being synthesized by each of the six clones whereas the parent line 4LP-B4, although synthesizing predominantly λ and γ chains, also was synthesizing μ and α heavy chains as well as κ light chains. Thus it is likely that the antitetanus antibody synthesized is of the IgG-λ isotype.

Recloning of the six clones producing antibody revealed a marked increase in the frequency of colonies, seeded at 1 cell per well, that produced antitetanus antibody. It has also served as an approximate measure of the clonotypic nature of the parent lines. For example, it can be seen (Table 2) that with recloning of clone 1E5, 15% of wells contained colonies that were secreting antibody. With clone 2F11, 87% of such wells contained detectable antitetanus antibody.

Table 2: Cloning efficiencies of clones 1E5 and 2F11 on HFF monolayers

Parent clone	Passage	Cells seeded per well	Wells with lymphoblastoid colonies (%)	Wells with colonies producing antitetanus antibody (%)
1E5	2	1000	100	100
		100	100	N.D.
		10	89	N.D.
		1	12	15
2F11	3	1000	100	100
		100	100	100
		10	89	100
		1	10	89

We have established clones of lymphoblastoid cell lines, which are evidently stable producers of antibody to tetanus toxoid. However, the true monoclonal nature of either these first or second generation clones must still be verified by thorough analysis of the secreted antibody. Monoclonal human IgG antibody to tetanus toxoid (toxin), synthesized in vitro and obtained in sufficiently large quantities, may prove useful clinically for passive immunization.

Acknowledgements

This work is from the Department of Medicine, Massachusetts General Hospital, and the Harvard Medical School, Boston, MA. Supported in part by USPHS Grant CA10126-11 and HL19259. V.R. Zurawski, Jr., is a recipient of NIH Fellowship Award 5F32AI05338-02.

References

1. Collins, J.J., Black, P.H., Strosberg, A.D., Haber, E., Bloch K.J.: Transformation by simian virus 40 of spleen cells from a hyperimmune rabbit: evidence for synthesis of immunoglobulin by the transformed cells. Proc.

Nat. Acad. Sci. USA 71, 260-262 (1974).

2. Strosberg, A.D., Collins, J.J., Black, P.H., Malamud, D., Wilbert, S., Bloch, K.J., Haber, E.: Transformation by simian virus 40 of spleen cells from a hyperimmune rabbit: demonstration of production of specific antibody to the immunizing antigen. Proc. Nat. Acad. Sci. USA 71, 263-264 (1974).

3. Black, P.H., Zurawski, V.R., Jr.: In vitro production of antibody by viral transformed cell lines. In: Antibodies in Human Diagnosis and Therapy. Haber, E. Krause, R.M. (eds.). New York: Raven Press, 1977, pp. 239-253.

4. Kohler, G., Milstein, C.: Continuous cultures of fused cells secreting antibody of predefined specificity. Nature 256, 495-497 (1975).

5. Fahey, J.L., Buell, D.N., Sox, H.C.: Proliferation and differentiation of lymphoid cells: studies with human lymphoid cell lines and immunoglobulin synthesis. Ann. N.Y. Acad. Sci. 190, 221-234 (1971).

6. Zurawski, V.R., Jr., Haber, E., Black, P.H.: Production of antibody to tetanus toxoid by continuous human lymphoblastoid cell lines. Science 199, 1439-1441 (1978).

7. Steinitz, M., Klein, G., Koskimies, S., Hakel, O.: EB virus induced B lymphocyte cell lines producing specific antibody. Nature 269, 420-422 (1977).

8. Grogan, E.A., Enders, J.F., Miller, G.: Trypsinized placental cell cultures for the propagation of viruses and as "feeder layers." J. Virol. 5, 406-409 (1970).

9. This sample of tetanus toxin was the generous gift of William C. Latham of the State Biologic Laboratories, Jamaica Plain, Massachusetts, USA.

Establishment of Specific Antibody Producing Human Lines by Antigen Preselection and EBV-Transformation

M. Steinitz, S. Koskimies, G. Klein, O. Mäkelä

A. Introduction

Human lymphocytes can be immortalized ("transformed") by laboratory (B95-8) isolates of Epstein Barr virus (EBV) (1). Following primary infection, the virus induces EBNA, the EBV-determined nuclear antigen, in B-lymphocytes only. Immortalized cell lines emerge in 10-20 days. Such lines, designated as LCL (lymphoblastoid cell lines) carry multiple copies of the viral genome and regularly express EBNA in 100% of the cells (2).

Only cells with EBV receptors on their surface (3) can be transformed (3, 4). It has been suggested that all B-cells carry EBV-receptor. In addition we have shown (5) that the transformation of different fractions of human peripheral blood could be directly related to the number of surface Ig positive cells.

All EBV-immortalized lines expressed B-cell characteristics. Normal B-cell-derived lymphoblastoid lines carry EBV receptors and complement receptors, they are often positive for surface immunoglobulin and also secrete immuno-globulin into the medium (2). They seem to represent their original EBV-infected B-cell.

We have departed from the working hypothesis that preselection of B-lymphocytes with appropriate antigen combining specificities and subsequent EBV-induced immortalization should lead to permanent cell lines that produce specific antibodies against the corresponding antigen. We recently published a preliminary report on that topic (6).

B. Materials and Methods

I. Separation and Selection of Lymphocytes

1. Lymphocytes: Fresh blood was obtained from three healthy members of our laboratory staff in Helsinki (E, L and S), that were previously found to have relatively high antibody titers against the synthetic hapten NNP (4 hydroxy-3.5-dinitro phenacetic acid). Their peripheral lymphocytes showed ($\sim$ 0.3%) specific NNP rosetting. Lymphocytes were isolated on Ficoll Isopaque density gradient as described by Böyum (7).

2. Hapten: The hapten used for selection of lymphocytes was NNP-azide. It was prepared according to Brownstone et al. (8). NIP shows > 90% cross reactivity with NNP and could therefore be used instead of NNP in some tests.

3. Coupling of NNP to red blood cells: 40 mg NNP-azide were dissolved in 1 ml of N-N-dimethylformamide (C_3H_7NO, Merck). The solution was diluted 1/10 in 0.12 M $NaHCO_3$, Na_2CO_3 buffer, pH 9.25, to give 4 mg NNP azide/ml. 1 volume of the latter mixed with 1 volume of (v/v) 20% autologous red blood cells in carbonate buffer was incubated in room temperature for 60 min, washed and finally adjusted to 20% red cells in PBS (8).

4. Rosetting of cells with hapten coupled erythrocytes: The method described by Osoba was used (9). To prepare specific rosettes with the hapten coupled erythrocytes (NNP-RBC), 5×10^6 lymphocytes/ml were mixed with 0.01 ml 20% coupled red cells in 4°C. The mixture was spinned down 100 g, 15 min. The pellet was further incubated at 4°C for 30 min. Separation of the rosettes from the non-rosetting lymphocytes was achieved by spinning down the mixture on Ficoll Isopaque (density 1.077) for 20 min at 400 g.

Rosette formation with cells from established lines was performed in the same way.

II. Epstein Barr Virus Infection

The transforming prototype B95-8 substrain of EBV (1) was used to immortalize the lymphocytes. 4 ml supernatant of the mycoplasma-free virus producer line was added to peripheral lymphocytes at 37°C for 10 hrs. The cells were fed regularly and their growth was followed. Within 2-4 weeks, permanently growing cell lines emerged.

III. Medium and Culture Conditions

RPMI-1640 (Gibco) was supplemented with glutamine $(4 \times 10^{-2}$ M$)$ streptomycin (100 µg/ml), penicillin (100 u/ml), Hepes buffer $(10^{-2}$ M$)$ and 10% heat inactivated fetal calf serum (FCS, Gibco). The established cell lines were cultured in 30 ml plastic flasks (Falcon). The cells were diluted twice a week to about 1×10^5 cells/ml and kept at 37°C in an incubator with 5% CO_2 in air. They are preserved at frozen storage in 10% DMSO at -80-90°C. Cloning of the established cell lines was performed as we described earlier (10).

Human embryo lung fibroblasts (5×10^5) were seeded into Falcon petri dishes. When the fibroblasts covered the bottom of the dish, they were exposed to 6000 r (Siemens roentgen unit, 220 kV, 15 mA, filtration 1 mm aluminium) and covered with 2.5 ml of 0.45% agarose (Sea plaque agarose mci, biomedical) in RPMI 1640 with 20% FCS, and 0.1 µg/ml sodium pyruvate. Twenty hours later 2.5 ml 0.35% agarose containing 2×10^4 cells in the same medium, was layered on the top of the feeder layer. The cultures were allowed to solidify and incubated at 37°C, 5% CO_2 in air in a high humidity box. The growth of the colonies was inhibited when the top layer contained more than 0.35% agarose. To achieve an accurate percentage of agarose, a mixture was prepared and layered on the top of the feeder layer as follows: 45 mg agarose in 1 ml BSS was autoclaved and mixed with 11.2 ml warm medium; 1.9 ml aliquots were mixed with 0.1 ml cell suspensions with the required cell concentration.

IV. Characterization of the Cell Lines Supernatants

1. Inhibition of haptenated T_4 plaque formation: Inhibition of haptenated T_4 phage induced plaque formation in E. coli (10) was used to quantitate the anti-NNP activity of the culture supernatants. Supernatant aliquots of 0.3 ml were filtered through 0.22 µ filter, mixed with 0.3 ml haptenated T_4 suspensions containing 200 plaque forming units and incubated for 4 hours at 37°C. 5 ml 0.7% fluid agar, containing about 10^8 E. coli were added, poured onto a solid 2% agar plate, and incubated overnight at 37°C. The titers are given as the reciprocal dilution causing 50% inactivation of the plaque formation activity of the haptenated phage.

2. Agglutination test: Two-fold supernatant dilutions in BSS were admixed in 0.1

ml aliquots to 0.02 ml of 1% NNP-RBC in round bottom wells of a microtest plate. After 60 min, the highest agglutinating dilution was determined.

3. Characterization of anti-NNP antibodies: NNP-RBC were exposed to test or control culture supernatants and stained by indirect fluorescence. In each sample, 0.1 ml of a 2% erythrocyte suspension was mixed with 1 ml supernatant and incubated for 60 min. After 3 washings, the cells were divided into 5 tubes and mixed with 0.05 ml 1/10 diluted FITC conjugated reagent. All reagents were from Dakopatts, Denmark. They included 1) rabbit anti-human IgG, IgA, IgM, kappa and lambda; 2) rabbit and human IgM, μ chain specific; 3) rabbit anti-IgM Fc fragment; 4) rabbit anti-kappa (Bence Jones) and 5) rabbit anti-lambda (Bence Jones). The following cell lines were included as positive and negative staining controls: Surface IgM, kappa positive BJAB (11), IgM, lambda positive Ramos (12) and IgG, lambda positive Rael (13).

4. Anti-NNP plaque test: The test was carried out according to Cunningham and Szenberg (14), with NNP-RBC. Plaques were counted under an inverted microscope after 90 min.

C. Results

I. Establishment of Anti-NNP Producing Cell Lines

10^8 lymphocytes derived from donors with high anti-NNP titers were rosetted at 4°C with autologous NNP-coupled erythrocytes. The autologous combination was prefered, to avoid NNP-unrelated rosette formation. Rosetting cells were separated from the rest of the lymphocytes by the Ficoll Isopaque method. Pelleted rosettes with erythrocytes, interface (non-rosetting) lymphocytes and unselected lymphocytes of the same donors were infected with transforming EBV. Polyclonal lymphoblastoid cell lines emerged and grew with an approximate doubling time of 24 hrs. They were 100% EBNA positive (15) and surface Ig positive demonstrated by direct staining.

II. Properties of the Lines Established by Transforming NNP Receptor Selected Cells.

Surface receptors for NNP were assessed by rosetting the cells in the cold with NNP-coated SRBC. At the single cell level, NNP antibody release was tested by the Cunningham-Szenberg plaque assay. Figure 1 shows a typical NNP specific plaque. The cell lines derived from the EBV-transformed NNP-rosetting fraction contained 13-18% NNP-rosetting and a somewhat lower frequency of plaque forming cells (Table 1A). In contrast, the lines derived from the EBV-transformed, unselected lymphocytes or from the EBV-transformed, NNP rosette deprived lymphocytes gave no NNP specific rosettes or plaques. Reselection of the EP line by another round of NNP-rosetting and Ficoll separation led to a subline that contained 86% NNP-specific RFC and 69% PPC. This implies that PFC and RFC were identical or largely overlapping. Plaque formation was completely inhibited by NaN_3, confirming that the plaques were due to active antibody release.

III. Antibody Tests on Culture Supernatants

Supernatants of the established, EBV-transformed lines were tested for anti-NNP activity by 1) agglutination of NNP-RBC and 2) inactivation of haptenated T_4 phage. Table 1B shows the results obtained with unconcentrated supernatants, harvested from logarithmically growing cultures ($0.4 - 0.8 \times 10^6$ cells/ml).

158

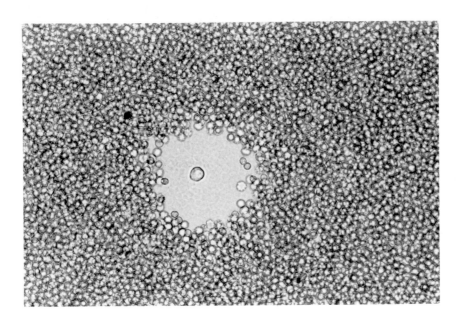

Fig. 1. A specific plaque formed by a cell from the EBV established cell line
EP. The erythrocytes were coupled with NNP and guinea pig serum was used as
a source for complement (X312).

Antibody activity was detected only in the lines derived from the NNP-selected
cells. Non-coupled erythrocytes were not agglutinated. The double selected EP
line showed a four-fold increase in agglutinating activity compared to the original
EP line. This corresponds to the increase of specific NNP-RFC formation (Table
1A). Haptenated phage inactivation activity was abolished by prior treatment of
the cell supernatants with 2-mercapto-ethanol (2-ME). This agrees with the fact
that sucrose gradient fractionation localized all activity to the 19S region (results
not presented).

The polyclonal, NNP-selected EP lymphocyte derived culture maintained its
antibody secretion during at least 14 weeks after EBV infection (Table 2).

IV. Anti-NNP Antibody Class

NNP-RBC were incubated with culture supernatants and subsequently stained with
FITC coupled reagents against different heavy and light chains. The results are
shown in Table 3. All positive reactions appeared as brilliant membrane staining
of all erythrocytes. Negative tests showed no trace of any staining.

All three NNP selected cell lines produced IgM kappa antibodies against NNP. In
contrast, the sera of the three original donors contained both IgM and IgG anti-
NNP with both lambda and kappa light chains.

V. Cloning of NNP Antibody Producing Lines

As a prerequisite for further immunochemical characterization of the NNP-anti-

159

Table 1. NNP rosette and plaque formation by EBV-transformed lines derived from NNP-rosetting lymphocytes (A) and agglutination and haptenated phage inactivation titre in unconcentrated culture supernatants (B).

Cell lines	Source of EBV-transformed explant	A			B		
		RFC[1] (%)	PFC[2] (%)	PFC with NaN3 (%)	NNP agglutination titre 3 §	NIP-capT4 inactivation titre 4 §§	NIP-capT4 inactivation after 2 ME treatment
EP	NNP-selected lymphocytes	15	11	0	256	N.D.	N.D.
LP	NNP-selected lymphocytes	13	N.D.	N.D.	64	86	< 3
SP	NNP-selected lymphocytes	18	N.D.	N.D.	64	981	< 3
E	Unselected lymphocytes	0	0	0	0	N.D.	N.D.
L	NNP rosette deprived lymphocytes	0	0	0	0	N.D.	N.D.
S	Unselected lymphocytes	0	0	0	0	N.D.	N.D.
EP reselected	NNP reselected EP cell line	86	69	0	1024	4070	< 3

1. RFC - rosette forming cells.
2. PFC - plaque forming cells.
RFC were performed with NNP-RBC in $4^{\circ}C$.
No rosettes or plaques were formed with uncoupled erythrocytes or when complement was omitted from Cunningham-Szenberg test.
3. Reciprocal titres.
4. 50% inhibition titres of NIP-cap-T4 phage on E. coli.
§ Non-coupled erythrocytes did not agglutinate.
§§ Non-haptenated phage was not inactivated.

Table 2. Anti-NNP activity of EP (NNP-selected) culture supernatants during 14 weeks after EBV infection.

Weeks after EBV infection	NNP-RBC agglutination (titre)	NIP-cap-T4 inactivation (titre)
6	512	2570
7	512	2980
8	512	3000
9	2048	9800
10	1024	5150
12	1024	5500
14	1024	2974
Control-FCS	0	8

body product, one of the NNP-selected lines (EP) was cloned. Cells were seeded in agarose on a fibroblast feeder layer. 150 colonies were isolated but only 30 grew into lines. 9 lines were positive for NNP. Data are shown for 7 in Table 4. The other 21 lines were totally negative for RFC, for PFC, for agglutination of NNP-RBC and for NNP-cap-T4 inactivation. The specific clones had high percentage of RFC and PFC. The anti-DNP activity detected in the polyclonal EP supernatant was absent from the supernatants of the clones. 4.5 - 16.5 μg/ml human IgM were detected in the supernatants of the specific clones. The anti-NNP antibodies of all the positive clones were IgM kappa.

160

Table 3. Indirect fluorescent staining of NNP-RBC after exposure to culture supernatants or sera (see Materials and Methods).

Supernatants[1]	Reagent[2]				
	Ig	IgM	IgG	Lambda	Kappa
EP	+	+	-	-	+
LP	+	+	-	-	+
SP	+	+	-	-	+
E	-	-	-	-	-
L	-	-	-	-	-
S	-	-	-	-	-
EP-reselected	+	+	-	-	+
Sera					
Serum E	+	+	+	+	+
Serum L	+	+	+	+	+
Serum S	+	+	+	+	+

1. Designation of cell lines - as in Table 1.
2. The staining was done with FITC coupled antisera, as explained in Materials and Methods.

Table 4. Anti-NNP activity of EP clones.

Clone number[1]	RFC[2] (%)	PFC[3] (%)	NNP-RBC agglutination	Heptenated T$_4$ phage inactivation titer					Human IgM in medium[4] μg/ml
				NNP	DNP	cloxa-cillin	ABA-MIP	NP	
3	60	70	640	947	22	< 1	< 1	< 3	5.6
21	N.D.[5]	N.D.	160	N.D.	N.D.	N.D.	N.D.	N.D.	N.D.
22	52	86	640	N.D.	N.D.	N.D.	N.D.	N.D.	N.D.
24	62	82	640	997	9	< 1	< 1	< 3	7.3
42	63	67	640	876	8	< 1	< 1	< 3	5.9
22	77	77	640	993	8	< 1	< 1	3	4.5
35	31	82	640	1050	21	< 1	< 1	3	16.5
6	0	0	0	10	13	< 1	< 1	< 3	N.D.
43	0	0	0	11	18	4	< 1	< 3	< 0.8
EP	15	11	320	2200	1800				
Medium			0	14	20				

1. The numbers refer to clones derived from the polyclonal culture EP (see Table 1).
2. Rosette formation was done with NNP-RBC.
3. Plaque formation was done with NNP-RBC.
4. Culture medium was first concentrated x 10 and then, using Mancini technique, the IgM was assessed.
5. N.D. - not done.

D. Concluding Remarks

EBV-transformation of specific antibody forming human lymphocytes requires preselection of B-cells with the appropriate, antigen binding specificity. This could be achieved in the present study, by using three donors with high antibody titers against the hapten NNP. All three donors were laboratory technicians with extensive previous contacts with the hapten. Following preselection by NNP-coupled erythrocytes rosetting and Ficoll Isopaque separation, a small number of lymphocytes could be immortalized by transforming EBV (B95-8 substrain). The

161

specific cell lines produced anti-NNP antibodies all being IgM kappa type. 10-20% of the cells were RFC and PFC with NNP-RBC. The polyclonal lines established were cloned in agarose. Thus monoclonal antibodies could be produced.

The availability of cloned antibody forming lymphoblastoid cells should facilitate the studies on the idiotypic and other characteristics of human antibodies directed against the well-defined haptens.

The availability of a method to establish specific antibody forming human lymphoid lines are obvious. In principle, there should be no limitation on this possibility, as long as antigen binding B-lymphocytes can be preselected and subsequently EBV-transformed. Human antibodies can be expected to discriminate human tissue and human antigens better than heterologous antibodies. If practical use is concentrated, the avoidance of unwanted allergic reactions may be a great advantage. The disadvantage inherent to the small quantities of the immunoglobulin might be overcome after fusion with myeloma cells.

Acknowledgements. This research was supported in part by Contract No. NO1 CP 33316 within the Virus Cancer Program of the National Cancer Institute and the Swedish Cancer Society. Michael Steinitz is the recipient of a fellowship from the U.S. Cancer Research Institute.

References

1. Miller, G., Lipman, M.: Release of infectious Epstein-Barr virus by transformed marmoset leukocytes. Proc. Natl. Acad. Sci (USA) 70, 190-194 (1973).
2. Nilsson, K., Pontén, J.: Classification and biological nature of established human haematopoietic cell lines. Int. J. Cancer 15, 321-341 (1975).
3. Yefenof, E., Bakacs, T., Einhorn, L., Ernberg, I., Klein, G.: Epstein-Barr virus (EBV) receptors, complement receptor, and EBV infectibility of different lymphocyte fractions of human peripheral blood. I. Complement receptor distribution and complement binding by separated lymphocyte subpopulations. Cell. Immunol. 35, 34-42 (1978).
4. Einhorn, L., Steinitz, M., Yefenof, E., Ernberg, I., Bakacs, T., Klein, G.: Epstein Barr virus (EBV) receptors, complement and EBV-infectibility of different lymphocyte fractions of human peripheral blood. II. EBV studies. Cell. Immunol. 35, 43-58 (1978).
5. Steinitz, M., Bakács, T., Klein, G.: Interaction of the B95-8 and P3HR-1 substrains of Epstein Barr virus (EBV) with peripheral human lymphocytes. Submitted for publication (1978).
6. Steinitz, M., Klein, G., Koskimies, S., Mäkelä, O.: EB virus-induced B lymphocyte cell lines producing specific antibody. Nature 269, 420-422 (1977).
7. Böyum, A.: A one stage procedure for isolation of granulocytes and lymphocytes from human blood. Scand. J. Clin. Lab. Invest. 21, suppl. 97, 51-76 (1968).
8. Brownstone, A., Mitchison, N.A., Pitt-Rivers, R.: Chemical and serological studies with an iodine containing synthetic immunological determinant 4-hydroxy-3-iodo-5-nitrophenylacetic acid (NIP) and related compounds. Immunology 10, 465-479 (1966).
9. Osoba, D.: Some physical and radiobiological properties of immunologically reactive mouse spleen cells. J. Exp. Med. 132, 368-383 (1970).
10. Steinitz, M., Klein, G.: Further studies on the differences in serum dependence in EBV negative lymphoma lines and their in vitro EBV converted, virus-genome carrying sublines. Europ. J. Cancer 13, 1269-1275 (1977).
11. Menezes, J., Leibold, W., Klein, G., Clements, G.: Establishment and characterization of an Epstein Barr virus (EBV) negative lymphoblastoid B

cell line (BJAB) from an exceptional EBV genome negative African Burkitt's lymphoma. Biomedicine 22, 276-284 (1975).

12. Klein, G., Giovanella, B., Westman, A., Stehlin, J.G., Mumford, D.: An EBV-genome-negative cell line established from an American Burkitt lymphoma; Receptor characteristics. EBV infectibility and permanent conversion into EBV-positive sublines by in vitro infection. Intervirology 5, 319-334 (1975).

13. Klein, E., Nilsson, K., Yefenof, E.: An established Burkitt's lymphoma line with cell membrane IgG. Clin. Immunol. Immunopathol. 3, 575-583 (1975).

14. Cunningham, A.J., Szenberg, A.: Further improvements in the plaque technique for detecting single antibody-forming cells. Immunology 14, 599-600 (1968).

15. Reedman, B.M., Klein, G.: Cellular localization of an Epstein Barr virus (EBV) associated complement-fixing antigen in producer and non-producer lymphoblastoid cell lines. Int. J. Cancer 11, 499-520 (1973).

Human Normal and Leukemia Cell Surface Antigens. Mouse Monoclonal Antibodies as Probes

R. Levy, J. Dilley, L.A. Lampson

A. Introduction

The definition of human cell surface antigens is an area in which monoclonal anti-bodies should make an especially important contribution. As in other systems, we are faced with a complex mixture of antigens to be analyzed and an exceedingly complex conventional antibody response. However, with human systems, we are further limited because of our inability to employ the tools of classical breeding genetics and deliberate alloimmunization.

Libraries of homogenous xenoantibodies to the human cell surface should contain the reagents necessary to define differentiation antigens, polymorphic antigens and, if they exist, tumor specific antigens. The only limitation of this approach is the antibody repertoire of the immunized animal. Our job, then, is to generate such libraries and to search through them for the antibodies of special interest.

We have employed two different approaches for the generation of monoclonal anti-bodies to human cell surface antigens: the spleen fragment culture technique of Klinman (1) and the somatic cell hybridization technique of Kohler and Milstein (2). Data will be presented on antibodies derived by both of these techniques.

From our point to view, the choice between these techniques depends upon the goal of the experiment. If the intent is to examine the mouse repertoire of evokable antibody responses to human cell surface antigens, and to quickly define as many antigens as possible, then we prefer the technique of fragment culture. However, if large quantities or stable sources of standard reagents are to be produced, then somatic cell hybridization is preferred. Obtaining antibody producing somatic cell hybrids depends upon immunization, and therefore, until data to the contrary appears, we assume that both techniques tap the same potential library of antibodies.

B. Technique

I. Fragment Culture System

Details of the basic technique (1) and our adaption of it for the study of human cell surface antigens (3) are covered in previous publications. Limiting dilutions of immune spleen cells are transferred with antigen to syngeneic irradiated recipi-ents, and shortly thereafter $1mm^3$ fragments of the recipient spleen are placed into culture. Culture fluids are changed and collected every three to five days and assayed for antibody activity by a cell binding assay (4). The responses are anti-gen dependent, donor cell dependent and, if a low enough donor cell number is used, they show a restricted isoelectric focusing pattern. These cultures usually secrete antibody for several weeks and sometimes for as long as two or three months, during which time ng–μg quantities of antibody are produced. This is usually enough to perform several hundred binding assays or up to a thousand micro-complement cytotox-icity assays.

II. Somatic Cell Hybridization

Our technique combines elements of those initially described by Kohler and Milstein (2) and by Gefter, et al (5). We have not critically examined many of the steps and therefore, our procedure may not be optimal.

Animals, either singly immunized or hyperimmunized, are given an I.V. boost of 10^6-10^7 human cells three days prior to use. P3/NS1/1-Ag4 (6) or P3/X63-Ag8 (2) cells are taken from a growing culture of less than 10^6 cells/ml. Spleen cells are mixed with NS-1 or P3/X63 cells in ratios varying from 1/4 to 4/1 - typically 5×10^7 cells of each. Cells are washed in protein-free Dulbecco's Modified Eagle's Medium (DME) and centrifuged into a common pellet. The pellet is gently re-suspended in 2ml of 35% polyethylene glycol 1540 (Baker) and centrifuged gradually bringing the speed up to 400xg for a total of six minutes. The PEG is removed and the pellet is re-suspended in 2 ml of protein-free DME containing HAT. After ten minutes, the cells are re-centrifuged and re-suspended in a large volume of DME-HAT containing 15% fetal calf serum and plated into microculture wells at a density of approximately 10^6 cells/well. Hybrids can usually be seen by microscope by day 6 of culture. Their growth after that time, if sluggish, can be enhanced by the addition of a drop of fresh medium containing 3×10^4 fresh spleen cells. Subcloning of antibody producing culture is then performed by dilution into microtiter wells containing spleen cells as feeders. The overall efficiency of hybridization has varied greatly, but has been as high as one hybrid per 10^6 input spleen cells. The positivity rate for production of antibody to human cell surface antigens for hybrids made with P3/X63 is approximately 10%, and for hybrids made with NS-1 is approximately 40%.

C. Results

Mice were immunized with human chronic lymphocytic leukemia (CLL) cells obtained from a single patient (V.S.). Individual monoclonal antibodies were obtained both by spleen fragment culture technique (Table I) and by somatic cell hybridization with P3/X63 or NS-1 (Table II). In each case the antibodies obtained were analyzed for their binding activity against a panel of target cells. Although the two test panels are not identical, certain generalizations can be made about the results. In both cases the most frequent reactivity pattern observed (approximately 50% of the total response) was positivity against all the test cells of the panel. These pan-human antibodies were screened, and found not to be directed against human β2 micro-globulin or human Ig κ, λ, or μ chains. Furthermore, they do not all detect the same antigens, since they all show different preferences of reactivity against the various test cells. The relatedness of the cell surface molecules they recognize is now being studied by immunoprecipitation and gel analysis.

The remainder of the antibodies produced by both techniques reacted with the respective test panels to give a number of different reactivity patterns. These patterns can be examined for specificities of special interest. For example, are there any antibodies that are specific for CLL cells as opposed to normal cells, i.e. leukemia specific? No, with the possible exception of the antibodies with pattern #1 in Table I, reacting only with the immunizing V.S. cells. These antibodies may detect individual specific antigens which may or may not relate to leukemia. Since the immunizing cells were B leukemia cells, are there any B cell antigens detected? None of the antibodies in either set reacted with all the B cells and none of the T cells, however, some of the antibodies (patterns 1,2,4,6 in Table I and pattern 2 in Table II) reacted with only some of the B cells. These may be candidates for detecting B cell alloantigens. Similarly, one can look for candidates which may detect other human polymorphic antigens such as HLA. None of these particular antibodies seem to detect HLA.

Table I. Fragment culture antibodies. Animals immunized against V.S.

Reactivity Pattern	Incidence	Normal	CLL				
		PBL pool	V.S.	P.H.	A.C.	V.M.	R.V.
1	5	−	+	−	−	−	−
2	2	−	+	+	−	−	−
3	3	+	+	−	−	−	−
4	1	−	+	+	−	+	−
5	1	+	+	−	+	+	−
6	1	−	+	+	+	+	−
7	1	+	+	+	−	−	+
8	1	+	+	+	−	+	−
9	1	+	+	+	+	−	+
10	5	+	+	+	+	+	−
11	1	+	+	+	−	+	+
12	27	+	+	+	+	+	+
Total	49						

Table II. Hybridoma antibodies. Animal immunized against V.S.

		Normal PBL			CLL			B-cell			T-cell	
Reactivity Pattern	Incidence	Pool	R	J	V.S.	P.H.	A.C.	8866	8396	S.B.	H-2SB2	8402
1	1	+	+	-	+	-	-	-	-	-	-	-
2	1	-	-	-	+	-	-	+	-	-	-	-
3	1	-	-	-	+	-	-	+	-	-	-	+
4	1	-	-	-	+	-	-	+	-	-	+	-
5	1	-	-	-	-	+	+	+	+	-	+	-
6	1	+	+	-	+	-	-	+	-	-	+	-
7	1	-	+	-	+	-	+	-	+	+	+	-
8	1	+	-	-	+	-	+	+	-	+	+	-
9	1	+	+	-	+	-	+	+	-	+	+	+
10	1	+	+	+	+	+	+	+	+	-	+	-
11	2	+	+	+	+	+	+	+	+	+	+	-
12	10	+	+	+	+	+	+	+	+	+	+	+
Total	22											

In general, this "shotgun" approach for the generation of possbily interesting antibodies has not been especially fruitful in our hands. Instead, we have turned to a different strategy, which involves deciding in advance which human cell surface antigens one wants to generate antibodies against, then devising immunization and screening partners to rapidly search for the desired specificity. For instance, in an attempt to derive mouse monoclonal antibodies against human B cell or T cell antigens, a series of human cell lines were used which were genetically identical, but differed with respect to T or B class. Mice were immunized to the B cell line and monoclonal antibodies were found which discriminated between the immunizing B cell and its genetically identical T cell and vice versa (7). In another example, human cell lines were employed which had been immunoselected for the loss of HLA alloantigens (8). Mice were immunized against the wild type cell line and monoclonal antibodies were found at a rate of approximately 1% which could discriminate between the immunizing wild type cell and the HLA loss mutant. These discriminator antibodies were subsequently demonstrated to detect bona fide HLA2-28 and HLA B27 (9).

Employing a similar strategy, we have recently returned to the question of human leukemia antigens. We have immunized mice with leukemia cells from children with acute lymphoblastic leukemia (ALL). A set of hybridoma antibodies were derived and screened for reactivity against the immunizing ALL blast cells as well as several control cells, including normal PBL taken from the same child in remission and normal PBL from the patient's father and mother. Fig. 1 shows the results of a binding assay of two different hybridoma antibodies showing the desired discrimination.

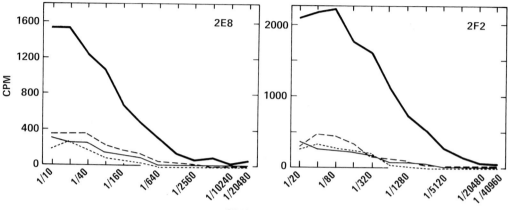

FLUID DILUTION

Fig. 1. Binding assay of hybridoma antibodies against the immunizing ALL blast cells (bold line) and the control normal PBL from the same child and his parents (dashed and fine lines).

In each case, reactivity against the blast cells was much greater than with the remission or parental normal cells. However, in both cases the difference was only quantitative. They did not detect antigens totally absent from the normal cells To date, we have only screened 22 monoclonal antibodies reactive with the immunizing leukemia cells. Clearly, if the frequency of mouse anti-HLA antibodies is any kind of indicator, our search for leukemia specific antigens has just begun.

R. Levy is a Howard Hughes investigator. This work was supported by the American Cancer Society grant #IM114 and the U.S.P.H.S. grant #CA21223-01. L. Lampson is supported by an NIAID post-doctoral fellowship

References

1. Klinman, N.R.: Antibody with homogeneous antigen binding produced by splenic foci in organ culture. Immunochem. 6, 757-759 (1969)
2. Kohler, G. and Milstein, C.: Continuous cultures of fused cells secreting antibody of predefined specificty. Nature 256, 495-497 (1975)
3. Levy, R. and Dilley, J.: The *in vitro* antibody response to cell surface antigens. II. Monoclonal antibodies to human leukemia cells. J. Immunol. 119, 394-400 (1977)
4. Levy, R. and Dilley, J.: The *in vitro* antibody response to cell surface antigens. I. The xenogeneic response to human leukemia cells J. Immunol. 119, 387-393 (1977)
5. Gefter, M.L., Margulies, D.H. and Scharff, M.D.: A smiple method for polyethylene glycol-promoted hybridization of mouse myeloma cells. Somatic Cell Genetics 3, 231-295 (1977)
6. Kohler, G., Howe, S.C. and Milstein, C.: Fusion between immunoglobulin-secreting and non-secreting myeloma cell lines. Eur. J. Immunol. 6, 231-236 (1977)
7. Lampson, L.A., Royston, I. and Levy, R.: Homogeneous antibodies directed against human cell surface antigens: I. The mouse spleen fragment culture response to T and B cell lines derived from the same individual. J. Supramol. Struct. 6, 441-448 (1977)
8. Pious, D., Hawley, P. and Forrest, G.: Isolation and characterization of HLA-A variants in cultured human lymphoid cells. Proc. Nat. Acad. Sci. 70, 1397-1400 (1973)
9. Lampson, L.A., Levy, R., Grumet, F.C., Ness, D. and Pious, D.: *In vitro* production of murine antibody to a human histocompatibility alloantigen. Nature 271, 461-462 (1978)

Mouse-Human Hybridomas. The Conversion of Non-Secreting Human B Cells into Ig Secretors

R. Levy, J. Dilley, K. Sikora, R. Kucherlapati

A. Introduction

When antibody secreting hybridomas are isolated from immune spleen cell suspensions, which normal cells are the target for hybridization?

A number of immunization protocols have been designed to maximize the plaque-forming cells in the donor spleen on the assumption that it is the plasma cell which gives rise to secreting hybrids. However, it is also possible that cell types at other stages of B cell differentiation are appropriate hybridization targets. Clearly, cells which have recently responded to antigens are much more likely targets than unstimulated cells. The frequency of specific antibody producing hybridomas (between 10 and 50% in our own experiments) is astoundingly high when one considers the frequency of plaque formers, antigen binders, or secondary B cell precursors among the total B cell population of the immune spleen.

One approach to this question is to determine which types of B cells are capable of forming Ig secreting hybridomas. For instance, homogenous B cell populations taken from various stages of differentiation can be hybridized with myeloma cells and screened for their ability to secrete Ig. Human B leukemia and lymphoma cells should provide a ready source of such homogeneous populations. We have performed a series of hybridizations between non-Ig secreting human B leukemia cells and mouse myeloma cells. Greater than 50% of the hybrids isolated have switched on secretion of the human Ig. Human Ig secretion has been a stable property of several of the mouse-human hybrids over months of continuous culture (1).

B. Technique

Human B leukemia cells were obtained from the blood of six different patients; five with chronic lymphocytic leukemia, and one with lymphosarcoma cell leukemia. These cells were hybridized to P3/X63/Ag8 (2) or P3/NS1/1-Ag4 (3), using polyethylene glycol according to the procedure outlined in our companion paper in this volume.

Hybrids were screened for secretion of human Ig κ or λ light chains and μ heavy chains as well as for mouse κ and γ chains by radioimmunoassay (1).

C. Results

The human leukemic cells used in these studies all contained surface Ig which was restricted to a single light chain type, either κ or λ, in each case (Table I). However, no Ig secretion by these cells was evident, either in the serum of the patients, or in short term culture _in vitro_. Table I contains the results of hybridization of the human B leukemia cells from the six patients with P3/X63 or NS-1.

Table I. Human immunoglobulin secretion by somatic cell hybrids.

Patients	Disease	Cell surface Ig light chain type	Hybrids isolated	Hybrids secreting human Ig	
				κ	λ
M	CLL[+]	κ	27	19	0
S	CLL	κ	21	14	0
C	CLL	κ	5	2	0
E	CLL	λ	11	0	4
W	LSCL[*]	λ	8	0	5
		Total	72	44	

+ Chronic lymphocytic leukemia
* Lymphosarcoma cell leukemia

Forty-four out of seventy-two total hybrids which were isolated secreted significant amounts of human Ig - between 1-10 µg/ml of spent culture medium. In each case, the light chain type of the secreted human Ig was identical to that on the surface of the input human leukemia cells. We presume, therefore, that the individual hybrids derived from a single patient all secreted the identical human Ig chains, but this remains to be proven. Most of the hybrids also secreted human µ chains in addition to the human light chain. All of the hybrids made with P3/X63 secreted the mouse κ and γ chains, as well. A detailed analysis of one of the hybrids (S2F3) was performed. This particular hybrid has now been in continuous culture for eight months, still producing human κ, human µ, mouse κ, and mouse γ, in a ratio of 3:1:3:3, respectively.

The secreted products of this clone were biosythetically labeled with ^{14}C-leucine and analyzed by immunoabsorption and sodium dodecyl sulfate-gel electrophoresis. The conclusion of this analysis was that the predominate molecule secreted by the hybrid contained two mouse γ chains, one mouse light chain, and one human light chain ($κ_M γ_M γ_M κ_H$). Other molecular species included ($κ_M µ_H µ_H κ_H$)5 and free $κ_H$. No molecules containing mouse γ and human µ were produced.

Immunofluorescence, both membrane and cytoplasmic, was performed on these hybrid cells. We found that a high percentage (>75%) of the cells produced the human Ig heavy and light chains.

Chromosome analysis of S2F3 and a series of subclones thereof has been started. The initial results are shown in Table II.

From this table, it can be seen that only chromsomes 6 and 11 were found in greater than 70% of the cells of all three subclones. Chromosomes 3,12 and 15 were present in greater than 50% of the cells of all three lines. Previous reports have implicated human chromosomes 2,6,11 and 12 as candidates for containing Ig genes (4,5). From our limited analysis above, it would appear that chromosome 2 is not necessary for human Ig light or heavy chain expression. A number of other subclones which have lost production of either one or both human Ig chains are now being analyzed.

171

Table II. Human chromosomes in percent

Cell line	No. of Cells	3	4	5	6	7	9	10	11	12	X	13	14	15	17	18	20	21	22
S2F3D12	28	89	54	29	89	57		54	71	71		57	7	50	4	7	7	11	7
S2F369	25	64		48	88	56		16	76	76	48	32	4	56	44		16	4	
S2F3	30	73	13	43	83	30	17	33	87	57	47	40	7	50	23	7	40	23	27

A number of other types of human lymphomas are derived from B cells (6). That is, they contain cytoplasmic and/or membrane Ig. It will now be of interest to determine which of these other types of non-secreting malignant B cells can serve as targets for the production of Ig secreting hybridomas.

R. Levy is a Howard Hughes Investigator. This work was supported by the American Cancer Society grant #IM114 and the U.S.P.H.S. grant #CA21223-01.

References

1. Levy, R., and Dilley, J.: Rescue of immunoglobulin secretion from human neoplastic lymphoid cells by somatic cell hybridization. Proc. Nat. Acad. Sci. In press (1978)
2. Kohler, G. and Milstein, C.: Continuous cultures of fused cells secreting antibody of predefined specificity. Nature 256, 495-497 (1975)
3. Kohler, G., Howe, S.C. and Milstein, C.: Fusion between immunoglobulin-secreting and non-secreting myeloma cell lines. Eur. J. Immunol. 6, 231-236 (1977)
4. Smith, M., Hirschhorn, K. Schuster, J. and Gold, P.: The study of human immunoglobulins in hybrid cells: tentative assignment of the genes responsible for human heavy chain immunoglobulin production to chromosome 2. Third International Workshop on Human Gene Mapping. Baltimore Conf. 1975. Birth defects original article, Series XII, ed. D. Bergsma (1976)
5. Smith, M. and Hirschhorn, K.: Production of human immunoglobulin in heavy and light chains in man-mouse somatic cell hybrids. Fed. Proc. 36, abs #4906, 1196 (1977)
6. Weissman, I.L., Warnke, R., Butcher, E.C., Rouse, R. and Levy, R.: The lymphoid system, Its normal architecture and the potential for understanding the system through study of lymphoproliferative diseases. Human Path. 9, 25-45 (1978)

172

Induction of IgM Secretion by Fusing Murine B-Lymphoma with Myeloma Cells

R. Laskov, K.J. Kimm, R. Asofsky

A. Introduction

The technique of cell fusion has been recently used as a tool for studying the regulation of immunoglobulin (Ig) synthesis in lymphoid cells. Studies of myeloma x myeloma cell hybrids have demonstrated that these hybrids continue to synthesize the parental types of the Ig polypeptide chains. Activation of a "silent" Ig gene, or modification of the expression of already "turned on" genes were never found in these hybrids (1, 2). Fusion between a less mature lymphoid cell with a plasma cell seems to be of interest, since it may elucidate some of the mechanisms involved in the differentiation of B-lymphocytes to plasma cells. One of the questions is, for example, whether hybrids between lymphocytes and plasma cells will be able to secrete the IgM represented on the cell surface of the parent lymphocyte. Since normal populations of lymphocytes are heterogenous and may contain plasma cells, it is not certain which of the cell types fuse with myeloma cells. This difficulty may be overcome by the use of B-lymphoma cells which are presumed to be of monoclonal origin and are homogenous with regard to cell type.

In the present work, we have fused three different murine B-lymphomas with myelomas and found that many of the hybrids did produce and secrete large quantities of IgM.

B. Methods

1. Cells

The B-lymphomas appeared as spontaneous tumors in aged BALB/c mice. These tumors were collected and maintained for several years by Dr. Ruth Merwin of the National Cancer Institute (3). Some of the tumors were recently adapted to continuous growth in culture and characterized to be of B-origin: the cells bear IgM and Ia on their outer surface and lack the Thy 1.2 antigen (4, 5). Two clones of the MPC-11 myeloma, kindly provided by Dr. Scharff, were used for fusion: clone 4T00.1 producing IgG, and clone 4T00.1L1 which produces only light chains. Both clones were 6-thioguanine and Ouabain resistant (1).

2. Fusion and testing procedures

This was done using polyethylene glycol (PEG 1000, Sigma), essentially accoring to Margulies et al (1). The seeding of the cells after fusion was done either to microwell plates (96 wells) or to Linbro plates (24 wells). Clones became visible after about 2 weeks, and the plates were fed with the selective medium lacking aminopterin at 3-4 weeks after fusion. Feeding with Dulbecco's medium without NCTC-109 and HT resulted in a very slow growth of the hybrid clones.

The clones were tested for membrane and cytoplasmic Ig by immunofluorescence, and for secretion of Ig by radial and double immunodiffusion in agar, and by polyacrylamide gel electrophoresis of biosynthetically labeled Ig (1).

C. Results

The frequency of recoverable clones in the fusion experiments varied between 1 to >10 hybrid clones/10^6 input lymphoma cells depending on the B-lymphoma used for fusion. Chromosomal analysis showed the parent lymphomas and myelomas to contain mean chromosomal numbers of 70±4 and 62±2 respectively. Individual hybrid clones contained mean chromosomal numbers of between 81±2 and 93±2. Figure 1 shows an example of an analysis of the Ig secreted by 3 hybrid clones, using the Ouchterlony technique. While two of the clones secreted IgM only, one secreted both IgM and IgG_{2b} [clone A6(D)].

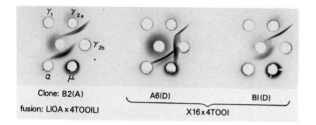

Fig. 1 Analysis of Ig secretion of three hybrid clones by double diffusion in agar. Culture medium from cloned hybrid cells was concentrated 2-4x and applied to the central wells. Affinity purified class specific antisera were placed in the outer wells as specified for clone B2(A). The parent L10A and X16 lymphoma cell lines did not secrete detectable amounts of Ig by this method. The 4T00.1 myeloma secreted IgG_{2b} (not shown). The precipitation line between the anti γ_{2a} and γ_{2b} is due to cross absorption of these antisera with the respective purified myeloma proteins.

Table 1 summarizes a survey of Ig secreted by the various clones obtained in 3 fusion experiments.

Table 1.   Secretion of Ig Polypeptide Chains by B-Lymphoma x Myeloma Hybrids.

Exp. No.	Fusion Lymphoma x Myeloma	Ig Secretion Parents	No. Clones Analyzed	No.Clones Secreting/ Total Clones		
				μ	γ_{2b}	$\mu+\gamma$
1	L10A x 4T00.1L1	$(\mu-)$ x $(\gamma_{2b}-)$	9	2/9	0/9	0/9
2	X16 x 4T00.1	$(\mu-)$ x $(\gamma_{2b}+)$	6	3/6	5/6	2/6
3	K46 x 4T00.1	$(\mu-)$ x $(\gamma_{2b}+)$	3	3/3	3/3	3/3

It can be seen that IgM synthesis and secretion was "induced" in many of the fused hybrids. The synthesis of the γ_{2b} chain was not rescued by fusing a γ_{2b} non-producer myeloma cell with B-lymphoma (Exp. #1). Fusion of B-lymphoma with a γ_{2b} producing myeloma did not suppress the synthesis of γ_{2b} chain in the hybrids (Exp. #2 and 3). The secreted IgM in stationary phase cultures reached a level of up to

100 μg/ml by radial immunodiffusion. This high level is comparable with levels previously determined for IgG secretion by the MPC-11 cultured myeloma cells (6). Many of the hybrids revealed cytoplasmic IgM and also membrane IgM by immunofluorescence. The synthesis and secretion of large quantities of IgM by the B-lymphoma x myeloma hybrids was confirmed by analysis of biosynthetically labeled Ig molecules on SDS polyacrylamide gel electrophoresis (not shown). This analysis showed that the hybrids secreted IgM mainly in its pentameric form.

D. Conclusions

Studies from other laboratories (1,2) have demonstrated that myeloma x myeloma cell hybrids continue to synthesize the parental types of Ig.

In the present work fusion of B-lymphoma with myeloma cells resulted in alterations in the expression of the IgM genes leading to 1) an amplification of IgM synthesis and 2) secretion of IgM into the medium. These apparent changes may be due to: 1) Contribution by the myeloma cells of the necessary machinery for the synthesis and secretion of large quantities of IgM; 2) Myeloma cells contain a diffusable substance which "induces" the immature B-lymphoma cells to differentiate and form a mature IgM secreting cell, i.e., a transdominant control mechanism is taking place in the final maturation step of B-lymphocytes.

Our results support the contention that antibody producing hybridomas resulting from fusion of myeloma cells with immunized spleen cells (2) are not only due to fusion of myeloma and antibody secreting cells, but also to fusion of myeloma with a less mature splenic B-cell.

Acknowledgements: We are greatly indebted to Mr. Charles B. Evans, who performed the Ouchterlony analyses.

References

1. Margulies, D.H., Cieplinsky, W., Dharmgrongartama, B., Gefter, M.L., Morrison, S.L., Kelly, T., Scharff, M.D.: Regulation of immunoglobulin expression in mouse myeloma cells. Cold Spring Harbor Symp. Quant. Biol. Vol. XLI, 1976, p. 781.
2. Milstein, C., Adetugbo, K., Cowan, N.J., Kohler, G.,Secher, D.S., Wilde, C.D.: Somatic cell genetics of antibody-secreting cells: Studies of clonal diversification and analysis of cell fusion. Cold Spring Harbor Symp. Quant. Biol. Vol. XLI, 1976, p. 793.
3. Merwin, R., unpublished.
4. Kim, K.J., Kanellopoulos, C., Merwin, R., Sachs, D., Asofsky, R.: Characterization of six BALB/c B-cell tumor lines. In preparation.
5. McKeever, P.E., Kim, K.J., Logan, W.J., and Asofsky, R.: Spontaneous BALB/c lymphomas synthesize monomeric IgM and IgM half molecules. In preparation.
6. Laskov, R., and Scharff, M.D., unpublished.

Cloned T-Cell Lines with Specific Cytolytic Activity

M. Nabholz, H.D. Engers, D. Collavo, M. North

A. Introduction

The search for hybrids derived from somatic crosses between specific
cytolytic T-cells (CTL) and cell lines, which express the specific
activity of the parental lymphocytes in our and many other laboratories has, sofar, not been successful (see this volume). There are a
number of possible reasons for this failure, but it is difficult to
obtain evidence distinguishing between the different possibilities
without CTL populations which are absolutely pure.

Attempts to obtain lymphocyte lines with specific cytolytic activity
by transforming CTL populations with viruses or by screening of spontaneous thymomas (D. Zagury, personal communication) have also been
unsuccessful.

Alloreactive T-cells can, on the other hand, be maintained in culture,
apparently indefinitely, by periodic stimulation with allogeneic cells
(1); and recent experiments indicate that it is possible to obtain clones from these populations (see Fathman, Hengartner and Elliot, this
volume). But while these populations retain their specificity as measured by their proliferative response to allogeneic stimulators they
lose their capacity to generate CTL activity, probably because a non-
cytolytic alloreactive T-cell class outgrows the CTL.

Recently, however, Gillis and Smith (2) have reported that it is possible to maintain CTL in continuous proliferation *in vitro* for periods
of more than four months without loss of cytolytic activity by addition of supernatant from Concanavalin A (Con A) stimulated spleen
cells to the medium. We have confirmed these results and have derived
sublines from such populations which have most likely originated from
single cells. These cloned populations maintained varying levels of
cytolytic activity.

B. Materials and Methods

I. Abbreviations

The following abbreviations will be used : B6 for C57Bl/6, Con A for
Concanavalin A, CTL for cytolytic T-lymphocyte, FBS for fetal bovine
serum, LTC for CTL populations maintained by culture in Con A supernatant (long-term CTL), MLC for mixed leukocyte culture, PEC for peritoneal exudate cells, PHA for phytohemagglutinin, SN for supernatant
from Con A stimulated spleen cells, SNMx for MLC medium containing
x % SN, T-cells for thymus derived lymphocytes.

176

II. Animals

C57Bl/6J (B6, H-2^b), CBA/CaJ (H-2^k) and DBA/2J (H-2^d) mice were obtained from the animal colony of the Swiss Institute for Experimental Cancer Research. This colony has been established from stocks obtained from Jackson Laboratories, Bar Harbor, Maine. AKR (H-2^k) mice were obtained from Gl. Bomholtgard, Ry, Denmark. SJL (H-2^s) mice were a gift from the Institut für Tierforschung, Füllinsdorf, Switzerland. (AKR x DBA/2)F$_1$ mice were bred in Lausanne. Sprague Dawley rats were a gift of Sandoz Co., Basel, Switzerland.

III. Cell lines

The following cell lines were used as targets for CTL : AKR-A (AKR lymphoma, H-2^k), EL-4 (B6 lymphoma, H-2^b) and P815-X-2 (DBA/2 mastocytoma, H-2^d) were maintained in culture. MB1-2 (Moloney virus induced lymphoma from B6, H-2^b) was maintained in ascites form by passage *in vivo*.

IV. Medium and Con A Supernatant

The medium used for all cultures is the MLC-medium described previously (3). It has as a base Dulbecco's modification of Eagle's minimum essential medium (DMEM) which is supplemented with additional arginin, aspartic acid and folic acid, as well as glutamin, penicillin and streptomycin, 2-mercaptoethanol (5 x 10^{-5}M), N-2-hydroxyethyl piperazine-N-2-ethane sulfonic acid (HEPES) and 4-5 % fetal bovine serum (FBS).

For supernatant production mouse or rat spleen cells are prepared by gentle homogenization of large spleen fragments in DMEM or MLC medium. The cells are spun down, washed once and resuspended to a concentration of 3 to 5 x 10^6 cells/ml in MLC medium containing, usually, 5 μg Con A (Pharmacia) per ml, and cultured in 50 ml aliquots in 250 ml plastic bottles (Falcon, No 3024), lying flat, for 44 to 50 hours. The supernatants are harvested by decantation, and cleared by a low speed centrifugation (500 g, 15 min), and a second, high speed spin (27'000 g, 20 min). They are then passed through 0.45 u average pore size filters (Millipore, type HA) by pressure filtration and stored in 10 or 20 ml aliquots at -20°C. In this condition they can be stored without significant loss of growth promoting activity for at least a month.

The activity of the supernatants is tested by culturing 10^4 - 1.5 x 10^4 cells of one of the long-term cytotoxic T-cell populations or clones derived from them in 1 ml of serial dilutions of the supernatant. The cultures are set up in 24 well clusters (Costar, No 3524) and the cells in each well counted by an electronic cell counter (Coulter Electronics) four or five days later.

For maintenance and growth of the long-term CTL populations supernatant and medium are mixed, usually to give a final concentration of 50 (SNM 50) or 17 (SNM 17) % of SN.

V. MLC cultures and production of long-term cytolytic T-cell populations

Primary, secondary and tertiary MLC cultures were set up as described previously (4) in MLC medium containing 5 % FBS. Cultures were restimulated not less than 10 days after the previous stimulation.

Approximately 10 days after the last allogeneic stimulation between 10^6 and 2×10^6 cells from the MLC were transferred into 10 ml SNM 50 (mouse SN) in a 100 mm tissue culture dish (Nunc, No 1415). These cultures were harvested after four to six days and from then on the cells were maintained by weekly transfers of 2.5×10^5 cells to 5 ml of SNM 50 in 30 ml plastic tissue culture flasks (Falcon, No 3013, Nunc, No 1460) or 60 mm tissue culture plates. Later, after about 40 days of culture, they were subcultured every three to five days in SNM 17 (containing rat or mouse SN) by seeding 1.5×10^4 to 3×10^4 cells per ml in tissue culture dishes. The long-term cytotoxic cells (LCT) show a progressive tendency to attach to the plastic substrate and have to be squirted off the plastic plate by vigorous pipetting.

VI. Cloning

1.   Cloning by limiting dilution

Ninety six-well plates with flat bottom wells (Falcon, No 3040) were prepared by seeding into each well peritoneal exudate cells (PEC) from normal mice, in 200 μl of MLC medium per well (for optimal PEC numbers see below). The plates were incubated at 37°C and irradiated with 2000 rad. On the next day the medium was sucked off and 100 μl of SNM 20, containing the required concentration of LCT was added to each well. Two days later the cultures were fed by adding approximately 100 μl SNM 20, and subsequently medium was changed every 2 to 4 days by sucking off most of the medium and replacing it with SNM 20. Wells were inspected periodically for growth and 10 days after seeding, cells from wells with growth were transferred to wells of a 24 cluster plate (Costar, No 3524) in SNM 17.

2.   Cloning in soft agar

The 30 mm wells of 6 well-clusters (Costar, No 3506) were prepared by seeding them with a given number (see below) of PEC from normal mice in MLC medium. They were irradiated 1 day later and the medium replaced with a mixture containing 0.3 % agar (Bacto-Agar, Difco Laboratories), 10 % FBS and 20 % SN in DMEM supplemented as described above. On the next day the bottom layer was overlaid with 1 ml of the same mixture containing various numbers of LCT. In general individual colonies were transferred after 6 to 8 days with a pasteur pipette into flat bottom wells of a 96 well plate containing 5×10^4 irradiated PEC well in SNM 17. The cells growing in these wells were, at various times, transferred to the wells of 24 well-clusters containing SNM 17, and were then maintained as described above. In some instances colonies were picked after 12 to 15 days and transferred directly to 24 clusters without feeder layer.

VII. Assay of cytolytic activity

The LCT were harvested from the culture vessels, spun down and resuspended to the required concentration. Their activity against various ^{51}Chromium labelled target cells was assayed mostly in tubes (3) and sometimes in round bottom microtiter plates (5). In some experiments PHA (PHA-M, Difco Laboratories) was added to the effector-target cell mixtures to give a final concentration of 1:200 of the reconstituted stock solution, which gave maximum lytic activities when used with

178

primary or secondary MLC cells. Specific ^{51}Cr release was calculated as described previously (3) and lytic units determined from the dose response curves, one lytic unit being defined as the number of effector cells required to give, as indicated in the results section, 50 % or 25 % specific ^{51}Cr release from 10^4 target cells.

VIII. Chromosome preparations

Cells were prepared for chromosome analysis by adding to a culture set up 2 to 3 days earlier Colchicine (Colchinéos Laboratoires Houdé, Paris) to give a final concentration of 0.2 μg/ml. Four hours later the cells were harvested, spun down and resuspended in hypotonic (1 %) Sodium citrate. They were centrifuged again immediately and processed according to the method of Ford (6).

C. Results

I. Variables affecting the growth promoting activity of the Con supernatant

The LCT described here were established in medium containing 50 % supernatant from mouse spleen Con A cultures. When we started to test the growth promoting activity of each supernatant, we found that most SN give the same or slightly better growth at a concentration of 25 %. But at higher dilutions growth promoting activity invariably falls off. Testing different concentrations of Con A (up to 5 μg/ml) for supernatant production (in cultures with 10^7 spleen cells per ml) we found that the highest concentration tested gave the strongest activity.

SN from different mouse strains (CBA, B6, DBA/2) gave similar activities. Rat spleen cells yield somewhat stronger SN than mouse cells, at least when tested on already established LCT, and the two populations described below have now been maintained for more than 150 days in SNM 17 containing rat SN. Spleen cell concentrations of 2.5 x 10^6 and 5 x 10^6 rat spleen cells per ml yielded similar growth promoting activities per ml. Supernatant from Con A cultures harvested after 24 hours gave weaker SN.

II. General description of LCT

In all experiments described here (AKR x DBA/2) F_1 cells were used as responders and stimulated with cells from B6 mice or a mixture of cells from B6 and SJL mice. After transfer into SNM the *in vitro* primed cells begin to grow vigorously with a doubling time of approximately 30 hours. Several LCT populations initiated from cells stimulated only once in MLC gradually lost their cytolytic activity (e.g. see Fig. 1), while cultures started from secondary or tertiary MLC maintained their activity for at least 6 weeks (Fig. 1). Two of these populations we have now maintained for more than 200 days in supernatant supplemented medium : The responder cells giving rise to the population D were stimulated with allogeneic cells 21 and 9 days, those giving rise to H 32, 21 and 9 days before transfer to Con A supernatant. In the following the day of transfer to SNM is referred to as day O. All the cells in these populations and in the clones derived from them (see below) are stained by

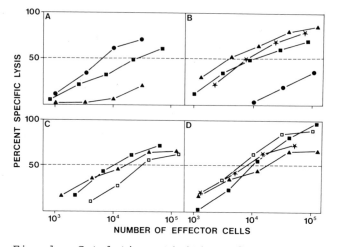

Fig. 1. Cytolytic activities of MLC responder cells one (●), two (□), four (■), six (▲) and ten (*) weeks after transfer to Con A supernatant. Responder cells : (AKR x DBA/2) F_1 stimulated once (A), twice (B) or three times (C,D) with SJL and B6 cells (A,B,C) or B6 cells (D). Target : EL-4.

a rabbit anti-mouse T lymphocyte antigen (MTLA) serum (7). No cells expressing surface immunoglobulin could be detected in the long-term cultures. Further antigenic analysis of these cells has not yet been carried out, one of the reasons being that they are more resistent to complement-dependent antibody mediated lysis than all other normal or tumor cell populations tested. Morphologically both the H and D populations were heterogeneous with regard to both shape and size of the cells. Most of the cells adhere to plastic. The cells also produce a varying but considerable degree of debris, the origin of this being unclear. In an exponentially growing culture, one finds usually less than 10 % dead cells but after reaching stationary phase, the cells die very rapidly. The growth rate depends on the batch of supernatant used and the doubling time varies from 20 to 48 hours. Growth strictly depends on the presence of supernatant. Cells desintegrate within 1 to 2 days in its absence. Con A cannot replace the supernatant. During earlier passages the cells grew to a maximum density of about 3×10^5 cells per ml but concentrations of up to 5 to 8×10^5/ml have been reached lately.

Chromosome preparations made from H after 146 and 203 days of culture in SNM showed that all of 30 well spread metaphase plates scored on each occasion contained a large metacentric chromosome. Metacentric chromosomes were not seen in spreads from two clones derived from H after 94 days of culture nor in the D population and its sublines. At the time of the first chromosome preparation we had also observed that this population was becoming morphologically much more homogeneous. It now consists entirely of relatively small adherent cells of irregular shape. It seems thus that a fast growing clone with a rearranged chromosome has taken over this population.

III. Derivation of LCT sublines

Cloning of the LCT was attempted after approximately 100 days of culture in SNM. Preliminary experiments had shown that when less than 10 to

20 cells were seeded in the flat bottom wells of a 96 well-plate, the cells would rarely survive the first few days. When, however, irradiated peritoneal exudate cells from either DBA/2 or B6 mice were added as a feeder layer all wells seeded with approximately 10 cells gave rise to growing cultures. The first sign of growth consisted in a destruction of the PEC feeder layer, whether the PEC were allogeneic or semi-syngeneic with the LCT. Irradiated PEC never gave rise to growing cultures without addition of LCT. Irradiated spleen cells were less efficient in supporting growth than PEC. These preliminary experiments suggested that 10^5 PEC per well gave optimal results, and the experiment reported in Table 1a shows that less than this number gives declining frequencies of growing cultures under limiting dilution conditions.

Table 1. Cloning by limiting dilution

A. Effect of PEC-Feederlayer

Cells/well plated (LCT H)	No of wells plated	PEC/well[*]				
		10^5	5×10^4	2.5×10^4	1.25×10^4	0
20	4	4[**]	4	3	3	0
5	4	3	2	2	1	
1	8	2	2	1	0	
0.5	12	2	0	0	0	
0.25	20	1	2	0	0	

[*] PEC irradiated with 2000 rad

[**] No of wells with growth 13 days after plating

B. Limiting Dilution of LCT

Cells/well plated (LCT D)	No of wells[o]	
	Plated	With growth[oo]
3	24	22
1	40	11
0.3	41	1
0.1	49	1

[o] 10^5 irradiated DBA/2 PEC/well

[oo] No of wells with growth 11 days after plating

Although the results of the limiting dilution experiment reported in Table 1b are insufficient to give a precise estimate of the proportion of growing cultures at different densities, or to verify that the distribution of negative cultures could be fitted to a poisson distribution, they indicate that the plating efficiency in this experiment was approximately 30 %.

Two sublines of H (designated HA-2, HA-3) have been established from plates seeded with less than 0.5 cells/well.

Simultaneously with the limiting dilution experiments the D population was cloned in soft agar : The plating efficiency in this experiment was about 5 %, regardless of the numbers of LCT (Fig. 2). It was independent

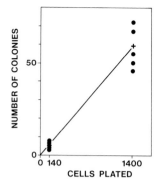

Fig. 2. Cloning of LCT population D in soft agar. Each point represents the colonies counted in one well. Wells contained different numbers of irradiated PEC (see text).

of the number of PEC added as feeder layer (between 5×10^4 and 8×10^5/ well), and in other experiments the difference in the plating efficiency in presence or absence of feederlayers was marginal. In some experiments the plating efficiency declined slightly when decreasing numbers of LCT were plated. The colonies grow vigorously for at least 12 to 14 days and almost all of them could eventually be seen with the naked eye. The morphological heterogeneity of the LCT populations is reflected in morphological differences between the colonies, while the cells within each colony are much more homogeneous : some colonies consist of tight clusters of large or medium sized cells, in others the relatively small cells tend to migrate away from the center to form disperse colonies.

Colonies were transferred with pasteur pipettes, usually into wells of a 96-well plate containing irradiated PEC from DBA/2 mice. From more than half of the picked colonies sublines have been established. They are designated DA-1, DA-4, etc. They differ among each other in cell size and shape, reflecting the differences between the colonies. The same applies to the sublines of H.

The one subline, DA-7, with strong cytolytic activity (see below), consisted of medium sized cells with fairly regular oval or round shape, quite distinct from the homogeneous population of cells in the H line. A considerable degree of heterogeneity with regard to size has reappeared in DA-7 since it was established.

Chromosome preparations were made from some of the sublines (HA-2, HA-3, DA-5, DA-6, DA-7). In all of them the karyotypes of most cells were diploid or pseudodiploid, except for subline DA-6, all metaphases of which contained an approximately tetraploid chromosome set.

IV. Cytolytic activity of LCT populations H and D, and their sublines

The cytolytic activities of the LCT populations H and D have been measured periodically, usually against EL-4 target cells, in a standard ^{51}Chromium release assay (Fig. 3). The activity of the cultures underwent fairly drastic fluctuations. We don't yet know the sources

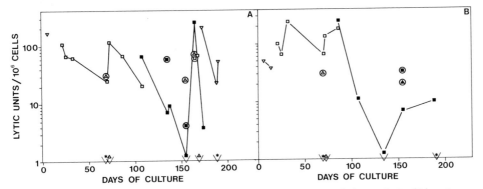

Fig. 3. Lytic activity of LCT-populations H (A) and D (B) at various times after transfer to Con A supernatant (Day O). Plotted are lytic units for 50 % (open symbols) or 25 % (closed symbols) specific ^{51}Cr release. Targets : EL-4 (□,■), P815 (△,▲), MB1-2 (▽,▼), AKR-A (*, 25 % lysis). Circles around symbols indicate addition of PHA to assay. The symbol V signifies that lytic activity was less than one unit.

of these variations. Cells assayed were always obtained from cultures in exponential growth phase. But one experiment in which growth and lytic activity on various days after subculturing at different densities was measured shows that this is not sufficient to obtain consistent activities : The activities began to decline after three days even in cultures which were still growing exponentially (Fig. 4).

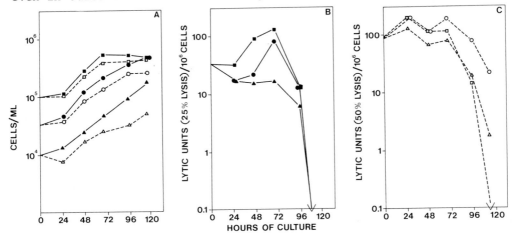

Fig. 4. Growth and lytic activity of H and DA-7. Cells were plated at the indicated densities (O hr) in 60 mm plates, and at various times thereafter plates were harvested, cells counted and their lytic activity against EL-4 measured. A : Growth, B : Lytic activity of H, C : Lytic activity of DA-7. V signifies that activity was below O.1 unit.

The activity of both populations H and D remained specific in the sense that they were active against the H-2^{b} targets MB1-2 and EL4 but not against semi-syngeneic targets P815 (H-2^{d}) or AKR-A (H-2^{k}). Lysis

183

against the latter two cell lines was obtained when PHA was added to
the assay (Fig. 5), in order to circumvent the requirement for specific
determinant recognition by the effector cells (8). On some occasions
when the specific lysis against EL-4 was very weak, addition of PHA

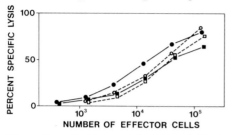

Fig. 5. Lytic activity of H (■,□) and D (●,○) against EL-4 (■,●), and
against P815 in the presence of PHA (□,○). There was no significant
lysis of P815 in the absence of PHA. Day 70 after transfer to Con A
supernatant.

increased the activity against this target considerably, but there was
no consistent relationship between specific and PHA boosted lysis.

Testing, at various times, the cytolytic activities of nine sublines,
two derived from H, and seven from D, we found that all, except one
subline of H, showed some specific activity (Fig. 6), but, except for
DA-7, it was below the activity of the uncloned populations, when
tested simultaneously. Addition of PHA to the assay increased the
activity of the weak sublines against EL-4 somewhat, but had no effect
on DA-7. None of the sublines had any activity against AKR-A or P815
in the absence of PHA. DA-7 has maintained its relatively high activity
for more than 3 months after cloning. When we compared the cytolytic

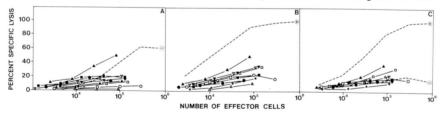

Fig. 6. Lytic activity of sublines derived from H and D against EL-4
without (A) and in the presence (B) of PHA, and against AKR-A with PHA(C).
No significant lysis was observed against AKR-A or P815 in the absence
of PHA. Sublines : : DA-7,▲ : DA-6, ● : HA-2,△ For comparison the
lytic activity of C3H cells after secondary stimulation with DBA/2 was
assayed against the same targets in presence (---⊕) or absence (---⊖)
of PHA.

activities of DA-7 and H, at a time when all cells of the latter line
carried the metacentric chromosome (see above), against the two B6 derived target cells EL-4 and MB1-2 (Fig. 7), we found a difference in
their specificity : H gave, in several experiments, only low levels
of lysis against EL-4 but high activity against MB1-2, while DA-7, in
the same experiments showed approximately similar activity against both
targets.

184

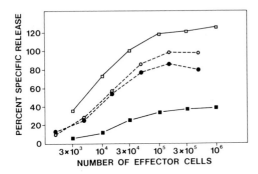

Fig. 7. Lytic activity of H (□,■) and DA-7 (O,●) against two targets derived from B6 mice : EL-4 (■,●) and MBl-2 (□,O). Day 174 after transfer to Con A supernatant.

D. Discussion

Our results confirm those of Gillis and Smith (2). While our LCT were derived from (AKR x DBA/2) mice, theirs came from B6 animals. Whether LCT can be established from other strains remains to be seen. One difference between our lines and those of Gillis and Smith is that while our starting populations were obtained by repeated *in vitro* stimulations with normal allogeneic spleen cells, theirs had been *in vivo* primed, and *in vitro* restimulated with an allogeneic cell line established from a Friend virus leukemia.

While the two LCT described here have maintained their cytolytic activity for over 200 days after transfer to medium containing supernatant from Con A stimulated spleen cells the lytic activity has fluctuated a great deal. To some extent, these fluctuations probably reflect the evolving cellular composition of the populations but other factors very likely play a significant role. Among those may be the supernatant : while we found little variation in the growth promoting activity of supernatants prepared under standard conditions it may well be that other components, which are essential for maintaining the expression of cytolytic activity by the LCT vary independently of the growth promoting activity.

Our results show that it is possible to obtain sublines from the LCT populations, which are most likely clones. We have not tried to clone LCT earlier than after 90 days of culture in supernatant. We know that cytolytic cells obtained from secondary or tertiary MLC can form small colonies in soft agar containing Con A supernatant, but the colonies formed almost never exceed 20 cells and die after about 10 days in agar. (Collavo, Engers, Nabholz, manuscript in preparation).

While we have not proved that the sublines derived from the LCT originated from single cells, there are a number of observations suggesting that this is so : (1) the plating efficiency in soft agar is, under optimal conditions, independent of the number of cells plated, down to 140 cells per 30 mm well. (2) While the uncloned populations are morphologically very heterogeneous, the colonies and sublines consist of one of

185

the cell types found in the uncloned populations, although during subsequent subcultures some sublines become more heterogeneous again. (3) The karyotype of one of the sublines (DA-6) was close to tetraploid. This line had some, albeit very weak, specific cytolytic activity. The finding that the H population has been overgrown by a single type of cells which carries a metacentric chromosome, but still shows strong cytolytic activity against MB1-2, also argues against the possibility that more than one cell type is required to maintain growth and specific activity of the LCT.

While all these considerations strongly suggest that the sublines described here are true clones further work is required to prove this. The apparent difference in the specificity of H and DA-7, detected by comparing their lytic activities against two different targets derived from B6 mice, will be useful in this respect. Experiments to identify the determinants responsible for this difference are in progress.

While it is thought that normal mammalian cells can undergo only a limited number of divisions, there are now two systems in which murine T lymphocytes can be maintained in culture for apparently indefinite periods. In populations maintained by periodic stimulation with allogeneic cells (1) no evidence for any sort of abnormality has been found. There are, on the other hand, some indications that the long-term populations described here consist of cells, which may have undergone some sort of transforming event : (1) besides the already mentioned cases of chromosomal abnormalities, some cells with 41 chromosomes have been found in several sublines. (2) The morphological heterogeneity of the cells described here is in great contrast with the homogeneity we have observed in populations maintained by allogeneic stimulations. In addition we found that some initially homogeneous sublines have regenerated a restricted degree of heterogeneity.

Whether the LCT have undergone a transformation event or not may be relevant to the observation that the mean of the activities of the sublines derived from H and D is much less than the average activity of uncloned populations, regardless of whether it is measured in the absence or presence of PHA. It is possible that our cloning procedures select for lines with low activity. If these cells are transformed, variants with low activity may arise frequently in a population due to epigenetic changes in the expression of one or more of the genes involved in controlling cytolytic activity. Subcloning of the sublines should provide some information bearing on this question. An alternative explanation which seems implausible, but cannot be ruled out, is that each subline recognizes determinants which are expressed always only on a fraction of the target cells. The cell lines described here may become instrumental for the solution of some of riddles around specific determinant recognition by T-cells. But to establish their usefulness further work on the stability of their specificity and their cytolytic activity is required.

Acknowledgments. We thank G. Trinchieri and H.R. MacDonald for helpful criticisms. J. Duc and S. Saule provided excellent and patient typing assistance. P. Dubied prepared the figures. We gratefully acknowledge the gifts of animals by the Institut für Tierforschung, Füllinsdorf and Sandoz Co., Basel.

This work was supported by grants from the Swiss National Science Foundation and the Consiglio Nazionale per la Ricerca.

186

References

1. Dennert, G., Raschke, W.: Continuously proliferating allospecific T-cells, lifespan and antigen receptors. Eur. J. Immunol. 7, 352-359 (1977)
2. Gillis, S., Smith, K.A.: Long term culture of tumour-specific cytotoxic T-cells. Nature 268, 154-156 (1977)
3. Cerottini, J.C., Engers, H.D., MacDonald, H.R., Brunner, K.T.: Generation of cytotoxic T lymphocytes *in vitro*. J. Exp. Med. 140, 703-717 (1974)
4. Fathman, C.G., Collavo, D., Davies, S., Nabholz, M.: *In vitro* secondary MLR. J. Immunol. 118, 1232-1238 (1977)
5. Nabholz, M., Vives, J., Young, H.M., Meo, T., Miggiano, V., Rijnbeek, A., Shreffler, D.C.: Cell-mediated cell lysis *in vitro*: genetic control of killer cell production and target specificities in the mouse. Eur. J. Immunol. 4, 378-387 (1974)
6. Ford, C.E.: Radiation Chimeras: The use of chromosome markers. In: Tissue Grafting and Radiation. Micklam, H.S., Loutid, J.F. (eds). Academic Press, Appendix I, 1966, pp. 197-203
7. Sauser, D., Anckers, C., Bron, C.: Isolation of mouse thymus-derived lymphocyte specific surface antigens. J. Immunol. 113, 617-624 (1974)
8. Bevan, M.J., Cohn, M.: Cytotoxic effects of antigen- and mitogen-induced T-cells on various targets. J. Immunol. 114, 559-565 (1975)

Characterization and Functional Analysis of T Cell Hybrids

R. Grützmann, G.J. Hämmerling

Introduction

Cell hybridization of selected T lymphocyte populations with T tumor
cells could provide a useful tool for the establishment of permanent
tissue culture lines which display the immunological characteristics
of the T lymphocytes such as antigen reactivity and surface characteristics. The rationale for employing T lymphomas for production of T
hybrids is the possibility that expression of particular T lymphocyte
functions may only be allowed by hybridization with tumor cells the
differentiation stage of which is similar to the one of the T lymphocyte. This notion is supported by the observation that hybridization
of murine T cells with the myeloma P3-X63-Ag8 results in the formation
of hybrids which do not express T cell specific markers such as Thy-1
while the immunoglobulin chains of the myeloma are still secreted
(Grützmann and Hämmerling, unpublished). Therefore, the AKR derived
thymoma BW5147 (hypoxanthine guanine phosphoribosyl transferase negative, a gift of Dr. R. Hyman, La Jolla) was used.

Production and Characteristics of T Cell Hybrids

Lymphocytes were hybridized with BW5147 cells in the presence of polyethylene glycol 4000 exactly as described previously (1). Under these
conditions a particular subclone of BW5147 (clone 4) fused with an efficiency of 1 to 3 hybrid lines per 10^5 lymphocytes. It should be noted
that other sublines of BW5147 had a 100 fold lower fusion efficiency.
This illustrates that it may be worthwhile to screen subclones of T tumors for high fusion frequency.

A large number of hybrids were produced by fusion of BW5147 clone 4
with various lymphocyte populations such as thymocytes, lymph node
cells, splenic B and T cells, Con A blasts, MLC blasts etc.(1). The
surface characteristics and chromosome numbers of a representative number of hybrids obtained with C57BL/6 spleen cells is depicted in Table
1. The presence of H-2, Thy-1 and Ia was determined by quantitative
absorption of respective alloantisera and subsequent microcytotoxic
assay of the absorbed sera on appropriate target cells. It can be seen,
that most hybrids express the H-2 and Thy-1 of both parent cells. Neither
Ia nor Ig determinants could be found suggesting preferential hybridization with T lymphocytes. During the course of these quantitative absorption studies it was observed that BW5147 cells as well as the resulting hybrids express only about 2% of the amount of H-2K and H-2D
region antigens compared to spleen cells (data not shown) indicating
that the tumor parent determines the amount of H-2 on the cell surface
of a hybrid cell. It is not yet clear whether this phenomenon is due
to regulatory mechanisms or to particular membrane properties of BW5147
cells. The estimation of chromosome numbers in hybrids revealed massive
loss of chromosomes. It is interesting to see that in spite of this

Table 1. Surface Analysis of Hybrids Derived from Fusions of BW5147 with C57BL/6 Spleen Cells

Hybrid	Thy-1 1.1	1.2	H-2 k	b	Ig	B cell Ia	Chromosomes (Range)
T48-3	+	+	+	+	−	−	58-65
T48-7	+	+	+	+	−	−	48-64
T48-11	+	+	+	+	−	−	41-44
T48-12	+	+	+	+	−	−	42-46
T44-2	+	+	+	+	−	−	34-37
T56-3	+	+	+	+	−	−	ND
T48-5	+	+	+	−	−	−	41-48
T56-2	+	+	−	+	−	−	45-48
T56-1	+	−	−	+	−	−	42-47
BW5147	+	−	+	−	−	−	38-40
C57BL/6 spleen	−	+	−	+	+	+	ND

Chromosome numbers were determined 10-18 weeks after fusion. The presence or absence of H-2 and Ia was determined by quantitative absorption of alloantisera with hybrid cells. For detection of Ia antigens, absorbed A.TH anti-A.TL serum was assayed against C57BL/6 spleen cells. The absence of surface Ig was concluded from the negative reactions of radiolabeled anti-Ig with hybrid cells.

substantial chromosome loss most hybrids still express both parental H-2 and Thy-1 antigens suggesting preferential retention of the respective chromosomes Nos. 17 and 6. Similar observations have been reported by Goldsby et al. (2). It is conceivable that loss of chromosomes may also result in loss of specific T cell functions.

Functional Analysis of T Hybrids

A variety of approaches were used for the detection of immunological function in T hybrids.

1. Haptenated phage inactivation assay. Splenic T cells from C57BL/6 mice sensitized with NP-CG have been shown to contain T cell receptors with specificity for the hapten NP (4-hydroxy-3-nitrophenacetyl). The presence of these T receptors can be assayed by their ability to inactivate NP conjugated T4 phage (3). Utilizing this extremely sensitive assay 1400 culture supernatants from T hybrids generated with NP-CG immunized C57BL/6 splenic T or spleen cells were analyzed for shedded or secreted NP specific material. Positive reaction was only observed occasionally and in these instances the activity disappeared after further culture of hybrids.

2. Antigen binding. 300 hybrid lines obtained with NP-CG sensitized splenic T or spleen cells were investigated for antigen binding using ^{125}I labelled antigen and autoradiographic techniques as described

previously (4). This technique allows the detection of less than 100 antigen molecules bound per cell. No positive hybrids could be found.

3. In vitro antibody responses. 280 supernatants from hybrids produced with spleen cells suppressed or immune for sheep red blood cells (SRBC) were analysed in Mishell-Dutton cultures for suppression or enhancement of primary and secondary SRBC responses. Two active culture supernatants were observed which specifically suppressed SRBC responses. However, these hybrid cultures lost rapidly their activity and it was not possible to isolate reactive subclones by soft agar cloning. It is, therefore, not clear whether the observed suppressive activity was in fact derived from specific T hybrids or from surviving not hybridized T cells.

4. Killer activity. Hybrids were generated with MLC blasts from primary and secondary mixed lymphocyte cultures which are known to contain a high percentage of antigen specific cytotoxic T cells (CTL). Among 370 hybrids with MLC blasts no hybrids with cytolytic activity were found. Since it is known that CTL can kill non specifically any target if they are brought together by agglutinating substances such as PHA (5) it was likely that during hybridization the CTL would lyse also syngeneic tumor cells due to the presence of fusion reagents. This possibility was verified in the experiment depicted in Table 2. It can be seen from the controls in which the CTL were replaced by EL4 cells that both PEG 4000 and Sendai Virus are of considerable toxicity for BW5147 by themselves. However, when $H-2^k$ anti-$H-2^b$ CTL were added lysis was markedly increased suggesting that the tumor cells are lysed non specifically in the presence of fusion reagents. Similar results have been described by Köhler et al. (6).

Table 2. Non Specific Lysis of BW5147 ($H-2^k$) Cells by B10.BR ($H-2^k$) anti-C57BL/6 ($H-2^b$) Cytotoxic T Cells (CTL) in the Presence of Fusion Reagents

| Targets | Reagent | % Cr Release with | |
		CTL	EL4
EL4 ($H-2^b$)	-	38.8	0
BW5147	-	7.6	0
BW5147	PHA	44.1	0
BW5147	PEG 4000	48.5	26.5
BW5147	Sendai Virus	24.5	13.0

It has been reported that the lytic activity of CTL is reversably blocked if either medium without calcium ions (7) or if medium with 2-deoxy-glucose (8) is used. Therefore, it was attempted to rescue killer function by hybridization under these reversible blocking conditions. While no hybrids could be obtained in the absence of Ca^{++}, 264 hybrids were obtained with secondary MLC blasts in the presence of 100 mM and 200 mM 2-deoxy-glucose. However, none of these hybrids displayed cytolytic activity.

190

Concluding Remarks

The failure to isolate antigen reactive T hybrids may have several reasons. Since the frequency of antigen reactive T cells is low either more hybrids have to be analysed or, preferably, hybridizations should be performed with highly enriched specific T cell populations. Another problem consists in the choice of the T tumors. It appears possible that for the rescue of particular T cell functions such as helper, suppressor or killer activity different T cell tumors have to be used for hybridization. Finally, since loss of chromosomes may also lead to loss of function such tumors should be employed which form stable hybrids.

Acknowledgements. We thank Ms. M. Palmen for expert technical help and Ms. A. Böhm for preparing the manuscript. This work was supported by the Deutsche Forschungsgemeinschaft through Sonderforschungsbereich 74.

References

1. Hämmerling, G.J.: T-lymphocyte tissue culture lines produced by cell hybridization. Eur. J. Immunol. 7, 743-746 (1977)
2. Goldsby, R.A., Osborne, B.A., Simpson, E., Herzenberg, L.A.: Hybrid cell lines with T cell characteristics. Nature, 267, 707-708 (1977)
3. Krawinkel, U., Rajewsky, K.: Specific enrichment of antigen binding receptors from sensitized murine lymphocytes. Eur. J. Immunol. 6, 529-536 (1976)
4. Hämmerling, G.J., McDevitt, H.O.: Antigen binding T and B lymphocytes. I. Differences in cellular specificity and influence of metabolic activity on interaction of antigen with T and B cells. J. Immunol. 112, 1726-1733 (1974)
5. Forman, J., Möller, G.: Generation of cytotoxic lymphocytes in mixed lymphocyte reactions. I. Specificity of the effector cells. J. Exp. Med. 138, 672-685 (1973)
6. Köhler, G., Lefkovits, I., Elliot, B., Coutinho, A.: Derivation of hybrids between a thymoma line and spleen cells activated in a mixed leucocyte reaction. Eur. J. Immunol. 7, 758-761 (1977)
7. Golstein, P., Smith, E.T.: The lethal hit stage of mouse T and non-T cell-mediated cytolysis: differences in cation requirements and characterization of an analytical "cation pulse" method. Eur. J. Immunol. 6, 31-37 (1976)
8. MacDonald, H.R.: Energy metabolism and T-cell-mediated cytolysis. II. Selective inhibition of cytolysis by 2-deoxy-D-glucose. J. Exp. Med. 146, 710-719 (1977)

Expression of Thy 1.2 Antigen on Hybrids of B Cells and a T Lymphoma

G.M. Iverson, R.A. Goldsby, L.A. Herzenberg

We and others reported that hybrids of spleen cells and the lymphoma line BW5147 express the Thy1.2 antigen of the splenic donor (1-3). Whether only T cells were the splenic parent of these hybrids could not be concluded from the hybrid phenotypes. We report here that either Ig$^+$ (B) cells or Ig$^-$ (T and other) cells give hybrids with BW5147 that express the Thy1.2 allele in the genome of the spleen cell whether or not that allele was expressed on the spleen cell parent.

Our specific purpose was to find out if any hybrids resulting from the fusion of B cells with BW5147 express the Thy-1 antigen coded for in the B cell. This could be accomplished on non-cloned populations. Therefore, we did not clone the hybrids but rather tested them as populations of hybrid cells.

Spleen cells from (BALB/c x SJL)F$_1$ mice (Thy1.2) were stained with rhodamine-labeled rabbit anti-mouse immunoglobulin (R-anti-MIg) that had been adsorbed with a 10% volume of normal BALB/c thymocytes. The brightly stained cells, Ig$^+$ B cells, were separated from the non-stained cells, Ig$^-$, T cell-enriched populations, by means of the fluorescence-activated cell sorter (FACS) (4) (see Fig. 1). Each separated population was fused, with the aid of polyethylene glycol after the method of Pontecorvo (5) as adapted by Goldsby (1), with BW5147 (Thy1.1) at a ratio of 4:1 (spleen cells to BW5147). After fusion the cells were plated at 1.2 x 10^5, 0.3 x 10^5, and 0.17 x 10^5 cells/ml and cultured in HAT medium. Approximately 4 wks later each well that had visible cell growth was transferred to a new culture vessel and kept as a unique hybrid population.

The fusion of BW5147 with purified B cells did not result in fewer hybrid populations than fusion of BW5147 with the T cell-enriched population (see Table 1). Therefore, we conclude that BW5147 does not preferentially hybridize with T cells.

Table 1. Hybrids between BW5147 and separated T or B spleen cells

	No. of wells with hybrids / Total no. of wells plated
BW x Ig$^+$	37/72
BW x Ig$^-$	21/72

The hybrid populations were tested for the expression of the Thy-1 allele of each parental type -- Thy1.1 from BW5147 and Thy1.2 from the (BALB/c x SJL)F$_1$ spleen cells. Each hybrid population was tested by direct cytotoxicity using anti-Thy1.1 and anti-Thy1.2, produced in congenic strains of mice, plus rabbit complement. All of the hybrid populations were lysed with anti-Thy1.1 plus complement. Approximately 80% of thy hybrid populations showed significant cytotoxic indices with anti-Thy1.2 plus complement (see Fig. 2).

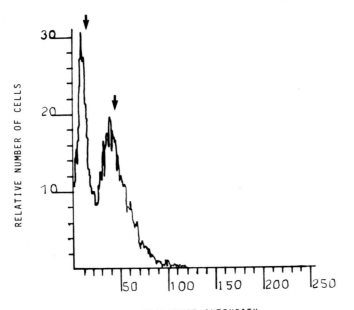

Fig. 1. FACS fluorescence profile of (BALB/c x SJL)F₁ spleen cells stained with rhodamine-labeled R-anti-MIg. The arrows indicate the gates used to separate the Ig⁺ (the brightest 38%) from the Ig⁻ (the dullest 40%).

Fig. 2. The expression of Thy1.2 on hybrids between BW5147 (Thy1.1) and separated Ig⁺ and Ig⁻ spleen cells of Thy1.2⁺ mice

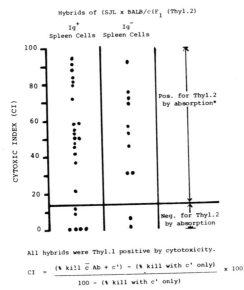

All hybrids were Thy1.1 positive by cytotoxicity.

$$CI = \frac{(\% \text{ kill } \bar{c} \text{ Ab} + c') - (\% \text{ kill with } c' \text{ only})}{100 - (\% \text{ kill with } c' \text{ only})} \times 100$$

*Tested back on Thy1.2 thymocytes. No absorption by BW5147 was found.

193

The background cytotoxicity of rabbit complement alone was constant for any given hybrid population but was quite variable from one population to another. The background cytotoxicity varied from as low as 10% with some hybrid populations to as high as 50% with others. This was true for six different lots of rabbit complement.

To ensure the specificity of the direct cytotoxcity test, all of the hybrids were tested for their ability to specifically adsorb anti-Thy1.2 activity against BALB/c thymocytes. Normal BALB/c thymocytes absorbed and BW5147 did not absorb the anti-Thy1.2 activity. Absorption of anti-Thy1.2 was observed for a given hybrid population only if a cytotoxic index (CI) above 20% had been found for that population. None of the hybrid populations had detectable surface Ig by immunofluorescence and FACS analysis nor secreted Ig by radioimmune assay.

From these experiments we conclude that BW5147 can and does fuse with both splenic B and T cells. In about 80% of the hybrid populations the Thy-1 antigen of the normal B or T cell donor is expressed. Also, the surface Ig characteristics of the splenic B cell are lost in the hybrid with BW5147.

This research supported, in part, by grants from the National Institutes of Health, GM-20832, GM-17367, and CA-04681.

References

1. Goldsby, R.A., Osborne, B.A., Simpson, E., Herzenberg, L.A.: Hybrid cell lines with T-cell characteristics. Nature, 267, 707-708 (1977)
2. Hämmerling, G.J.: T lymphocyte tissue culture lines produced by cell hybridization. Eur. J. Immunol. 7, 743-746 (1977)
3. Köhler, G., Lefkovits, I., Elliott, B., Coutinho, A.: Derivation of hybrids between a thymoma line and spleen cells activated in a mixed leukocyte reaction. Eur. J. Immunol. 7, 758-761 (1977)
4. Herzenberg, L.A., Herzenberg, L.A.: Analysis and separation using the fluorescence-activated cell sorter (FACS). In: Handbook of Experimental Immunology 3rd Edition, D.M. Weir (ed.). Blackwell Scientific Pub., Oxford, England, Chap. 22 (1978)
5. Pontecorvo, G.: Production of indefinitely multiplying mammalian somatic cell hybrids by polyethylene glycol (PEG) treatment. Somatic Cell Genetics 1, 397-400 (1976)

T Cell Hybrids with T Cell Functions

E. Simpson, S. Kontiainen, L.A. Herzenberg, E. Bohrer, A.Torano, P. Vogt,
P. Beverley, W. Fitzpatrick, M. Feldmann

A. Introduction

Different T cell functions, e.g. help, suppression and cytotoxicity are associated
with different T cell subsets, bearing distinctive differentiation markers. Thus
effector T cells mediating help are Ly $1^+2^-3^-$ and may bear Ia determinants, those
mediating suppression are Ly $1^-2^+3^+$, IJ^+, and cytotoxic cells are Ly $1^-2^+3^+$ and do
not bear Ia determinants (1). At least the cytotoxic effectors are Ly 6.2 positive,
but they arise from Ly 6.2 negative precursors (2). Thus expression of certain
differentiation markers (Ly 123) is associated with acquisition of the ability to
perform a particular function, even before encounter with antigen, whilst others
are expressed as a result of further differentiation, including clonal expansion,
following response to antigen (e.g. Ly 6.2 and possibly Ia antigens).

The successful hybridization of primed B cells with a B cell tumour to produce
functional antibody secreting hybrids appears to require that each of the two populations be at the same stage of differentiation (3). It is possible that the
same restrictions may apply to production of functional T cell hybrids between
normal primed T cell populations and a T cell tumour. There may be additional restrictions on the formation of hybrids between T cells capable of carrying out
certain functions, such as cytotoxicity, since it could be that the very act of
fusion between a cytotoxic cell and a tumour cell would lead to the killing of the
tumour cell (4) and thus prevent hybrid formation. However, there is no reason to
assume that T cells with other functions e.g. help and suppression cannot be successfully fused with an appropriate T cell tumour, and function be expressed.

We have used the AKR thymoma BW5147 to make somatic cell hybrids with T cell populations induced *in vitro* to express (a) cytotoxicity, (b) antigen specific suppression and (c) antigen specific help. In each case, we have obtained hybrids
which carry the Thy 1 marker of each parental type and for suppression and help we
have obtained T cell hybrids which express the function shown by the original
primed normal parental T cell population. Typing of other cell surface markers has
yielded some rather surprising and unexpected results: some hybrids express apparently 'inappropriate' differentiation antigens and fail to express differentiation
antigens which would be appropriate to the function their supernatants show.

Finding of immunologically functional material in the supernatants of hybrids growing *in vitro* has allowed us to do preliminary analyses of some immunochemical characteristics of the functional moieties. These and further investigations should
throw substantial light on the question of the nature of the T cell receptor(s).

B. MATERIALS AND METHODS

I. Animals and Antigens

Mice. CBA/Ca, C57BL/10, B10.BR, AKR, A.TH, A.TL, SJL, (B6xBALB/c)F$_1$, B6.Ly1.1, and

(A.Thy1.1x CBA) F$_1$ mice were obtained from the Division of Comparative Medicine, CRC and from the Imperial Cancer Research Fund Breeding Unit. The antigens used were keyhole limpet haemocyanin (KLH, a gift from Dr M. Rittenberg, Portland, Oregon), a copolymer of L-glutamic acid ^{60}-L-alanine ^{30}-L-tyrosine ^{10}(GAT, a gift from Dr P. Maurer, Philadelphia) and poly(L-tyrosine-L-glutamine)-poly(D-L alanine) -poly(L-lysine) [(T,G(-A--L) a gift from Dr Edna Moses, the Weizmann Institute, Rehovot]. Trinitrophenylated KLH and GAT had 8 and 4 groups of TNP per 100,000 daltons. Trinitrophenylated polyacrylamide beads (TNP-PAA), a thymus independent antigen, were prepared by Dr Marilyn Baltz at University College London. The preparation of TNP antigens was as described earlier (5).

II. T Cell Tumour Line

In every case the tumour line used for the fusions has been BW5147 (BW). This was obtained from Dr Robert Hyman, La Jolla via Prof. L.A. Herzenberg of Stanford in September 1976. It was HGPRT negative, thus allowing for suppression of its growth in HAT medium (see below) following fusion with a source of normal T cells: non-fused T cells die out in a matter of days, and only hybrids between these cells and BW grow in HAT. BW and all fusion hybrids were maintained in Dulbecco's Modification of Eagle's Medium DMEM, Flow Cat. No. 12-332-54(1-001M) with 10% foetal calf serum (FCS) obtained from Gibco (Scotland) and screened for its ability to support mixed lymphocyte responses for the generation of cytotoxic cells, i.e. our FCS was not specially screened for fusion work. Cells were incubated at 37°C in a humidified 5% CO$_2$ atmosphere.

III. Normal Primed T Cell Populations

Spleen cells were used as a source of normal T cells: they were primed *in vitro* in the following ways:- (1) for cytotoxicity, either 25x10^6 AKR or CBA spleen cells were co-cultured in tissue culture flasks in a humidified 10% CO$_2$ atmosphere at 37°C for 4 or 5 days with equal numbers of 2000R irradiated C57BL/10 spleen cells in bicarbonate buffered RPMI medium containing 10% FCS, 5x10^{-5}M 2ME, 10mM of Hepes, glutamine, penicillin and streptomycin; (2) for suppression, CBA spleen cells were cultured in the inserts of Marbrook Diener flasks for 4 days with 100 µg of KLH/ml in Eagles Minimum Essential Medium (MEM, Gibco, Cat. No. F15), containing 3% FCS and Hepes at 4.36g/litre. The same medium, but with bicarbonate added at 3.7g/litre instead of Hepes, was used in the outer chamber (6); and (3) for help, B10 spleen cells were cultured as above in Marbrook Diener flasks, but with (T,G)-A--L at 1 µg/ml (7).

IV. Cell Hybridization

The cell hybridizations, as previously described (3,8) were performed with polyethylene glycol. (PEG, BDH MW 1500, 50% in BSS, pH 7.8: to prepare, 25 gm PEG crystals were autoclaved, and 25 mls of warmed balanced salt solution (BSS) with no foetal calf serum were added at 56°. The solution can be used immediately or can be stored at +4°C for several months). 10^8 *in vitro* primed cells
        and 10^7 BW5147 cells were washed twice in serum free BSS and pelleted together at 400 g. 0.5ml of PEG was added slowly (0.1ml per 10 seconds) over a period of 1-2 minutes as the cells were gently shaken into suspension, and 0.5ml of serum free BSS was then added at the same rate. A further 5ml of BSS were then added dropwise before slowly filling the tube to 20ml with BSS. The cells were spun at 400g, the supernatant discarded and the cells resuspended in 100ml of DMEM with 20% FCS. The cell suspension was dispensed in 2ml aliquots into 48 wells of two 24 well Linbro trays (Cat. No. Flow FM 1624TC) and incubated at 37°C in a humidified atmosphere of 5% CO$_2$. 24 hours after the fusion 1.0ml of medium was removed from each well and replaced by 1.0ml HAT (DMEM plus 20% FCS, 1x10^{-4}M hypoxanthine, 1.6x10^{-5}M thymidine and 4x10^{-7}M aminopterin) and this procedure was repeated 48 and 72 hours after fusion. On the sixth, eighth and tenth day the medium was

changed to HT (DMEM plus 20% FCS, 1×10^{-4}M hypoxanthine and 1.6×10^{-5}M thymidine). Thereafter (day 13 from fusion) the medium was changed to DMEM plus 10% FCS, and the contents of each Linbro well which started to grow within the next 1-2 weeks were subcultured in Linbro wells and then transferred to Nunc tissue culture flasks (50ml, Nunclon-Delta 1461). Aliquots of supernatants were tested for function when the cells were growing in the Linbro plates, and subsequently in the flasks, in the case of the suppressor and helper hybrids, whilst in the case of hybrids made with cytotoxic populations, functional testing for cytotoxicity using ^{51}Cr labelled target cells was done with hybrid cells as attackers following growth and transfer of hybrid lines to flasks.

V. Cell Cultures for Functional Assay of Hybridoma Products

Only supernatants of hybridoma cultures were initially tested for suppression and help. For the assay Marbrook Diener cultures incubated under the conditions described above were used. The assay for suppression was to add the hybridoma supernatants, at a 10% or 1% final concentration, to the mixture of KLH, GAT or (T,G)-A--L primed helper cells (^{HC}KLH, ^{HC}GAT, or ^{HC}(T,G)-A--L) primed *in vitro* as described previously (6,7,9), normal spleen cells, and haptenated antigens TNP-KLH, DNP-GAT or DNP-(T,G)-A--L. The anti-DNP antibody forming cells (AFC) were assayed on day 4 of the *in vitro* cooperative culture (7). Using unprimed B cells only IgM anti-DNP AFC were detected. All cultures were in triplicate. The assay is summarised as: spleen cells + haptenated antigen = background, helper cells + spleen cells + haptenated antigen = help, helper cells + spleen cells + haptenated antigen + suppressor cell S/N = suppression.

The results are expressed as numbers of anti-DNP AFC/culture $\pm$ SE. The degree of suppression was counted as % suppression $= 100 - \dfrac{(\text{suppression} - \text{background})}{(\text{help} - \text{background})} \times 100$

The assay for helper cell hybridoma products was to add instead of the helper cells the hybridoma supernatants at 1% or 0.1% final concentration to a mixture of normal spleen cells and antigen, 1 µg/ml DNP-(T,G)-A--L. Anti-DNP AFC were assayed on day4.

VI. Cell Surface Antigens

The following antisera were used to test for cell surface antigens Thy 1.1, Thy 1.2, Ly 1.1, Ly 1.2, Ly 2.1, Ly 6.2, H-2^k and Ia^k on parental and hybridoma cells: Thy 1 antisera were prepared conventionally by injecting CBA thymocytes into AKR mice (Thy 1.2) and AKR thymocytes into CBA mice (Thy 1.1) and obtaining hyperimmune serum or ascites fluid after 6 or more immunizations. Ly 1.1 antisera were obtained by hyperimmunizing (B6xBALB/c)F$_1$ mice with thymocytes from B6 Ly 1.1 mice. Ly 1.2 and Ly 2.1 antisera were the kind gift of Dr E.A. Boyse, Sloan Kettering Institute. Ly 6.2 antisera were obtained by hyperimmunizing (A. Thy 1.1xCBA)F$_1$ mice with thymocytes and spleen cells from AKR mice. Ia^k antisera were prepared by hyperimmunizing A.TH mice with A.TL spleen cells.

The possession of cell surface antigens Thy 1.1, Thy 1.2, H-2^k and Ia^k were tested by direct complement mediated cytotoxicity using ^{51}Cr labelled hybrids as target cells. In addition, a quantitative absorption technique was used for testing for the presence of Ly 1.1, Ly 1.2, Ly 2.1, Ly 6.2 and Ia^k on the cell surface of hybrids and their parental cells. Briefly, each of the antisera were first titrated by a micro-trypan blue exclusion test on appropriate targets, using thymocytes for Ly 1.1 1.2 and 2.1, spleen cells for Ia^k and lymph node cells for Ly 6.2, and a dilution of each antiserum was then chosen which still just showed plateau level killing: 50 µl aliquots of these dilutions of antisera were then absorbed for 1 hour at RT with 5 different numbers, usually 50×10^6, 25×10^6, 12.5×10^6, 6.25×10^6 and 3.12×10^6, of hybrid cells and appropriate control cells, and the supernatants then tested on the indicator target cells, using the micro-trypan blue exclusion test.

VII. Immunoabsorptions

One of the suppressor cell hybridoma products, S1.41 supernatant, SFS1.41, was subjected to detailed functional and immunochemical analyses. SFS1.41 was taken from bulk cultures of this hybrid 9 months after the fusion. It was absorbed and 'acid eluted', by using Sorensen's buffer pH 2.4, from the following immunoabsorbents: KLH (1mg of KLH/1ml beads), GAT (1mg of GAT/1ml beads), anti Ia^k (0.5ml of the anti Ia^k serum described above/5ml beads) anti Ia^s (ATL anti ATH serum kindly donated by Dr McKenzie, Canberra, Australia, 0.5ml of serum/5ml beads), anti $I-J^k$ (B10.HTT anti B10.S (9R) serum donated by Dr I. McKenzie, 0.5ml of serum/5ml beads anti $I-J^s$ (B10.S(9R) anti B10.HTT serum donated by Dr I. McKenzie, 0.5ml of serum/5 ml beads, and anti MIg (0.5ml of rabbit anti mouse Ig/5ml beads). The beads used were Sepharose 4B (Pharmacia, Sweden). The cyanogen bromide activation of beads and coupling of antigen or antisera was as described (10).

VIII. Anti suppressor Factor Antisera

The preparation of these antisera and their characteristics have been recently described (11). The antisera made in rabbits (R α SF) against KLH specific suppressor factor of CBA origin seems to recognise a constant region like determinant in all suppressor factors, while the antibody made in CBA mice (M α SF) recognises idiotypic determinant(s) unique to the KLH specific suppressor factor of CBA mice. These antisera were used at a 0.1% final concentration.

C. RESULTS

Table 1 summarizes the results in terms of the numbers of hybrids recovered from each type of hybridization, the cell surface markers shown by some of these hybrids and the number of functional hybrids obtained.

Table 1. Summary of Results   (* number positive/number tested; + by adsorption; o by cytotoxicity; NT = not tested)

Normal parent function	No of wells set up	No of wells with hybrids	Thy1.1 + Thy1.2	Ia^k	Cell surface markers H-2^k	Ly1.1	Ly1.2	Ly2.1	Ly6.2	specific	non-specific	specificity unknown
Cytotoxicity H-2^k anti H-2^b	144	60	11/12 ^{*o}	0/20 ^{o}	NT	NT	NT	NT	NT	0	0	
Suppression CBA-KLHSC	96	14	3/13 ^{o}	0/2 $^+$	2/2 ^{o}	1/2 $^+$	1/2 $^+$	0/2 $^+$	2/2 $^+$	2	2	4
Help B10 (T,G) -A--LHC	48	19	NT	NT	NT	NT	NT	NT	NT			12

198

I. Cytotoxicity

We have reported earlier that the majority of hybrids tested following fusion of BW with mixed lymphocyte culture cells had the Thy 1 markers of both the tumour parent (Thy 1.1) and the normal parent (Thy 1.2) (12). This indicated to us that they were probably T cell hybrids, although this type of evidence is merely circumstantial. Ia antigens were never found on the cell surface of these hybrids by direct cytotoxic testing. No functional activity was detected in any of the 60 hybrids tested, following either incubation with ^{51}Cr labelled H-2^b tumour target cells, EL-4, or with Con A blasts of H-2^b or H-2^k origin, either in the presence or in the absence of 10 µg/ml Con A as a non-specific 'glue'.

II. Suppression

The results shown here are those accummulated from two separate hybridization experiments. In the first, bacterial contamination wiped out all but two of the hybrids, but both of them proved to be of interest, the supernatants of one of them having antigen specific (KLH) suppressor activity (S1.41) whilst the supernatants of the other (S1.34) were non-specifically suppressive. In a subsequent hybridization 12 more hybrids were grown, of which six had suppressor activity, at least one of which is specific. Data showing the antigen specificity of suppression by S1.34 and S1.41 supernatant materials (SFS1.41 and SFS1.34) are shown in Table 2, in which the activities of SFS1.34 and of SFS1.41 are compared with that of SF prepared from CBA suppressor cells (9) induced by high doses of KLH (100 µg/ml) i.e. the same type of suppressor cells which were used as the normal 'parent' of S1.34 and S1.41. Supernatant material from neither hybrid 'helped' the anti-DNP response when HC were absent.

Table 2. Specificity of suppression by SFS1.34 and SFS1.41

STIMULUS		SUPPRESSION	RESPONSE (d.4, IgM)	Percent of
HC[1]	Antigen	SF (%)	Anti-DNP AFC/culture ±S.E.	Suppression
-	TNP-KLH	-	10 ± 8	-
HC$_{KLH}$	"	-	263 ± 19	0
"	"	CBA SF$_{KLH}$ 10	30 ± 10	92
"	"	SFS1.41 10	117 ± 33	58
"	"	" 1	50 ± 5	84
"	"	SFS1.34 10	70 ± 34	76
"	"	" 1	160 ± 16	61
-	"	SFS1.41 10	47 ± 12	-
-	"	SFS1.34 10	25 ± 10	-
-	DNP-(T,G) -A--L	-	156 ± 45	-
HC$_{(T,G)}$ -A--L	"	-	453 ± 106	0
"	"	SFS1.41 10	447 ± 98	0
"	"	SFS1.34 10	253 ± 53	67
-	DNP-PAA[2]	-	4660 ± 821	-
-	"	SFS1.41 10	6413 ± 696	-

[1] 3×10^5 B10.BR HC$_{KLH}$ or 1×10^5 B10.BR HC$_{TGAL}$ + 10^7 normal B10.BR spleen + 0.1 µg/ml of TNP-KLH or 1 µg/ml DNP-(T,G)-A--L ± SF.

[2] 10^7 normal B10.BR spleen + 0.3% final conc. DNP-PAA ± SFS1.41.

The effect of S1.41 on KLH specific help has been tested about 20 times over a period of 13 months with similar results. The specificity of S1.41 suppression has been tested on either GAT or (T,G)-A--L specific help 5 times over a period of 13 months.

The preliminary immunochemical characterization of SFS1.41 is shown in Table 3. The functional molecule(s) was removed not only by antigen (KLH), but also by anti-Ia^k and anti-I-J^k, and the activity could be eluted from the solid phase immuno-absorbents used. In contrast, neither anti-Ia^s nor I-J^s absorbed the activity, and nor did an irrelevant antigen, GAT or rabbit-anti-mouse immunoglobulin (data not shown). The presence of rabbit anti-suppressor factor antiserum and mouse anti-suppressor factor antiserum in the cultures abrogated the suppressive effect of SFS1.41.

Table 3. Molecular characteristics of SFS1.41

STIMULUS		SUPPRESSION	RESPONSE (d.4, IgM)	Percent of
$HC^{1)}$	Antigen	SF (10%)	Anti-DNP AFC/culture ± S.E.	Suppression
–	TNP-KLH	–	0	–
B10.	"	–	193 ± 40	0
HC_{KLH}				
"	"	SFS1.41	3 ± 3	98
"	"	" abs KLH	270 ± 8	0
"	"	" elu "	13 ± 11	92
"	"	" abs α Ia^k	210 ± 5	0
"	"	" elu "	0	100
"	"	" abs α I-J^k	70 ± 38	64
"	"	" elu "	0	100
"	"	" abs α I-J^s	7 ± 6	96
"	"	" elu "	193 ± 7	0
"	"	$CBASF_{KLH}$	0	100
–	TNP-KLH	–	470 ± 56	–
$CBAHC_{KLH}$	"	–	1630 ± 78	0
"	"	SFS1.41	973 ± 48	57
"	"	" +M α $SF^{2)}$	1463 ± 385	14
"	"	" +R α $SF^{2)}$	1823 ± 94	0

1) 3×10^5 HC + 10^7 normal B10 spleen + 0.1 µg/ml of TNP-KLH ± S1.41. Two other experiments gave similar results.

2) Rabbit anti-$CBASF_{KLH}$ (R α SF) and CBA α $CBASF_{KLH}$ (M α SF) were added to cultures at 0.1% final concentration together with S1.41 or S1.34. Two other experiments gave similar results.

The cell surface markers of selected suppressor hybrids were interesting and not altogether expected (Table 1). Firstly, by no means all of them expressed both Thy 1.1 and Thy 1.2 - 10 of the 13 tested were only Thy 1.1 positive: however, both S1.34 and S1.41 were Thy 1.1 and Thy 1.2 positive, and in view of the more detailed analysis of their functions, quantitative absorptions were carried out with these two cell lines, and with BW, to determine their Ly and Ia phenotypes (Table 1). Neither BW, nor either of these two hybrids were Ia positive using this test with ATH anti ATL antiserum. S1.34, the non-specific suppressor line, was neither Ly 1.1 (CBA genotype) Ly 1.2 (AKR genotype) nor Ly 2.1 (CBA and AKR genotype) positive, and nor did our sub-line of BW express any of these markers, in contrast to a report that it is Ly 1 positive (13). In contrast S1.41, the specific suppressor line, was Ly 1.1 and Ly 1.2 positive, thus expressing both parental alleles for this locus. The finding of an Ly 1^+2^- phenotype for S1.41 was unexpected, since KLH specific suppressor cells induced from normal spleens are Ly 1^-2^+(1)

200

There remains the possibility that as we phenotyped an uncloned hybrid line of S1.41, it contains more than one population of cells, but this seems unlikely since it has now been growing continuously for 13 months, without any alteration in function of the antigen specific suppressor supernatant material.

III. Help

The fusion performed with *in vitro* induced (T,G)-A--L specific helper cells is a recent one, and although 19 lines have been established of which 15 have been tested for function, there is no specificity data on whether the helper function displayed by any of them is antigen specific, but 12 gave supernatants which showed the same degree of help as 'conventional' (T,G)-A--L helper factor (14): 3 of these have been cloned producing 49 clones altogether, and in each case more than 50% of the clones of each hybrid have shown help for anti-DNP responses stimulated by DNP-(T,G)-A--L.

D. DISCUSSION

T-T cell hybrids, expressing T cell surface marker characteristics of both parental cell types are not difficult to obtain using BW as the T cell tumour line, and polyethylene glycol as the fusing agent. Approximately 1 cell per 5×10^6 *in vitro* primed T cells fuses with BW to produce a hybrid. Detecting function in such hybrids appears not to be easy in the case of cytotoxicity. There are possible explanations for this, although the negative findings do not prove the widely discussed hypothesis that cytotoxic cells kill the tumour parent, and so prevent formation of fusion hybrids (4,15).

In the experiments reported here, the 'rescue' and perpetuation of specific suppressor function is very much better established than that of helper function, where we have no specificity data at the moment. The need to examine each supernatant for functional activity in a fairly complicated biological assay has placed certain restrictions on the speed of progress. Nevertheless our most studied suppressor hybrid, S1.41, appears to be both specific in its action, and stable in as much as it has retained function and cell surface markers for over a year.

Immunochemical analyses of the suppressor factor found in the supernatant of S1.41 indicate that it is similar if not identical to 'conventionally' induced suppressor factor, and the possibility of obtaining very large quantities of this material for further analyses should throw light on the nature of suppressor factor. It is possible that this material may be the T cell receptor(s). Thy 1 phenotyping was initially done early but H-2 and Ly phenotyping was performed only after one year of growth. It may be significant that with the apparently inevitable loss of chromosomes by hybrids with time (13), H-2, Thy 1 and Ly 1 and 6 antigens and therefore the chromosomes coding for these loci appear to be preserved. This may reflect the need for antigens of these loci to be present on the cell surface, in order to permit T cell function.

Acknowledgments. We thank Deirdre Hayton, Pirkko Himberg and Pirjo Ranttila for excellent technical assistance, Dr M. Rittenberg for KLH, Dr P. Maurer for GAT, Dr Edna Mozes for (T,G)-A--L and Dr M. Baltz for DNP beads. These experiments have been supported by the Imperial Cancer Research Fund, USA Public Health Service Grant A1-13145-02, and the Medical Research Council of Great Britain. Sirrka Kontiainen and P. Vogt are supported by EMBO Fellowships.

References

1. Simpson, E., Beverley, P.C.L.: T cell subpopulations. Progress in Immunology III, 206-216 (1977).
2. Woody, J.N.: Ly-6 is a T cell differentiation antigen. Nature 269, 61-63 (1977).
3. Köhler, G., Milstein, C.: Derivation of specific antibody - producing tissue culture and tumour lines by cell fusion. Eur. J. Immunol. 6, 511-519 (1976).
4. Köhler, G., Lefkovits, I., Elliot, B., Coutinho, A.: Derivation of hybrids between a thymoma line and spleen cells activated in a mixed leucocyte reaction. Eur. J. Immunol. 7, 758-761 (1977).
5. Kontiainen, S., Feldmann, M.: Suppressor cell induction *in vitro*. IV. Target of antigen specific suppressor factor and its genetic relationships. J. Exp. Med. 147, 110-127 (1978).
6. Kontiainen, S., Feldmann, M.: Suppressor cell induction *in vitro*. I. Kinetics of induction of antigen specific suppressor cells. Eur. J. Immunol. 6, 296-301 (1976).
7. Kontiainen, S., Feldmann, M. Induction of specific helper cells *in vitro*. Nature 245, 285-286 (1973).
8. Pontecorvo, G.: Production of mammalian somatic cell hybrids by means of polyethylene glycol treatment. Somat. Cell Genet. 1, 397-400 (1975).
9. Kontiainen, S., Feldmann, M. Suppressor cell induction *in vitro*. III. Antigen specific suppression by supernatants of suppressor cells. Eur. J. Immunol. 7, 310-314 (1977).
10. March, S.C., Parikh, I., Cuatre Casas, P.: A simplified method for cyanogen bromide activation of agarose for affinity chromatography. Analyt. Biochem. 60, 149-152 (1974).
11. Kontiainen, S., Feldmann, M. Suppressor cell induction *in vitro*. V. Effects of anti-suppressor factor preparations. Submitted.
12. Goldsby, R.A., Osborne, B.A., Simpson, E., Herzenberg, L.A.: Hybrid cell lines with T cell characteristics. Nature 267, 707-708 (1977).
13. Hammerling, G.J.: T lymphocyte tissue culture lines produced by cell hybridization. Eur. J. Immunol. 7, 743-746 (1977).
14. Howie, S., Feldmann, M.: *In vitro* studies on H-2 linked unresponsiveness to synthetic polypeptides. III. Production of an antigen specific T helper cell factor to (T,G)-A--L. Eur. J. Immunol. 7, 417-421 (1977).

202

T Cell Hybrids with Specificity for Individual Antigens

N.H. Ruddle

A. Introduction

The purpose of the studies described here was to obtain functional
T cell hybrids which express specificity for particular antigens.
Such hybrids will aid in the purification and elucidation of the
molecular nature of the T cell receptor and products characteristic
of different functional T cell subsets.

B. General Discussion of T Cell Hybrid Systems

I. Sources of Primary T Cells

In order to obtain T cell hybrids with   specificity for particular
antigens, it is imperative that one initiate such crosses with
highly purified, enriched populations of T cells which express the
particular function in question. This problem is of greater moment
when one considers T cell hybrids rather than those formed with B
cells and myelomas due to difficulties inherent in screening T cells.
If one begins with highly purified populations, the hybrids, should,
at the very least, have the information for expression of particular
functions.

Three different sources of mouse primary T cells have been used in
our studies. The cell types which I have attempted to immortalize
by hybridization are: T cells which suppress the 5 day Igm plaque
forming cell response to sheep red blood cells (SRBC); T effector
cells for delayed hypersensitivity to ovalbumin (OVA); and TNP-CBA
self-killer cells. All hybrids have been generated with BW 5147(TGr)
as the input lymphoma parent. This line was kindly supplied by Dr.
Richard Goldsby. It is an AKR tumor which expresses H-2k, Thy 1.1,
and the slow (a) form of glucose _   phosphate isomerase.   We
have been unable to obtain hybrids with EL4.Bu or S49.1TB.2 obtained
from the Salk Institute. The three hybrid cell systems I have developed are described in Table 1.

II. Analysis of T Cell Hybrids

The analysis of T cell hybrids has included investigation of both
constituitive markers (e.g. GPI) and those which are more characteristic of differentiated functions, such as antigen binding. The
markers which have been examined, or are currently being investigated, are listed in Table 2.

Table 1. T Cell Hybrids Between BW 5147 (TGr) and Primary T Cells

Source of Primary T Cells

	C57Bl/6	C56Bl/6	CBA
Strain			
H-2	b	b	k
Thy 1.	1.2	1.2	1.2
GPI	fast b	fast b	fast b
Antigen	SRBC	OVA	TNP-CBA
Method of sensitization	SRBC in vitro 4 days	50ug OVA in CFA in vivo 8 days	In vitro culture 14 days; Re-stimulate 4 days
Function in vivo	Regulation of anti SRBC response	Delayed hypersensitivity	?
Assays in vitro	Suppression IgM anti SRBC Rosette formation	Incorporation of 3HdT Lymphotoxin	Cr 51 release
Cells			
Source	Spleen	Lymph node	Spleen
Purification	Nylon, LSM of RFC	Nylon, 1 g sediment	None: 80% blasts
Number of Hybrid Clones (Currently maintained)	7	20	18
Trivial Name	Hyb 29, 34	Hyb 20,21,23,33	Hyb 32

Table 2.   Analysis of T cell hybrids

Trait	Chromosome	Method
1. Expression of HPRT	X	Growth in HAT selection medium
2. Glucose phosphate isomerase	7	Starch gel electrophoresis
3. H-2	17	Quantitative absorption of cytotoxicity Direct cytotoxicity
4. Thy-1	9	Quantitative absorption of cytotoxicity Direct and indirect fluorescence Direct cytotoxicity
5. Chromosome counts		
6. Reactivity to specific antigen		Antigen binding Growth rate changes Lymphokine release
7. Function		Killing TNP conjugated spleen cells Suppression anti SRBC response Transfer of delayed hypersensitivity to OVA
8. Membrane analysis		

C.   Hybrids between BW 5147(TGr) and T Effector Cells for

   Delayed Hypersensitivity to Egg Albumin (OVA)

I.   Purification of OVA Sensitized T Cells

Mice sensitized with 50 ug egg albumin (OVA) in complete Freund's adjuvant undergo ear swelling when the specific antigen is reinjected. This is a manifestation of delayed hypersensitivity (DTH) (1). DNA synthesis of T cells from mice 8 days after sensitization is stimulated in the presence of OVA. T cells sensitized in this manner kill innocent bystander A9 cells in the presence of antigen by releasing a soluble factor called lymphotoxin (2,3). Cells have been purified by passing draining inguinal lymph node cells from sensitized mice through nylon wool and subjecting them to unit gravity sedimentation (4,5). A significant enrichment of cells which react to OVA with increased DNA synthesis and release of lymphotoxin has been achieved in fractions of cells which sediment most rapidly. The aim of the experiment described here was to immortalize this putative DTH effector cell by hybridization with BW 5147(TGr).

II. Hybridization

Two conditions of hybridization have been used. Most experiments
were carried out with a modification of the technique of Galfre et
al (6). Nylon-purified   unit gravity sedimented lymph node T
cells from OVA sensitized mice were mixed with an equal number of
BW 5147(TGr) (2-4 x 10^6 cells from each parent). The cells were
mixed and spun together (1000 rpm for 10 minutes) in a 15 ml centrifuge tube, and all the supernatant was removed. Room temperature
polyethylene glycol (PEG) Baker, MW 1540) was added at 40% (W/W) in
phosphate buffered saline in a volume of 0.3-0.4 ml over the space
of 45 seconds. The PEG was added with a wide mouth 1 ml pipette.
The latter was prepared by cutting the end of a plastic pipette with
a heated razor blade, rapidly flaming it and resterilizing by UV
light exposure for 30 minutes. The cells were slowly drawn up into
the pipette as the PEG solution was added. Cells were placed in a
37° water bath for an additional minute. One ml Dulbecco's Modified Eagle's medium with HEPES buffer (DME) (GIBCO) was added over
the course of 1 minute. Then 8 ml DME was added slowly over the
course of the next 3 minutes. The cells were spun at 1000 rpm for
10 minutes and resuspended in DME with 10% fetal calf serum. Approximately the same hybridization frequency has been obtained with
the second method, the pancake technique (7). This has ranged from
between one in 10^4 and one in 10^5 BW 5147 cells. Successful hybridization, as determined by expression of parental forms and an
intermediate form of GPI in vigorously growing cells, was obtained
with ratio of T cells to BW 5147 cells ranging from 7:1 to 1:1.
The limiting factor is the number of tumor cells.

III. Selection

Selection has been mainly by means of the dilution technique in the
HAT medium. Cells were dispensed with a wide mouth pipette immediately after PEG treatment into 96 well flatbottom tissue culture
plates. Generally, T cells and 6 x 10^4 BW 5147 cells were added in
0.2 ml to each well. After 24 hours, half the medium was removed
and 0.1 ml HAT medium was added (hypoxanthine 10^{-4}M, aminopterin
5 x 10^{-7}M, thymidine 10^{-5}M). The cells were fed with HAT every two
days. When the bottom of the wells became crowded, the cells were
transferred to larger wells in 2 ml medium and then to plastic
flasks in 10 ml medium. Cells were maintained in HAT for 3 weeks,
in HT for 1 week, and thereafter in DME with 10% fetal calf serum.

The second selection technique involved simultaneous cloning on 0.5%
agar plates in which the components of HAT medium had been incorporated (8). After PEG fusion, the cells were diluted in complete
medium and incubated overnight at 37°. Cells were resuspended at a
concentration of 1.0 x 10^6 BW 5147 cells per ml. One drop (0.2 ml)
was added to each hardened plate and spread with a glass bacterial
spreader. Cell growth was usually seen between 14 and 21 days.
The firm and well delineated colonies were then picked and suspended
in medium. This technique has distinct advantages in that the cells
are cloned early, recovered hybrids appear to be more stable and
maintenance and visualization of early hybrids is technically
simpler. Disadvantages include the poorer recovery of hybrids
(about 10 times less) and the greater expense of the medium.
Characteristic colonies which appear on such plates are depicted
in Figure 1.

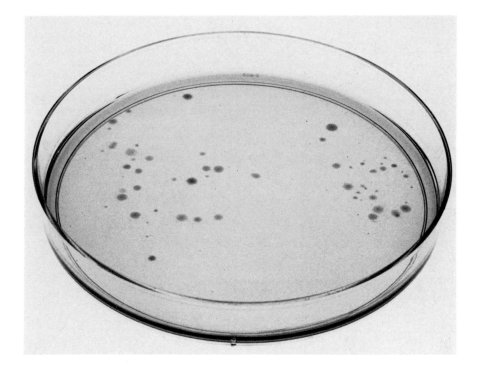

Figure 1.   BW 5147(TGr) colonies growing on 0.5% agar plates
             which contain thioguanine (2.5 ug/ml)

IV. Analysis

Hybrids were analyzed with a number of the methods described in
Table 1. All were positive for both a and b forms of GPI
(Figure 2).

All hybrids in this group expressed both parental forms of Thy 1
with the exception of Hyb 33 B8 which no longer expressed Thy 1.1
of the BW 5147 parent (Table 3).

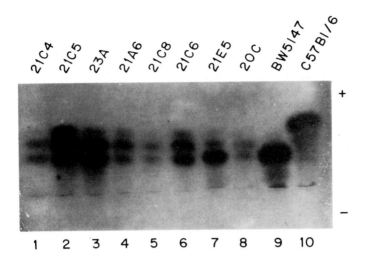

Figure 2. T Cell hybrids express both parental forms of
glucose phosphate isomerase

Table 3. Analysis of Thy 1 Expression OVA Hybrids

Laboratory	Serum	Method	Line	Results Thy 1.2	Thy 1.1
CJ	AKR anti C3H Fluorescein	Fluorescence	BW5147	−	+
			EL4Bu	+	−
			Hyb21C4	+	+
	C3H anti AKR rhodamine		Hyb23A	+	+
			Hyb33A4	+	+
			Hyb33A9	+	+
			Hyb33A11	+	+
			Hyb33B2	+	+
			Hyb33B7	+	+
			Hyb33B8	+	−
			Hyb33B11	+	+
NR	AKR anti C3H	Quantitative absorption	BW5147	−	+
			EL4Bu	+	−
			Hyb23A	+	+
			Hyb21C4	+	+

Hybrids have been tested for their ability to react with OVA. All
were negative for lymphotoxin production in the presence of specific
antigen. Furthermore there was no difference in growth rate or saturation density of cells in the presence and absence of OVA. Antigen binding studies have been slightly more encouraging. One percent of the cells of one line (Hyb 21C5) when tested for binding of
I^{125} labelled OVA were positive by autoradiography. The same cells
bound OVA-coupled sheep red blood cells at day 7 but not day 2,4,

or 9 after culture transfer. Only 0.7% of the cells bound the specific antigen. Future studies will concentrate on optimizing antigen presentation to such cells.

D. Hybrids between BW 5147(TGr) and T cells which Form Rosettes with Sheep Red Blood Cells

I. Purification of T Rosette Forming Cells (RFC)

Another set of hybrids have been prepared with Dr. Diane Eardley from populations of T cells which have been educated in vitro in a manner which induces suppressor cells. The system was chosen because of the previously noted proclivity of suppressor cells to bind antigen. Nylon purified splenic T cells after a 4 day exposure in vitro to sheep red blood cells (SRBC) suppress an IgM response to SRBC as assessed by a reduction in plaque forming cells. In the course of this education in vitro (9) and in vivo (10) a cell is seen which rosettes with SRBC. This cell is detected by cosedimenting T cells with SRBC in a refrigerated centrifuge. The supernatant is removed and the cells placed on ice for 30 minutes and then resuspended in 0.2% glutaraldehyde to stabilize the rosettes. The rosette forming cells (RFC) are Thy 1 positive Ig negative and express Ly 123 or Ly 23 (11). They are not sensitive to treatment with neuramidinase or anti Ig and complement. They are sensitive to trypsin.

T cell populations have been prepared for hybridization by a 4 day incubation of nylon purified cells with 1% SRBC in vitro as described above. The cells were allowed to form rosettes with SRBC, suspended in phosphate buffered saline and spun for 10 minutes at 1200 rpm through Lymphocyte Separation Medium (LSM Litton Bionetics) in a refrigerated centrifuge. The rosette forming cells (RFC) were recovered in a pellet, re-rosetted and spun through LSM a second time. The red cells were lysed with Gay's solution and the dead and red cells were pelleted through LSM. The population which was now enriched in RFC was collected at the interface.

II. Hybridization and Selection

Hybridization was essentially as described above. 2×10^6 cells were used from each parent. HAT selection was accomplished in culture wells and on agar plates.

III. Analysis

Five of five lines growing in HAT medium and one which grew on HAT agar plates were tested for, and expressed the parental and intermediate forms of GPI, which identified them as hybrids. All hybrid lines were tested for expression of Thy 1.1 and Thy 1.2 as above and were positive for both alleles. Hyb 29 lines were analyzed for their ability to form rosettes with SRBC and horse cells (HRBC). All Hyb 29 lines, BW 5147, and Hyb 21C5 (described in C above) were tested by Dr. Eardley for their ability to form rosettes after culture transfer. Only two, Hyb 29P and Hyb 29A10 formed rosettes with SRBC to a significant extent (greater than 0.3%), and both formed rosettes with SRBC and HRBC. Dual expression of receptors for SRBC and HRBC may reflect a cross reactivity at the level of the receptor, or may represent a hybrid molecule formed with

components provided by each parent. Neither 29P nor 29A10 formed
rosettes with human or chicken RBC. Rosette formation was culture cycle dependent. It did not appear on days 1 or 2 after
transfer, peaked at days 3 or 4, and then dropped off. Generally a
maximum of 10-20% of the cells formed rosettes in such 3 or 4 day
cultures. It was possible to enrich for fractions which contained
between 60 and 100% RFC by a 2 hour sedimentation at unit gravity
(4). The evidence favors the interpretation that rosette formation
is generated by an antigen specific receptor for SRBC. The characteristics of this interaction of hybrid and sheep cells are
similar to that of the RFC generated in vitro, the original input
parent. Both are trypsin sensitive and resistant to neuraminidase,
and optimal rosette conditions are the same with both cells.

E. Hybrids between BW 5147(TGr) and TNP-CBA Killer Cells

I. Generation of TNP-CBA Killer Cells

An additional group of hybrids has been generated with cells kindly
provided by Dr. Charles Janeway. Spleen cells from CBA mice were
cultured with irradiated TNP-CBA spleen cells for 15 days. At the
end of that time, fresh TNP-CBA cells were added for 4 days, the
cells were washed, passed through a cotton plug to remove dead
cells, and incubated in medium overnight at 37°. This technique
results in a population of which greater than 80% are blast cells
which kill TNP labelled autologous spleen cells in a 4 hr Cr^{51}
release assay.

II. Hybridization and Selection

TNP-CBA cells (7.5×10^7 cells) were fused with 1.23×10^7 BW 5147
(TGr) cells, and dispensed at 6×10^4 BW 5147 cells/well, and hybrids were selected in HAT. Growth was seen in more than 90% of
the wells.

III. Analysis

The contents of 72 wells were picked. Thirty two of the most rapidly growing "lines" were screened for Thy 1 and GPI. Twenty nine
were positive for both parental forms of GPI and Thy 1; two were
positive for GPI and negative for Thy 1.2 and one was negative for
GPIb and Thy 1.2 and was probably a revertant. These results do
not agree with those of Kohler et al (12) who found it difficult to
obtain hybrids with killer cells, and stated that such hybrids did
not express the Thy 1 allele of the killer cell. None of the hybrids in our study reacted in vitro to the specific antigen (TNP-
CBA) with alterations in DNA synthesis or changes in growth rate
or saturation density.

F. Concluding Remarks

I have emphasized the importance of initiating T cell hybrids with enriched highly purified populations of T cells. With this approach we have established three kinds of hybrids which have been provided information to react with OVA, SRBC, or TNP-CBA. Two lines bound SRBC and HRBC. This is believed to be a culture cycle dependent, antigen specific phenomenon. A small proportion of cells of another hybrid bound OVA. No effect of antigen preincubation has been seen with regard to changes in growth properties; nor has production of any lymphokines been observed.

References

1. Ruddle, N.H.: Delayed hypersensitivity to soluble antigens in mice. I. Analysis in vivo. Int.Arch.All. and Appl.Immunol. in press (1978)
2. Ruddle, N.H.: Delayed hypersensitivity to soluble antigens in mice. II. Analysis in vitro. Int. Arch. All. and Appl. Immunol. in press (1978)
3. Ruddle, N.H., Waksman, B.H.: Cytotoxicity mediated by soluble antigen and lymphocytes in delayed hypersensitivity. III. Analysis of mechanism. J.Exp.Med. 128, 1267 (1968)
4. Hecht, T.T., Ruddle, N.H., Ruddle, F.H.: Separation and analysis of differentiating B lymphocytes from mouse spleens. Cell Immunol. 2, 193 (1976).
5. Ruddle, N.H., Kelly, K. and Beezley, B.: Delayed hypersensitivity to soluble antigens in mice. III. Separation of sensitized cells by unit gravity sedimentation. in preparation.
6. Galfre, G., Howe, S.C., Milstein, C., Butcher, C.W., Howard, J.C.: Antibodies to major histocompatibility antigens produced by hybrid cell lines. Nature 266, 550 (1977)
7. O'Malley, K.A., Davidson, R.A.: A new dimension in suspension fusion techniques with polyethylene glycol. Somatic Cell Genet. 3, 441 (1977)
8. Pfahl, M., Kelleher, R.J., Bourgeois, S.: Steroid resistance in mouse lymphoma cell lines. Molec.and Cell. Endocrin. 10, (2) (1978)
9. Eardley, D.D., Gershon, R.K.: Induction of specific suppressor T cells in vitro. J. Immunol. 117, 313 (1976)
10. Cone R.E., Gershon, R.K., Askenase, P.W.: Nylon adherent antigen specific rosette forming T cells. J.Exp.Med. 146, 1390 (1977)
11. Eardley, D.D., Shen, F.W., Cone, R.E., Gershon, R.K.: The profile of subsets of T cells which bind antigen: Induction of specific rosette forming cells in vitro. Submitted to J. Immunol.
12. Kohler, G., Lefkovits, I., Elliott, B., Coutinho, A.: Derivation of hybrids between a thymoma line and spleen cells activated in a mixed leucocyte reaction. Eur. J. Immunol. 7, 758 (1977)

Specific Suppressor T Cell Hybridomas

M.Taniguchi, J.F.A.P. Miller

A. Introduction

Antigen-specific suppressive factors, bearing determinants coded by the I-J locus of the major histocompatibility complex, have been extracted from T cells of mice tolerant to human γ globulin (HGG) (1). The establishment of T cell hybridomas with specific suppressor function would provide an ideal tool to characterize the specific I-J bearing factor and its mechanism of action. Unfortunately, suppressor T cells (STC) with defined specificity form only a very minor proportion of the spleen population and the chances of fusing one such cell with a thymoma cell line are minimal. We were able to provide a richer source of STC by allowing Ig$^-$ spleen cells from HGG-tolerant mice to bind to dishes coated with HGG(2). This cell preparation was used for hybridization with tumor lines. In addition since STC bear I-J determinants (2), we used a monospecific anti-I-J serum together with fluorescein-conjugated antimouse immunoglobulin (MIg) to select, with the aid of a fluorescent activated cell sorter (FACS), after the fusion process, those normal and hybridized cells bearing I-J determinants. The separated cells were then cultured, cloned and characterized for surface markers and suppressive functions. The results show that hybridoma lines with specific suppressive properties can be established.

B. Methodology

The method used for the preparation and enrichment of antigen-specific STC has been given in detail elsewhere (2). A mixture of 5×10^6 enriched STC and 5×10^6 EL-4 lymphoma cells (Thy-1.2 and H-2^{b+}) in Dulbecco's modified Eagle's medium (DME) was centrifuged at 400 g. To the pellet was added 2 ml of polyethylene glycol (PEG) dimethylsulfoxide (DMSO) solution (reagent A:1 weight of PEG, MW 4000, plus 1.4 volume of 15% DMSO in DME). This was gently mixed with a broadened pipette. Immediately after this, the mixture was transferred into 2 ml of 50% (W/V) PEG (reagent B; 1 weight of PEG plus 1 volume of DME) and mixed well. The suspension was then gradually diluted to 16 ml of serum-free DME and further diluted with 180 ml of DME containing 13% fetal calf serum (FCS). Each of the above step was completed within 1 min. The mixture was then incubated at 37°C in 5% CO_2 in air. 3 hr after incubation the cells were washed 4 times with medium and cultured for 1-2 days in 10% FCS-DME at 37°C. 1-2 days after the fusion, cells were harvested and the membrane stained by a monospecific anti-I-J serum and fluorescein-conjugated rabbit anti-MIg. The stained cells were washed 4 times with HEPES-buffered DME (containing 20 mM HEPES). The I-J$^+$ cells were then separated by FACS-II (Becton-Dickinson Electronics Laboratory, Mountain View, California) and cultured at 37°C in 5% CO_2 in air in 0.3% agar containing 5% FCS-DME. After 12 days, colonies of fused cells were transferred and cultured in 10% FCS-DME to characterize their surface markers, chromosomes and suppressive activities. This was repeated at further intervals. Cell-free extracts of established lines were prepared by sonication as described elsewhere (1). They were tested in vivo on the secondary adoptive response of irradiated CBA mice given 5×10^6 HGG-primed and 5×10^6 DNP-primed spleen cells intravenously, 10^8 horse erythrocytes (HRBC) intravenously and 100 μg DNP.HGG intraperitoneally. Extract equivalent to 10^6 hybridoma cells was injected intravenously at the time of cell transfer and the anti-DNP and anti-HRBC plaque-forming cell (PFC) responses measured 7 days later.

C. Results

1. Separation of I-J Bearing Cells by FACS

Antigen-specific CBA (H-2^k) STC, enriched on antigen-coated dishes, were fused with C57BL (H-2^b) EL-4 tumor cells. One day later, the cells were reacted with anti-I-J^k serum or normal C57BL serum and stained with fluorescein-conjugated rabbit anti-MIg. The fluorescence distribution was analyzed by the FACS and the fluorescence profile obtained is shown in Fig. 1. A broad bright peak was observed with cells stained by anti-I-J^k (solid line) and computer analysis of this

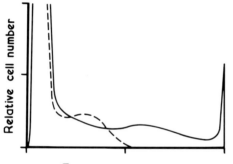

Fluorescence intensity

Fig. 1. Fluorescence distribution of hybridoma cells, labeled with anti-I-J^k and fluorescein-conjugated rabbit anti-MIg. Distributions are based on cumulative analyses of 40,000 viable cells. The solid line indicates the profile of cells incubated with a 1:8 dilution of anti-I-J^k serum. The broken line indicates the pattern of cells stained with a 1:10 dilution of normal mouse serum.

indicated that about 5% of the total population was specifically stained. Only a dull peak (broken line) was obtained with cells exposed to normal serum and this was likely to be nonspecific since no fluorescence could be seen by microscopy when hybridoma cells were reacted only with fluorescein-conjugated rabbit anti-MIg. The fluorescent I-J^{k+} cells were sorted by FACS and seeded into 0.3% agar plates with 5% FCS-DME. About 10 colonies per plate appeared 10-12 days later; 15-20 days later single colonies were picked up and cultured in 10% FCS-DME.

2. Detection of Gene Products on Cell Surface of Hybridoma Cells

Within 4 weeks after fusion less than 5% of established hybridoma cells stained after direct exposure to fluorescein-conjugated rabbit anti-MIg. From 60-90%, but never 100%, stained with the fluorescein reagent following preincubation with anti-H-2^b, anti-H-2^k, anti-I-J^k and anti-Thy-1.2 sera. EL-4 cells did not stain with anti-H-2^k or anti-I-J^k. By 3 months after fusion, most lines had lost H-2^k gene products (both I-J^k and H-2.33) (Table 1).

Table 1. H-2^k gene products on hybridoma cell lines*

| Cell lines | Percent of cells with following markers | | | | | | | | |
| | 6 weeks after cell fusion | | | | >3 months after cell fusion | | | | |
	Ig	H-2.23	I-J^k	Thy 1.2	Ig	H-2.23	H-2.33	I-J^k	Thy 1.2
72	<1	45	40	>90	<1	<5	58	<5	>90
73	<1	20	20	>90	<1	<5	66	<5	>90
77	<1	80	>90	>90	<1	16	45	10	>90
104	<1	80	>90	>90	<1	<5	47	<5	>90
110	<1	70	80	>90	<1	<5	55	<5	>90
111	<1	ND	30	>90	<1	<5	55	5	>90
119	<1	80	80	>90	<1	<5	50	<5	80
125	<1	80	70	>90	<1	20	61	15	85

* 4 weeks after cell fusion, >90% of cells of the parent line (no. 22), which gave rise to the lines listed above, had both H-2.23 and H-2.33 antigens. Markers on EL-4: Ig, <1%; H-2.23, <5%; H-2.33, >90%; I-J^k; <5%, Thy 1.2, 73%.

3. Karyotype of Hybridoma Cells

Soon after fusion most cells had from 60 to 80 chromosomes but this number rapidly dropped in most established clones to near 40 by 3 months (Table 2).

Table 2. Karyotype of hybridoma cells

| Cell Lines* | Weeks after fusion | Number of metaphases with following chromosome number | | | | | | | | |
		30-35	36-39	40	41-50	51-60	61-70	71-80	81-90	>90
EL-4	-	0	3	11	2	0	0	0	0	0
22	4	0	0	0	1	3	7	16	6	2
72	14	1	7	7	24	0	0	0	0	0
73	14	0	3	17	9	1	0	0	0	0
77	24	0	4	18	7	1	0	5	3	0
119	14	0	18	6	6	0	1	0	0	0
125	14	0	9	19	9	2	1	0	0	0

Lines 72, 73, 119 and 125 were originally derived from clones established from parent line 22 and were initially near-tetraploid.

4. Suppressive Activity of Hybridoma Cells

18 hybridoma cell lines containing I-J^{k+} cells were sonicated and the extracts were tested for specific suppressive activity in the in vivo adoptive transfer test.

Some of the results obtained are shown in Table 3. Extracts from 6 lines produced significant specific suppression ranging from 61 to 80%, 5 produced non-specific suppression (Table 3), and 7 had no activity. The activity of the lines

Table 3. Suppressive activity of extracts from hybridoma cell lines

Cell Lines	$\%I\text{-}J^{k+}$ bearing cells	Indirect anti-HRBC PFC/ spleen at 7 days	Anti DNP PFC/spleen at 7 days Direct	Indirect	% specific or nonspecific (NS) suppression
none	–	6450 (1.18)*	7030 (1.07)	104,370 (1.14)	–
EL-4	<5	5450 (1.22)	5440 (1.34)	95,860 (1.08)	8NS
72	30	7550 (1.40)	15,140 (1.19)	37,940 (1.12)	64
73	20	9730 (1.30)	10,820 (1.23)	21,340 (1.40)	80
77	>90	4970 (1.31)	3470 (1.30)	33,980 (1.13)	67
106	50	3040 (1.50)	4420 (1.39)	37,170 (1.31)	64NS
109	10	1040 (1.70)	3950 (1.55)	38,670 (1.32)	63NS
111	30	5630 (1.45)	5930 (1.23)	34,280 (1.19)	67
112	>90	480 (2.36)	1000 (1.68)	13,550 (2.26)	87NS
119	80	7490 (1.18)	1860 (1.18)	28,230 (1.14)	73
121	90	120 (3.52)	2970 (1.58)	18,460 (1.97)	82NS
125	70	6510 (1.36)	6050 (1.23)	40,310 (1.29)	61
127	>90	1210 (1.22)	70 (1.44)	1430 (1.32)	99NS

* Geometric means and SE; 5-6 mice per group.

did not correlate with their content of $I\text{-}J^{k+}$ cells: e.g. some lines with 90% $I\text{-}J^{k+}$ cells had no activity, whereas others with only 20% such cells were specifically suppressive. Specific suppression generally affected the indirect anti-DNP PFC response more markedly than the direct. It was no longer evident when extracts from most lines were tested after 3 months (Table 4). This loss paralleled that

Table 4. Loss of specific suppressive activity with time after cell fusion

Cell Lines	Percent suppression of DNP response at following times after fusion 6 weeks	10 weeks	>3 months
72	64	24	0
73	80	14	0
77	67	59	46
111	67	65	0
119	73	66	0
125	61	ND	ND

observed for $H\text{-}2^k$ gene products and the change in karyotype. Sonicates from line 77 were still specifically suppressive 6 months after fusion (Table 4).

D. Discussion and Summary:

The fusion of antigen-specific $I\text{-}J^k$ STC and the thymoma cell line, EL-4, was facilitated considerably by the use of two procedures: the method we developed to enrich antigen-specific, I-J$^+$, Ig$^-$ cells from HGG-tolerant spleen (2) and the selection, subsequent to fusion, of I-J$^+$ cells by appropriate fluorescent reagents and the FACS. The lines established after FACS separation were hybridomas of CBA T cells (H-2^k) and EL-4 (H-2^b), since they carried no Ig determinants but expressed both H-2.23 and H-2.33 as well as I-J^k surface markers. In addition, the karyotype of the majority of the cells observed soon after fusion was near-tetraploid. In vivo assays of sonicates from hybridomas containing I-J^{k+} cells enabled the lines to be classified into 3 groups with different suppressive activities: antigen-specific, nonantigen-specific and nonsuppressive. Specific suppression affected the indirect response more markedly than the direct, as we observed with material from STC of HGG tolerant mice (1). The activities of the lines did not correlate with their content of I-J^{k+} cells. Why this should be is not clear but one possible explanation may be that the expression of the I-J gene product varies with the cell cycle.

Cell surface markers, karyotype and suppressive activity were examined at progressively greater intervals after cell fusion. Most lines selectively lost the characteristic properties derived from the normal CBA parent. Thus only <5-10% of cells 3 months after fusion had H-2^k or I-J^k gene products. Likewise no significant suppression was observed with extracts of cells of 4 out of 5 hybridomas from which specifically suppressive material had been obtained several weeks before. Sonicates from line 77, however, retained specific suppressive activity 6 months after cell fusion. In parallel with the loss of surface markers and suppressive activity of most lines was a change in karyotype from near-tetraploid to near-diploid. A similar loss in chromosomes has been reported in other studies (3). Conceivably, the loss of suppressive function may result from extinction, by the tumor genome, of the expression of the gene products of the normal cell. Alternatively, as seems to have been the case here, there may be a selective loss of chromosomes derived from the normal parent. The state of differentiation of the tumor line may well determine subsequent behaviour, i.e. the extent to which the normal T cell effector function may be preserved. Preliminary results obtained with the HAT-sensitive L5178Y line indicate that some of the hybrids still express H-2^k and I-J^k antigens 18 weeks after fusion. We therefore think it profitable to screen a variety of thymoma cell lines to determine which may be optimal in hybridization studies for the production of stable lines which continue to express particular T cell functions.

E. References

1. Taniguchi, M., Miller, J.F.A.P.: Specific suppression of the immune response by a factor obtained from spleen cells of mice tolerant to human γ globulin. J. Immunol. 120, 21-26, 1978.
2. Taniguchi, M., Miller, J.F.A.P.: Enrichment of specific suppressor T cells and characterization of their surface markers. J. Exp. Med. 146, 1450-1454, 1977.
3. Hammerling, G.T.: T lymphocyte tissue culture lines produced by cell hybridization. Eur. J. Immunol. 7, 743-747, 1977.

Selective Expression of Loci in the I–J Region on T Cell Hybrids

B.A. Osborne, R.A. Goldsby, L.A. Herzenberg

Gene products of the major histocompatibility complex (MHC) in the mouse are responsible for a variety of effector and cooperator functions of immunocompetent lymphoid cells. Specifically, loci mapping to the I-region have been linked to suppression (1), MLR stimulator cells (2) and GVH reactions (3). Ia antigens, which are thought to be the products of the Ir genes, have been found on B lymphocytes, macrophages, sperm and a variety of functionally different subsets of T cells (4). The selective expression of I-region molecules on T cell subsets has led many to postulate that these molecules are important in T-B collaboration (5).

Until recently it has been impossible to examine the antigens on these T cell subsets as they represent only a small fraction of the total number of lymphoid cells. Attempts at enrichment of specific T cells have been only partially successful.

Recently it has been demonstrated that cell hybridization offers the opportunity to preserve and immortalize individual B cells as well as T cells. Köhler and Milstein (6) were able not only to immortalize the B cell by fusion with a myeloma but were also able to immortalize its differentiated function, the production of immunoglobulin. Recent work from our laboratory and others (7,8) has shown that by fusion with a T-lymphoma, T cells may also be immortalized. It has therefore become possible to isolate clonal populations of T cells and to characterize surface alloantigens of such clones.

The production of clonal lines of T cell hybrids has been discussed elsewhere (7,9). The clones reported in this paper have come from two different hybridizations. The "921" clones were made with spleen cells from DNP-KLH primed and boosted (BALB/c x SJL)F$_1$ mice hybridized to BW5147 (a thioguanine, ouabain resistant thymoma cell line, kindly provided by Dr. R. Hyman of the Salk Institute). The "822" clones were made in an identical fashion except that spleen cells were from DNP-KLH primed and boosted A/J mice.

Three different lots of antisera directed against the I-J^k subregion have been employed in this study. Two of these lots (designated 035 and 055) were made by repeated injections of lymphoid cells of B10.A(5R) into B10.A(3R). The third lot (069) was made by repeated injections of B10.A(5R) lymphoid cells into (BALB.B x B10.A(3R))F$_1$ mice.

The cytotoxic activity of antisera against the hybrid clones was assayed by a micro-procedure using 2000 cells in Terasaki plates with selected lots of rabbit complement. Unless indicated, all cell lines were grown to a density of 3-4 x 10^6 cells/ml for testing. Quantitative absorptions of cytotoxicity were done to show specificity. Either normal spleen cells or hybrid clones were used for absorptions. The absorptions were done with varying cell numbers and 50 µl of antisera at a 1:10 dilution.

We noticed that the killing of 921 2b with anti-I-J sera varied markedly from test to test. Upon detailed examination, we found that the
absolute amount of cytotoxicity was dependent upon the culture cycle
of the test cells. Specifically, clone 921 2b, at a density of
2×10^6 cells/ml, was 20% killed with 035 at a 1:10 dilution whereas
24 hr later when the cells were approaching stationary phase of the
culture cycle and were at a density of 3×10^6 cells/ml, 90% of the
cells were killed with the same concentration of 035. Similar results
were found on the three subsequent occasions it was tested. This
phenomenon was also noted with other antisera (e.g., anti-Thy-1 and
anti-Ly-1 and -Ly-2). Thus we conclude that cells in stationary phase
of the culture cycle are more susceptible to complement-dependent
lysis than those in log phase.

It is known that mouse alloantisera often contain murine leukemia
virus activity (10). The concern that lymphoid hybrid lines are quite
likely to bear viral determinants on their surface led us to examine
the anti-I-J antisera for antiviral activity. Direct testing by a
plate binding radioimmunoassay (kindly performed by Dr. J. Ledbetter)
demonstrated that 035 had very little activity against AKR virus
whereas 055 had considerably more activity. Further, by immunoprecipitation of ^{125}I-labeled disrupted AKR virus followed by SDS-PAGE
analysis, 055 and 069 had high levels of activity against GP-70.
These antibodies were effectively removed by absorption with purified
AKR virus. The absorptions and SDS-PAGE analysis was kindly done by
Dr. R. Nowinski, Fred Hutchinson Cancer Center.

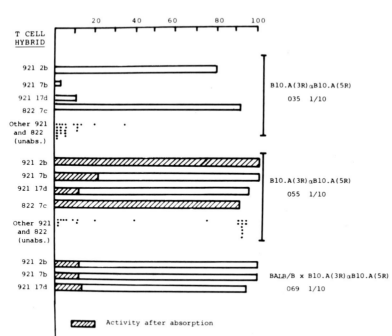

Fig. 1. Expression of loci in the I-J region. Both 055 and 069 were
used before and after absorption with 50 µg of AKR virus/30 µl
of neat antisera.

218

An early examination of 24 clones from the 921 series revealed that 921 2b was killed with 035, whereas 921 2b, 927 7b and 921 17d were killed with 055 and 069. Absorptions with appropriate spleen cells indicated that lysis with 035 on 921 2b was specific for I-J region determinants. However, absorptions with 055 and 069 never showed convincing specificity of lysis. When the antiviral antibody was detected in 055 and 069 clones, 921 2b, 7b and 17d were retested using unabsorbed and virus-absorbed antisera. The results are shown in Fig. 1. Serum 055 after virus absorption kills only 921 2b and indeed appears identical to 035 in our assay. However, 069 after absorption no longer kills 921 2b, 7b or 17d, indicating that the only reactivity present in this particular antiserum detected on these T cell hybrids is antiviral in nature.

Recently a similar screening of the 822 hybrids has been done (see Fig. 1). As can be seen, only one 822 clone, 822 7c, was found to be positive with 035. Many clones were found to be positive with 055 before absorption with virus. However, only 822 7c was shown to be killed when virus-absorbed 055 was used. Therefore we conclude that determinants coded for by loci mapping to the I-J region can be detected on T cell hybrids; however, it is essential to be aware of and to remove antiviral antibodies from the antisera used in such studies.

Convincing evidence that the determinants detected on 921 2b are coded for in the I-J region was obtained by absorptions with spleen cells from four congenic pairs of mice. Table 1 shows that a B10.A(3R) (I-J^b) anti-B10.A(5R) (I-J^k) antiserum can be absorbed with B10.A(5R) but not B10.A(3R). A series of other spleen cell populations was also used for absorption purposes. In all cases the absorption with I-J^k haplotype cell populations removed the activity against 921 2b whereas absorption with I-J^b or I-J^s did not remove the activity against this

* * *

TABLE 1

ABSORPTION OF I-J CYTOTOXIC ACTIVITY AGAINST HYBRID CELL LINE 921-2b[*]

Strain	I-J Haplotype	No. spleen cells (x 10^8) to half absorb anti-I-J[**]
3R	b	>1
5R	k	0.18
A.SW	s	>1
A/J	k	0.1
B10.S	s	>1
B10.A	k	0.1
BW5147	k	>1

Normal mouse serum + c' cytotoxicity was always ≤10%.

[*]921-2b grown to 3-4 x 10^6/ml (high density culture)

[**]B10.A(3R) α B10.A(5R) serum #035 was kindly provided by Dr. Hugh McDevitt, Stanford U. Absorptions were done with 50 λ/tube at a 1/10 dilution of antisera with varied cell concentrations.

219

clone. Thus we conclude that when there are no contaminating antiviral antibodies, it is possible to detect by direct cytotoxicity sepcificities coded for in the I-J region on T cell hybrids. It has been reported by others (11,12) that T cell hybrids made in a similar fashion to ours have specific suppressor activity. It will be interesting to determine if these T cell hybrids express I-J determinants and what correlation there is between immune functional activity and surface phenotype in T cell hybrids.

This work was supported, in part, by grants from the National Institutes of Health: AI-08917, CA-04681, and HD-01287.

References

1. Murphy, D.B., Herzenberg, L.A., Okumura, K., Herzenberg, L.A., McDevitt, H.O.: A new I subregion (I-J) marked by a locus (Ia-4) controlling surface determinants on suppressor T lymphocytes. J. Exp. Med. 144, 699-712 (1976)
2. Bach, F.H., Widner, M.B., Bach, M.L., Klein, J.: Serologically-defined and lymphocyte defined components of the major histocompatibility complex in mouse. J. Exp. Med. 136, 1430-1444 (1972)
3. Klein, J., Park, J.M.: Graft-versus-host reaction across different regions of the H-2 complex of the mouse. J. Exp. Med. 137, 1213-1225 (1973)
4. Hämmerling, G.J.: Tissue distribution of Ia antigens and their expression on lymphocyte subpopulations. Transpl. Rev. 30, 64-82 (1976)
5. Katz, D.H., Greaves, M., Dorf, M.E., Dimuzio, M., Benacerraf, B.: Cell interactions between histoincompatible T and B lymphocytes. VIII. Cooperative responses between lymphocytes are controlled by genes in the I-region of the H-2 complex. J. Exp. Med. 141, 263-268 (1975)
6. Köhler, G., Milstein, C.: Continuous cultures of fused cells secreting antibody of predefined specificity. Nature 256, 495-497 (1975)
7. Goldsby, R.A., Osborne, B.A., Simpson, E., Herzenberg, L.A.: Hybrid cell lines with T cell characteristics. Nature 267, 707-708 (1977)
8. Hämmerling, G.J.: T lymphocyte tissue culture lines produced by cell hybridization. Eur. J. Immunol. 7, 743-746 (1977)
9. Goldsby, R.A., Osborne, B.A., Murphy, D.B., Simpson,E., Schröder, J., Herzenberg, L.A.: Somatic cell hybrids wtih T cell characteristics. In: The Immune System: Genetics and Regulation, E.Sercarz, L.A. Herzenberg, C.F. Fox (eds). Academic Press, New York 1977, 265-271
10. Wettstein, P.T., Krammer, R., Nowinski, R.C., David, C.S., Frelinger, J.A., Shreffler, D.C.: A cautionary note regarding Ia and H-2 typing of murine lymphoid tumors. Immunogenetics 3, 507-516 (1976)
11. Simpson, E., et al.: in this volume
12. Tanaguchi, M., Miller, J.F.A.P.: in this volume

Attempts to Produce Cytolytically Active Cell Hybrids Using EL4 and MLC Spleen

R. DiPauli, G. DiPauli

A. Introduction

The highly successful fusions between myeloma cell lines and splenic antibody producing cells has encouraged many similar, though to date largely unsuccessful, attempts to fuse T lymphoma cell lines with a variety of effector T cell populations (1,2,3). The T lymphoma most commonly used has been the BW5147 line; however, this line is not known to carry the Ly 2 allo-antigen characteristic of cytotoxic T lymphocytes (CTL). We therefore chose the EL4 T lymphoma line which carries the Ly 2 allo-antigen to fuse with MLC spleen cells in the belief that this would enhance the probability of fusion products being CTL. This paper summarizes our experiences to date which lead us to believe that fusion between EL4 and MLC spleen is possible, but no functioning hybrids have yet emerged.

B. Methods and Results

Two fusion methods were developed and are summarized in Table 1.

Table 1.   T-Cell Fusions Methods

A	B
0.2-0.4 ml PEG (30-50%)	Mix cells to be fused and pellet by centrifugation
Overlay 0.2-0.4 ml mixture of cells to be fused on PEG (2×10^7-4×10^7 total cells)	Add 1-2 ml 10% PEG in DME
Centrifuge 3 mins. at 800-1000xg	Centrifuge 15 mins. at 800-1000xg
Leave pellet for 2 mins. at 37°C in PEG	Remove supernatant, suspend cells in 1 ml DME and keep at 37°C for 1 min.
Suspend cells in 5-10 ml DME	Suspend cells in 20 ml DME

After fusion procedure A and B the cells are treated identically: washed 2 times in DME and incubated overnight in DME + 10% FCS in tissue culture dishes. The following day the cells are distributed into microtiter plates with 96 wells after being suspended in DME supplemented with 20% FCS, non-essential amino acids, 3×10^{-6}M glycine, 2×10^{-6}M deoxycytidine and HAT.

These methods were applied to the fusion between EL4 and a primary
in vitro MLC spleen cell population. The MLC used (BALB/c x C57BL/6)Fl
responding to irradiated C57BL/H-2^k cells after 4 days of culture.
The results in terms of numbers of HAT resistant clones is summarized
in Table 2.

Table 2. EL4 BU.Oua x MLC Fusions

		#EL4 + #MLC Spleen per Fusion	
Method	PEG Conc. %	$10^7 + 1.7 \times 10^7$ # hybrids	$1.9 \times 10^6 + 1.9 \times 10^7$ # hybrids
A	35	16	3
	40	22	1
	45	12	4
	50	3	0
		$10^7 + 10^7$	$10^7 + 10^7$
B	10	20 (expt.1)	20 (expt.2)

An apparent fusion frequency of 2×10^{-6} with respect to the limiting
EL4 cells was obtained under the best conditions. Of potential
importance is the success of Method B which requires only 10% PEG
and may help reduce the cytotoxic effects of high concentrations for
spleen cells. However, as shown in Table 3 none of the 101 HAT
resistant clones were able

Table 3. Analysis of EL4 x MLC Fusions

MLC type (days after stimulation)	No. of hybrids analyzed	Direct killing on R1 (TL$^+$)	Con A killing assay on P815	4 days with irradiated CBA or A cells
(BALB/c x C57BL/6)Fl α C57/H-2^k (4)	101	0/101	0/16	0/10*

* 7/10 karyotyped: 6/7 mode 70-80 chromosomes; 1 mixed clone (40 and 65)

to kill ^{51}Cr-labelled targets under a variety of conditions. To verify that fusion had in fact occured karyotype analysis was performed
on 7 lines and in all but one case there was the expected doubling of
chromosome numbers. As a further test of hybrid formation we tested
for the presence of H-2^d antigen from the MLC donor cells by serologic
and cytotoxic T cell killing, but in no case was there any evidence
of H-2^d expression on the hybrids.

Since we have been unable to find any cytolytically active HAT resistant clones tested, and we have no direct proof of EL4-MLC fusion,
however, as discussed below we belive such fusions have in fact
occured.

C. Discussion

In principle, EL4 by EL4 hybrids should be HAT sensitive. However, the thymidine kinase gene responsible for BU resistance is an auto-somal recessive requiring therefore separate mutations in each gene to become BU resistant, and thus, it is possible to envisage intra-genic somatic recombination as a means of rescuing a functional thymidine kinase enzyme that would make such a cell HAT resistant. This sequence of events, though exceedingly rare, may at first sight explain our results. On closer analysis this seems an unlikely explanation because in other experiments (not shown) we have obtained a maximum fusion frequency of 10^{-5} with EL4-BU and WR1.2.3-TG$_6$ and the reversion frequency of our EL4 BU line is less than 5×10^{-8}. Thus, in our EL4 by MLC spleen fusions which give HAT resistant colonies at a frequency of 2×10^{-6}, one in every 5 of the possible EL4-EL4 fusion hybrids would have to undergo somatic intragenic recombination within the thymidine kinase gene in order to yield the present results.

We therefore reason that the methods described do in fact yield EL4 by MLC spleen fusion hybrids, but these are non-functional, and for some unknown, though maybe very important, reason express very little or no H-2^d antigens derived from the MLC donor cells. Our current working hypothesis is that by analogy with the myeloma fusions, the T cell tumor line may have to be a functional cell in the first place if we are to preserve the effector T cell functions present in MLC spleen.

Acknowledgements. We wish to thank M. Cohn and R. Langman for their interest and support. The excellent technical assistance of Mrs. M. Fairhurst was of great help. This work was supported by Grant Number RO1 CA19754 awarded by the National Cancer Institute to M. Cohn, and a Fellowship from the Deutsche Forschungsgemeinschaft to R.DP.

References

1. Goldsby, R.A., Osborne, B.A., Simpson, E., Herzenberg, L.A.: Hybrid cell lines with T-cell characteristics. Nature 267, 707-708 (1977).
2. Hammerling, G.J.: T lymphocyte tissue culture lines produced by cell hybridization. Eur. J. Immunol. 7, 743-746 (1977).
3. Köhler, G., Lefkovits, I., Elliot, B., Coutinho, A.: Derivation of hybrids between a thymoma line and spleen cells activated in a mixed leukocyte reaction. Eur. J. Immunol. 7, 758-761 (1977).

Phenotypic Expression in T-Lymphocyte x "Fibroblast" Cell Hybrids

N. Saravia, R.I. DeMars, F.J. Bollum, F.H. Bach

A. Introduction

T lymphocyte x fibroblast hybrids were constructed for use in genetic
and regulatory studies of T cell characteristics including immune
function. The demonstration that hybrid cell lines generated by the
fusion of T cells with a dedifferentiated fibroblast line continue
to express T lymphocyte facultative markers is critical to the
efficacy of this approach. In this communication we present the
results of phenotypic analysis of permanent cell lines derived from
Sendai virus mediated fusion of mixed leukocyte culture primed murine
T lymphocytes or unstimulated thymocytes and line 613 fibroblasts.
The latter is an L cell subline that lacks hypoxanthine-guanine
phosphoribosyl transferase (HPRT), thymidine kinase (TK) and adenine
phosphoribosyl transferase (APRT). Independent hybrids were examined
for the expression of T cell and thymocyte facultative markers as
well as the constitutive H-2 antigen specificities of the parental
cells.

I. Rationale for Using a Permanent Line with Multiple Selectable Markers

Given the uncertainties regarding the fusibility of T lymphocytes at
the outset of these experiments, and the possibility that expression
of T cell function is contingent upon gene products deriving from
loci on more than one chromosome, a permanent cell line known to be
fusible, and which would allow selection for multiple markers, was
chosen for hybridization with T cells. This approach was based on
the premise that selection for markers on several different chromosomes might favor recovery of hybrids that would (initially at
least) retain maximum numbers of chromosomes non-selectively, thereby
increasing the probability of rescuing non-selectable T lymphocyte
markers. Such a strategy also offered the practical advantage of
virtually eliminating the likelihood of revertants surviving selection, since at least two genetic reversions would be required. In
addition, hybrids isolated under conditions requiring complementation
of multiple enzyme deficiencies, can subsequently be exposed to the
appropriate substrate analog (e.g. β HGPRT-azaguanine; APRT-diamino-
purine; TK-bromodeoxyuridine) in order to select for cells that have
spontaneously lost the corresponding chromosome, and determine
whether certain phenotypic traits concordantly segregate with that
chromosome.

II. Use of Enriched T Lymphocyte Populations for Fusion

In order to increase the likelihood of obtaining hybrids which
derived from the fusion of T lymphocytes rather than some other cell
type with 613 cells, lymphocyte populations with minimal numbers of
B cells and macrophages were used for hybridization.

224

1. Thymocytes

Thymuses were taken from normal and cortisone acetate pretreated
(2.5 mg in 0.25 ml PBS intraperitoneally 48 hours prior to sacri-
fice) 4-6 week old B10.G mice. After removing attached lymph nodes,
the thymuses were pressed through wire mesh to express the thymocytes. The cells were washed 3 times in PBS and finally resuspended
in Hepes buffered F10 medium for fusion.

2. Primed Splenic T Lymphocytes

Whole spleen cell suspensions were prepared from the spleens of 8-12
week old B10.M mice. Preparative steps which preceded fusion have
been detailed elsewhere (1). The sequence of T lymphocyte enriching
procedures began with nylon wool depletion of adherent cells, followed by allogeneic stimulation of the nonadherent population in a 5
day mixed leukocyte culture (MLC) with irradiated B10.D2 spleen
cells and, subsequent size separation of responding cells by unit
gravity sedimentation. Pooled blast fractions and those containing
a mixture of small and large lymphocytes were used in separate
fusion experiments.

III.   Isolation of Hybrids

1. Cell Fusion

The T lymphocyte or thymocyte to 613 cell ratio was 4:1 for all
hybridizations. After a two minute exposure of adherent 613 cells
to 0.25% trypsin and resuspension in Hepes-F10 medium without serum
(0.9528 g Hepes/100 ml hypoxanthine-free F10) the cells of line 613
were mixed with T lymphocytes or thymocytes in a total volume of 1
ml. One thousand hemagglutinating units of β-propiolactone inactivated Sendai virus diluted to 1 ml with Hepes-F10 were added to the
cell suspension. This mixture was held on ice for 10 minutes to
promote agglutination, then placed in a $37^{\circ}C$ water bath for 5 minutes.
Fetal calf serum (0.5 ml) was added before incubating at $37^{\circ}C$ with
gentle rotation for 30 minutes. Afterwards, 20 ml of ice cold F10
with 15% fetal calf serum (FCS-F10) was slowly added to the fusion
mixture, which was subsequently dispensed into four 60 mm plastic
culture dishes to increase the probability of isolating independent
hybrids.

2.   Selection for Hybrids

Selection of hybrids in a medium containing aminopterin (5 x 10^{-7}M),
azaserine (1 x 10^{-6}M), hypoxanthine (1 x 10^{-4}M), thymidine (3 x 10^{-5}M),
and adenine (1 x 10^{-4}M), was initiated after two days of culture in
FCS-F10. De novo formation of nucleotides was blocked at two points
in the synthetic pathway to insure that the block would be maintained
even if cells became resistant to one of the inhibitors. While one
or the other of the purine nucleotide precursors is sufficient to
sustain the growth of hybrid cells generated by this cross, provision
of both hypoxanthine and adenine has often been found to result in
the retention of both HPRT and APRT in hybrid cells. Hybrids have
been maintained in selective F10 medium as described above.

IV. Criteria Defining Isolates as Hybrids

The hybrid nature of resulting cell clones was inferred from the
presence of line 613 marker chromosomes (several metacentric and a
single dicentric) in metaphase chromosome preparations, and the
stringency of selection against 613 in the modified HAT medium which
was such that growth of 613 would have required at least two independent genetic reversions. The enzyme deficiencies of 613 have not
been observed to revert during several years of culture. Therefore,
growth in selective medium provided strong evidence supportive of
the hybrid derivation of recovered clones. The range of total
numbers of chromosomes in the hybrids approximated the sum of one or
more of each of the parental genomes; extensive chromosome loss was
not apparent. Subsequent phenotypic characterizations reaffirmed
the conclusions drawn from chromosome analysis and the rigor of
selection.

B. Phenotypic Characterization

I. H-2 Antigen

Coexpression of parental histocompatibility antigens by somatic cell
hybrids has often been demonstrated and has been employed as a
criterion for establishing the hybrid nature of cells surviving
selection. Since the progeny of 613 x T lymphocyte fusion experiments were known to be hybrids as discussed above, the purpose of
testing the cells for H-2 antigen expression was to determine whether
the region of chromosome no. 17 comprising the major histocompatibility
complex (MHC) deriving from both parental cells had been retained.

Cytotoxic cells were generated in a 5 day MLC against the H-2 haplotypes of the parental cells from which the hybrid lines were derived,
as well as against unrelated specificities. The presence of the
appropriate target antigens on the hybrids was determined in a standard
cell mediated lympholysis (CML) assay (2). The results of a CML assay
designed to detect CD determinants characteristic of each of the cells
(H-2^f-B10.M; H-2^k-613) contributing to the hybrid are shown in Table
1. Cytotoxic cells sensitized against the B10.M H-2^f haplotype
selectively lyse the B10.M PHA blast targets, are able to lyse all
the hybrids tested, but do not significantly kill the 613 parent
line. B10.BR (H-2^k) sensitized killer cells (second line) also
specifically lyse the appropriate PHA blast target, all hybrid
lines, and 613, indicating that the hybrids do express MHC gene products characteristic of each parent cell. The lysis of hybrid cells,
and 613 by cytotoxic cells primed against an unrelated H-2 haplotype
(third line) was unfortunately uninformative. Looking at the PHA
blast controls, it is evident that the cells sensitized to B10.RIII
cross-kill B10.BR and B10.M, which individually express the target
antigens co-expressed in the hybrids. In addition, L cells and the
613 subline rarely form tumors in syngeneic mice, suggesting that new
antigenic specificities, which elicit an allograft response, are
expressed by these permanent cell lines. L cells do express the "L"
antigen of undefined origin (3), and the Moloney leukemia virus
envelope antigen (4), which may contribute determinants that crossreact with certain alloantigens (5). The recent finding that transformed cells express new alloantigen specificities may also have some
bearing on the interpretation of these data (6,7).

226

Table 1

Detection of H-2 CD Antigens on Primed T Lymphocyte x 613 Hybrid Cells

% CML[a]

Targets

Sensitization	Effector/Target	Primed B10.M x 613 hybrids							Parent line	PHA blasts			
		A-4	B-1	C-5	D-3	E-2	F-1	G-3	613	B10.BR	B10.M	B10.RIII	B10.D2
B10.D2 + B10.M$_x$	70:1	42.1	34.4	45.8	35.8	36.9	42.0	35.8	6.1	-2.1	40.2	-6.1	-10.9
B10.D2 + B10.BR$_x$	70:1	49.6	50.7	53.0	47.9	59.2	46.4	48.7	33.0	33.8	-1.2	-0.4	-10.0
B10.D2 + B10.RIII$_x$	70:1	55.5	52.1	57.9	47.1	55.2	54.8	61.3	46.3	33.3	15.7	57.0	-4.8
B10.D2 + B10.D2$_x$	70:1	-0.4	-1.7	0.1	-1.6	5.1	0.4	5.8	3.7	-11.3	-7.3	-11.4	-9.8

H-2 haplotypes: B10.BR (kkkkkk); B10.M (ffffff); B10.RIII (rrrrrr); B10.D2 (dddddd); 613 (C3H origin, kkkkkk).

[a] All Standard Deviation $\leq$ 7.3%.

II. Detection of T Lymphocyte Surface Markers by Radioiodination

Surface labeling techniques reveal distinctive components in murine T and B lymphocytes (8-10). Using lactoperoxidase-catalyzed radio-iodination of lymphocyte surface proteins Dunlap et al (10) have resolved the high molecular weight T cell marker described by Trowbridge and Bevan (8) into three discrete radioactive bands in the 170K to 200K molecular weight range on SDS polyacrylamide gels. None of these bands are evident in radiolabeled membrane extracts of macrophages, B cells or 613 cells while T lymphocyte x 613 hybrid cells consistently and stably (throughout 8 months of culture) express one or more of the three bands (1).

1. Resolution by Co-electrophoresis

To resolve the band patterns obtained by autoradiography, a precise comparison of the relative electrophoretic mobilities in SDS-poly-acrylamide was made by co-electrophoresing ^{125}I surface labeled hybrid and 613 membrane preparations with ^{131}I labeled spleen cell membrane extracts. The radioactive iodoprotein profiles of 613 and spleen cells are presented in Figure 1. No prominent peaks are evident in the areas of the T or B cell markers. While the contours of the hybrid peaks vary, the positions relative to the T cell triplet are constant. In each hybrid preparation, a prominent peak clearly co-migrates with the middle component of the T cell triplet. A radio-active peak approximating the position of, and perhaps representing a modified form of the 170K molecular weight structure in the T cells, is also apparent in the radioactivity profiles of the hybrid lines and is reproducibly shifted to lower molecular weight position. A typical hybrid profile is shown in Figure 2.

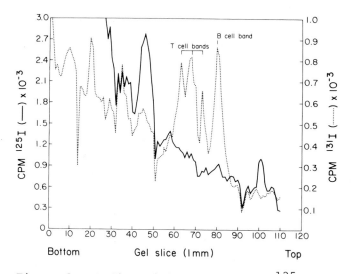

Figure 1. Radioactivity profile of ^{125}I labeled 613 fibroblast (solid line) and ^{131}I labeled spleen cell (dashed line) NP-40 membrane extracts co-electrophoresed on a 0.1% SDS, 5% acrylamide gel. The T cell bands, beginning with the most rapidly migrating, correspond to molecular weights of 170K, 187K, and 200K. Migration was from top to bottom.

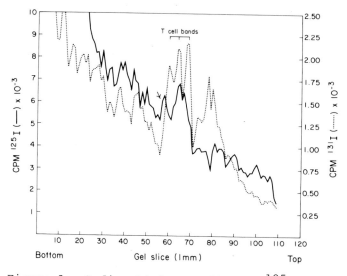

Figure 2. Radioactivity profile of ^{125}I labeled hybrid A-4 (solid line) and ^{131}I labeled spleen cell (dashed line) NP-40 membrane extracts co-electrophoresed on a 0.1% SDS, 5% acrylamide gel. Arrow indicates hybrid peak which may represent a modified form of the 170K T cell marker.

2. Counterselection

Azaguanine (AG) and diaminopurine (DAP) resistant variant hybrid clones isolated by counterselection continued to display both high molecular weight markers. Since most AG^R and DAP^R clones would have resulted from loss of the X chromosome and no. 8 respectively, linkage of these surface markers with either chromosome is unlikely, though the possibility of a translocation involving the loci for HGPRT and APRT cannot be excluded.

3. Radioimmunoprecipitation

In order to obtain evidence that the high molecular weight components expressed by the hybrid cells were of T cell origin, attempts were made to determine whether these were cross-reactive with the corresponding T cell surface proteins. Since Trowbridge and Bevan (8) had demonstrated that antiserum made against intact mouse lymphocytes would react with, and could be used to precipitate the radiolabeled cell markers from NP-40 extracts, rabbit anti-mouse lymphocyte globulin (ALG) was tested for its ability to specifically precipitate T and B markers from radioiodinated spleen extracts when co-precipitated with goat anti-rabbit globulin. The radioiodinated proteins characteristic of both T and B lymphocytes were effectively precipitated. The specificity of both antisera used in two step precipitation was affirmed by a control in which normal rabbit globulin rather than anti-lymphocyte globulin was used in the first step. ALG precipitation of hybrid radioiodinated membrane extracts revealed that the putative T cell markers expressed by the hybrid lines were both recovered in the immunoprecipitate. There was no evidence of these markers in the 613 immunoprecipitate.

III. Terminal Deoxynucleotidyl Transferase (TdT)

This enzyme has a highly restricted normal tissue distribution which involves the thymus (primarily cortisone sensitive thymocytes) and 1-4% of bone marrow cells. The latter are Thy-1 positive in the rat (11). No other lymphoid tissue examined to date displays detectable enzyme activity. However, transient expression of TdT has been demonstrated in about 1% of rat spleen cells between 2-4 weeks of age by the indirect fluorescent antibody technique (12). Because of its distribution TdT has been postulated as an internal differentiation antigen and a potentially useful cell marker (13). This was the basis of interest in testing the various T lymphocyte and thymocyte x 613 hybrids for terminal transferase expression.

Thirty-five different hybrid lines were screened for TdT using an indirect fluorescent antibody assay which employed immunoadsorbent column purified rabbit anti-calf thymus terminal transferase and fluorescein conjugated F(ab')$_2$ fragments of goat anti-rabbit Ig. Several lines showed bright cytoplasmic fluoresence while a few exhibited nuclear localization as well. Line 613 in contrast, demonstrated only faint cytoplasmic fluorescence. The pattern of positive fluorescence was not confined to any particular hybrid group; those derived from cortisone treated thymocyte populations as well as from MLC primed populations yielded lines with bright fluorescence. Tests for non-specific labeling by the FITC conjugated F(ab')$_2$ goat anti-rabbit γglobulin were negative. Selected lines are being tested for TdT activity by a polymerization assay utilizing an acid insoluble p(dA) initiator and ^{3}HdGTP as described by Barton et al (14).

C. Discussion and Concluding Remarks

The data presented demonstrate that the T-cell hybrids formed with a permanent fibroblast line can express certain T lymphocyte markers. At least one and perhaps two T cell specific membrane components are detected on all the hybrids examined using surface radioiodination. The identification of the radiolabeled surface molecules expressed in the hybrids as T cell markers is based upon their co-migration with labeled T cell components and the ability to immunoprecipitate them with antisera capable of precipitating the corresponding antigens from T cell preparations. Similar structures are either absent or present in greatly reduced amount on 613 cells, macrophages and B lymphocytes. Thus the evidence is consistent with the hybrid markers being of T cell derivation. Modification of the molecular structure of the 170K marker is suggested by its slightly faster migration on SDS gels. The immunoprecipitability of the "altered" molecule implies that the modification did not drastically affect its antigenic characteristics.

H-2 antigens represent a constitutive marker, i.e. property expressed by all cells of the same genotype. The significance of the expression of the respective normal cell products by hybrid lines is that generalized suppression of the normal genome as a consequence of hybridization can be ruled out. This information facilitates interpretation of the loss of differentiated markers. Expression of the T cell H-2 target antigen recognized by cytotoxic T lymphocytes on hybrids is of interest for another reason. Immunoselection by means of cell-mediated killing, or antibody-mediated lysis if serologically defined determinants are present, should make it possible to obtain hybrid subclones that have lost chromosome 17 and, to determine whether one or the other of the T cell markers is syntenic with H-2. Together with the salvage pathway marker chromosomes then, one-fifth of the normal mouse genome can already be probed using counterselective techniques.

The results obtained by indirect fluorescence for terminal deoxynucleotidyl transferase indicate that TdT or an antigenically cross-reactive protein is being produced by the hybrid cells. The high frequency of fluorescence positive cells among thymocyte and primed spleen cell derived hybrid lines might reflect some kind of preferential hybrid formation with relatively immature T lymphocytes, since TdT is characteristic of "progenitor" Thy-1 positive cells. Alternatively, fusion of T cells with a dedifferentiated cell could conceivably result in derepression of genes normally inactive beyond a certain stage of differentiation.

In conclusion, normal T lymphocyte x 613 somatic cell hybrids display several phenotypic characteristics deriving from the T cell genome, including T cell specific surface markers detectable by radiodination, and an antigen which is cross-reactive with the thymocyte "internal differentiation antigen" terminal deoxynucleotidyl transferase. Documentation of the stable expression of one or more T lymphocyte facultative markers after fusion with a non-T cell line is of particular interest since it has not been possible to demonstrate Thy-1 antigen on hybrids generated by the fusion of normal or malignant thymus derived lymphocytes with a variety of non-T cell lines. While the retention (or extinction) of some markers may not be predictive of the integrity of immune function, the phenotypic profile of these hybrid lines provides an opportunity to study genetically and biochemically some T cell specific expressions.

230

Acknowledgements. We thank Barbara Alter for her assistance with the CML assay, Brian Dunlap for guidance in cell surface analysis by radioiodination, and Karen Heim for her help in preparing this manuscript. NS was supported in part by NIH grant no. 5T01CA5295-03 awarded to Vanderbilt University Department of Microbiology, Nashville, Tennessee, for predoctoral training. This work was also supported by NIH grants GM-06983, GM-15422, CA-16836, AI-11576, AI-08439 and National Foundation-March of Dimes grants CRBS 246 and 6-76-213. This is paper no. 156 from the Immunobiology Research Center and paper no. 2228 from the Laboratory of Genetics, The University of Wisconsin, Madison, Wisconsin 53706.

References

1. Saravia, N., DeMars, R., Dunlap, B., Bach, F.H.: A T-cell marker in mouse fibroblast x T lymphocyte somatic cell hybrids. Scand. J. Immunol. 6, 1333-1337 (1977)
2. Peck, A.B., Bach, F.H.: Mouse cell-mediated lympholysis assay in serum free and mouse serum supplemented media: culture conditions and genetic factors. Scand. J. Immunol. 4, 53-62 (1975)
3. Leclerc, J.C., Levy, J.P., Varet, B., Oppenheim, S., Senik, A.: Antigenic analysis of L strain cells: A new murine leukemia-associated antigen, "L". Cancer Res. 30, 2073-2079 (1970)
4. Fenyo, E.A., Grundner, G., Klein, E.: Virus-associated surface antigens on L cells and Moloney lymphoma cells. J. Natl. Cancer Inst. 52, 743-751 (1974)
5. Martin, W.J.: Immune surveillance directed against derepressed cellular and viral alloantigens. Cell Immunol. 15, 1-10 (1975)
6. Pellegrino, M.A., Ferrone, S., Brautbar, C., Hayflick, L.: Changes in HL-A antigen profiles on SV40 transformed human fibroblasts. Exp. Cell Res. 97, 340-396 (1976)
7. Garrido, F., Schirrmacher, V., Festenstein, H.: H-2 like specificities of foreign haplotypes appearing on a mouse sarcoma after vaccinia virus infection. Nature 259, 228-230 (1976)
8. Trowbridge, I.S., Ralph, P., Bevan, M.J.: Differences in the surface proteins of mouse B and T cells. Proc. Natl. Acad. Sci. USA 72, 157-161 (1975)
9. Gahmberg, C.G., Häyry, P., Andersson, L.C.: Characterization of surface glycoproteins of mouse lymphoid cells. J. Cell. Biol. 68, 642-660 (1976)
10. Dunlap, B., Bach, F.H., Bach, M.L.: Cell surface changes in alloantigen activated T lymphocytes. Nature 271, 253-255 (1978)
11. Gregoire, K.E., Goldschneider, I., Barton, R.W., Bollum, F.J.: Intracellular distribution of terminal deoxynucleotidyl transferase in rat bone marrow and thymus. Proc. Natl. Acad. Sci. USA 74, 3993-3996 (1977)
12. Gregoire, K.E., Goldschneider, I., Barton, R.W., Bollum, F.J.: Federation Proceedings 36, 1301 (1977)
13. Bollum, F.J.: Antibody to terminal deoxynucleotidyl transferase. Proc. Natl. Acad. Sci. USA 72, 4119-4122 (1975)
14. Barton, R., Goldschneider, I., Bollum, F.J.: The distribution of terminal deoxynucleotidyl transferase (TdT) among subsets of thymocytes in the rat. J. Immunol. 116, 462-468 (1976)

A Quantitative Disc Radioimmunoassay for Antibodies Directed Against Membrane-Associated Antigens

L.A. Manson, E. Verastegui-Cerdan, R. Sporer

A. Introduction

The antigens coded for by the major histocompatibility locus are found to exist in cells associated with membranes. As a consequence a number of radioimmunoassays have been developed using whole cells as the immunoadsorbent (1, 2). There are several disadvantages in using whole cells, among them being non-reproducibility of creating monolayers of cells in microtiter plates with adhering cells and the need to centrifuge plates if lymphoid, non-adherent cells are being used.

We have reviewed (3) the immunogenic properties of a sub-cellular cell-free membrane preparation, the microsomal lipoproteins (MLP). The MLP of mouse spleen and of some lymphomas such as L-5178Y (a DBA/2 lymphoma) are excellent H-2, non-H-2 and TATA immunogens. By coupling an MLP (which contains no Ig) covalently to CNBr-activated filter paper discs, we obtain a stable immunoadsorbent that can be used in a quantitative binding assay. In this report we will present the essential elements of the assay, and how it is used in our laboratory for screening for hybridomas making anti-membrane antibodies.

B. Materials and Methods

I. Reagents.

We routinely use an ^{125}I-rabbit anti-mouse Fab reagent for screening and quantitation. This reagent is classless, since it is directed against the light chains of the antibody molecules. We have also used anti-heavy chain reagents in the procedure.

II. Membrane Preparations

The method for preparing L-5178Y MLP has been described (4). For absolute quantitation, we prepared washed membranes from sheep cells and bound these to the discs, and also prepared a highly purified anti-SRBC antibody preparation. This anti-SRBC antibody preparation was adsorbed and eluted twice from SRBC membranes and was thoroughly absorbed with horse red blood cell membranes. This antibody preparation was used as the primary standard (see Fig. 2).

III. Antisera

The anti-SRBC serum was made in Swiss mice. M24F is a C57BL/10-anti-L-5178Y serum. The mice were boostered every 4 weeks with 2.5×10^7 cells, and serum pool obtained 3 weeks after the sixth such injection. The pool had an end-point titer in the complement-dependent cytotoxic test of 1:2200. From Fig. 2 it was determined that M24F contains 3.9 mg of specific antibody per ml.

C. Discussion

The protocol for preparing antigen discs is described in Fig. 1 and the protocol for the assay is in Fig. 2. In our hands the discs

Preparation of Antigen Discs for Radioimmunoassay

I. Activation

2 g of 3/16" discs of #542 Whatman paper
↓ soaked in 40 ml distilled water for 1 h.
Add 40 ml of a fresh 5% CNBr solution.
Keep temperature at 19°C by means of ice
bath. Raise pH to 10.0 - 10.5 and maintain this pH with stirring with 1 M NaOH.

Reaction time = 3 minutes

↓ Pour into 300 ml of ice-cold 0.005 M
NaHCO₃. Discs are then washed
   5 times with 300 ml 0.005 M NaHCO₃
   4 times with 100 ml ice-cold acetone.
When dry, discs are stored in desiccator
at 0°C or in vials at -20°C.

II. Antigen Coupling

Membrane preparations (1 mg/ml)
in 0.1 M NaHCO₃ are incubated
overnight in cold with agitation
with activated discs (approx.
100 discs ml).
Discs are washed 2 X 10 ml 0.1 M
NaHCO₃. To quench the discs incubate in 10 ml 1 M ß-ethanolamine
pH 8.2 for 2 h at room temperature.
Discs are washed in IB⁺ and stored
in IB⁺ at 0°C.
   IB⁺: 0.01 M tris pH 8.0
        0.15 M NaCl
        0.25% bovine serum albumin
        0.02% sodium azide
        2% calf serum

Fig. 1.   The procedure for making antigen discs.

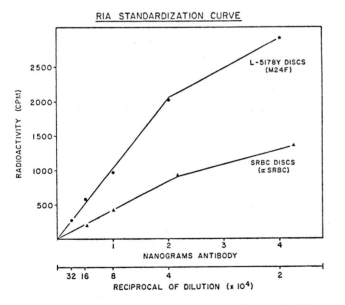

Fig. 2.   The protocol for carrying out the assay.   The standardiza-
tion curve is also shown in the figure.   Each point on the curves is
the average of triplicate discs, and the standard deviation was always
less than 10% among replicates.

233

remain stable and useable for 6 months. The background uptake of
iodinated reagents is of the order of 1 - 1.5% with non-Ig containing
MLP (L-5178Y). If spleen MLP is used, the anti-Fab reagent will
bind to the membrane Ig giving high backgrounds (> 10% of input).
We have recently found that an anti-γ_1 reagent can be used with spleen
antigen discs and still have acceptable backgrounds. The discs bind
approximately 2 µg of membrane protein.

With MLP, the number of antigen sites is limited, therefore, the assay
gives a straight line relationship only in the 0 - 2 ng range. We
have made mammary tumor virus discs (MMTV) using a purified preparation of the virus. With such discs we see a straight line relationship with as high as 10 ng of specific antibody.

When the assay is used to screen for hybridomas producing anti-H-2
or anti TATA antibodies, we do the initial screening at a 1:25 dilution. This is done by adding 2 λ of hybridoma supernatant to 50 λ
of IB$^+$. At higher concentrations of hybridoma supernatant, the specific binding of the ^{125}I-anti Fab is inhibited by constituents in the
growth medium. Our positive supernatant fluids, even in the initial
screening, will show positive binding at 1:100 dilutions. For quantitation, antibody binding is carried out for 24 hours, for rapid screening a 6 hour incubation is adequate.

D. Summary

A solid phase RIA has been developed for detecting antibodies directed against membrane-associated antigens. The conditions for quantitative analysis have been defined.

Acknowledgements. We are indebted to Dr. N. Klinman for the radioactive anti-antibody reagents. The excellent technical assistance
was provided by Ms. Anita Guarini. These studies were supported by
Research Grants from the National Cancer Institute, CA 10815, CA 07973
and RRO-5540.

References

1. Goldstein, L. T., Klinman, N. R., and Manson, L. A. : *Brief Com-
    munication:* A microtest radioimmunoassay for noncytotoxic tumor-
    specific antibody to cell-surface antigens. Journal of the Nation-
    al Cancer Institute 51, 1713-1715 (1973).
2. Tax, A. and Manson, L. A.: A radioimmunoassay for antibodies a-
    gainst surface membrane antigens using adhering cells. Journal of
    Immunological Methods 11, 281 - 285 (1976).
3. Manson, L. A.: Intracellular localization and immunogenic capaci-
    ties of phenotypic products of mouse histocompatibility genes.
    In: Biomembranes 8. Manson, L. A. (ed.) Plenum Publishing Corpor-
    ation, New York: 1976, p. 47 - 88.
4. Manson, L. A., Foschi, G. V. and Palm, J.: An association of
    transplantation antigens with microsomal lipoproteins of normal
    and malignant mouse tissues. Journal of Cellular and Comparative
    Physiology 61: 109 - 118 (1963).

234

Summary of Cloning of Differentiated Function Using Hybrid Cells and Comparison of the Mouse and Human Gene Maps for Homologous Markers

J. Minna

A. Introduction

There are several recent findings in somatic cell hybrid genetics that may prove of general use to cell biologists and to workers using hybrid cell technology to investigate the biology of the immune system. These findings are summarized in tabular form below.

I. Differentiated Functions May Be Cloned from Normal Tissues Using Hybrid Cell Technology and the Appropriate Tissue Culture Adapted Parent Cell Line

The cloning of hybrid cells producing antibodies of predefined specificity is a most interesting use of hybrid cells for study of differentiated functions. (1) It is important to realize that this phenomenon is probably a general one and of far reaching importance to biology. Many authors have studied fusions of cell lines often considered "undifferentiated" such as fibroblast cell lines and examples of both expression and extinction of differentiated functions have been seen. However, several studies have been performed by fusing cultured clonal lines of differentiated cells expressing specialized properties (usually derived from tumors) and normal cells taken directly from the animal that were differentiating along similar paths.

Table 1. Examples of cloning differentiated functions from normal cells in mammalian cell hybrids between tissue culture adapted tumor lines and normal tissues

Replicating Tumor Cell (Species)	Normal Tissue (Species)	Function Expressed	Reference
Plasmacytoma (Mouse)	Lymphoid cells (Mouse, Rat, Human)	Immunoglobulin of normal tissue parent	1,2, and this meeting
T Cell Lymphoma (Mouse)	Lymphoid Cells (Mouse)	Suppressor activity	this meeting
Erythroleukemia (Mouse)	Bone Marrow cells (Human, Chinese hamster)	Hemoglobin of normal tissue parent	3
Hepatoma (Mouse)	Fetal Liver (Human)	Albumin, transferrin, ceruloplasmin, alpha-1 antitrypsin	4
Neuroblastoma (Mouse)	Nervous System cells (Mouse, Chinese Hamster)	Neurotransmitter synthesis	5,6

The results of these experiments indicate that differentiated properties can be recovered from the normal tissue parent and cloned using the tumor cell parent as the replicative vehicle. In some cases the evidence shows expression of structural genes of the normal tissue parent. In others, it is possible that the structural genes for the function are from the tumor cell line and result from activiation of genes by the normal tissue parent. It appears likely in most cases, that at least certain chromosomes of the normal tissue parent have to be maintained for continued expression of the new function.

II. Results of Comparative Gene Mapping Studies

The results of recent comparative gene mapping studies using interspecific somatic cell hybrids preferentially segregating chromosomes of one parent, recombinational genetics, and chromosome banding analysis allow several conclusions to be reached concerning the organization of the mammalian gene map and its evolution. (See references 7-11 for a full discussion.) The conservation or disruption of the synteny of gene groups can be summarized as follows: syntenic genes on the X chromosome are highly conserved in all species examined; syntenic genes that are tightly linked (less than 1 map unit apart) including tandem duplications are highly conserved; species separated 10 million years or less in evolution usually have homologous genes assigned to chromosomes found to be homologous by chromosome banding studies (eg. primates and humans; rats and mice); syntenic human genes which are located on different arms of the same chromosome are unlikely to be syntenic in the mouse; syntenic mouse genes which are many map units apart (25-45 map units) are very unlikely to be syntenic in man; however, there are several autosomal synteny groups, some of which are 10-20 map units apart which appear to have their syntenic relationships conserved during the 80 million years of evolution separating humans and mice. Because of the great interest in the mouse and the human in the genetic analysis of the biology of the immune response Table 2 summaries and compares the mouse and human gene maps where homologous markers have been identified.

In Table 2, mouse enzyme nomenclature is used (12) and the human equivalent given in (). The McKusick Catelogue numbers provide a common reference ground for comparing humans with other species (13). The methodology for enzyme assays on starch gel or cellulose acetate electrophoresis are found in references (14,15). The mode of mouse gene assignment refers to sexual or recombinational genetics (F), and to somatic cell hybrid genetics (S). General references for human gene assignments can be found in (7,16) and for the mouse in (17,18). The provisional human assignments are underlined. Only some of the X-linked markers are shown.

The data in Table 2 indicate there are at least 9 different human autosomes with gene pairs which have remained together during the divergent evolution of man and mouse. Thus, both the X chromosome and the autosomes exhibit conservation of gene synteny relationships.

B. Concluding Remarks

The techniques of cell fusion are prowerful cell biologic tools for studying differentiation and evolution. They also provide unique biologic reagents aus the antibody producing hybridomas demonstrate. Several general principles are emerging from the hybrid cell studies. First, it should be possible to clone a variety of other highly specialized (i.e. "differentiated") molecules in cell hybrids. The approach appears to be the use of a clned tumor cell derived line capable of rapid replication and high cloning efficiency in vitro as one parent and a normal cell taken directly from the organ producing the molecules of interest as the other parent. The tissue culture line should be as closely related as possible in its pathway of differentiation to that of the organ producing the desired normal tissue product. Species barriers may be crossed and it remains to be determined how separate the pathways of differentiation of the tumor and normal cell parent may be. While, there are examples of gene activation in fusions between cells differing widely in their pathways of differentiation, these activations are for the present too complex to readily predict the outcome of any individual fusion.

236

The comparative gene mapping studies and summary of the mouse gene map should be of use to immunologists using hybridomas. The location of genes coding for immunologic functions such as light and heavy chain genes in one species should give clues to the location of homologous genes in other species. Of practical importance this comparative map should greatly aid in gene assignment, and in designing selective systems to hold in chromosomes carrying the structural genes (eg. specific antibody light and heavy chains) of interest.

Acknowledgments. I thank my collaborators in the somatic cell hybrid work: H. Coon, M. Nirenberg, X. Breakfield, A. Chalazonitis, L. Greene, W. Shain; and in the comparative gene mapping work: P. Lalley, U. Francke, W. Moss, G.A.P. Bruns, P. Gerald, J. Yavelow. I thank G. Darlington for allowing me to quote her work before publication.

References

1. Kohler, G. and Milstein, C.: Continuous cultures of fused cells secreting antibody of predefined specificity. Nature 256, 495-497 (1975)
2. Galfrè, G., Howe, S.C., Milstein, C., Butcher, G.W., Howard, J.D.: Antibodies to major histocompatibility antigens produced by hybrid cell lines. Nature 266, 550-552 (1977)
3. Deisseroth, A.: Isolation of hybrid cells that exhibit markers of erythroid differentiation. N. Engl. J. Med. 294, 148-152 (1976)
4. Darlington, G.: Linkage analysis of four human serum proteins using mouse x human cell hybrids. (personal communication)
5. Minna, J.D., Yavelow, J., Coon, H.G.: Expression of phenotypes in hybrid somatic cells derived from the nervous system. Genetics 79, (supp) 373-383 (1975)
6. Greene, L.A., Shain, W., Chalazonitis, A., Breakfield, X., Minna, J., Coon, H.G., Nirenberg, M.: Neuronal properties of hybrid neuroblastoma x sympathetic ganglion cells. Proc. Natl. Acad. Sci. U.S.A. 72, 4923-4927 (1975)
7. McKusick, V.A., Ruddle, F.H.: The status of the gene map of the human chromosomes. Science 196, 390-405 (1977)
8. Minna, J.D., Lalley, P.A., Francke, U.: Comparative mapping using somatic cell hybrids. In Vitro 12, 726-733 (1976)
9. Roderick, T.H., Pearson, P.L.: Report of the workshop on comparative gene mapping. (In reference 16)
10. Miller, D.A.: Evolution of primate chromosomes. Science 198, 1116-1124 (1977)
11. Womack, J.E., Sharp, M.: Comparative autosomal linkage in mammals: genetics of esterases in Mus musculus and Rattus norvegicus. Genetics 82, 665-675 (1976)
12. Committee on Standardized Genetic Nomenclature for Mice: J. Hered. 54,159-162 (1963)
13. McKusick, V.A.: Mendelian Inheritance in Man, 4th ed. Baltimore: Johns Hopkins Univ. Press, 1975
14. Harris, H., Hopkinson, D.A.: Handbook of Enzyme Electrophoresis in Human Genetics, 1st ed. New York: American Elsevier, 1976
15. Nichols, E.A., Ruddle, F.H.: A review of enzyme polymorphism, linkage and electrophoretic conditions for mouse and somatic cell hybrids in starch gels. J. Histochem. Cytochem. 21, 1066-1081 (1973)
16. Winnipeg Conference (1977). Fourth International Workshop on Human Gene Mapping. Bergsma, D. (ed.) Birth Defects: Orig. Artic. Ser., The Natl. Found., or Cytogen. Cell Gen. Basel-Munchen-Paris-London-New York-Sydney: S. Karger, (to be published in 1978)
17. Searle, A.G.: Mouse mutant gene list. Mouse News Lett. 56,4-23 (1977)
18. Womack, J.E.: Linkage map of the mouse. Mouse News Lett. 57, 6 (1977)

19. Francke, U., Lalley, P.A., Moss, W., Ivy, J., Minna, J.D.: Gene Mapping in Mus musculus by interspecific cell hybridization: assignment of the genes for tripeptidase-1 to chromosome 10, dipeptidase-2 to chromosome 18, acid phosphatase-1 to chromosome 12, and adenylate kinase-1 to chromosome 2. Cytogen. Cell Gen. 19, 57-84 (1977)

20. Hutton, J.J., Roderick, T.H.: Linkage analyses using biochemical variants in mice. III. Linkage relationships of eleven biochemical markers. Biochem. Genet. 4, 339-350 (1970)

21. Chapman, V.M., Ruddle, F.H., Roderick, T.H.: Linkage of isozyme loci in the mouse: phosphoglucomutase-2 (Pgm-2), mitochondrial NADP malate dehydrogenase (Mod-2), and dipeptidase-1 (Dip-1). Biochem. Genet. 5, 101-110 (1971)

22. Lalley, P.A., Francke, U., and Minna, J.D.: Homologous genes for enolase, phosphogluconate dehydrogenase, phosphoglucomutase, and adenylate kinase are syntenic on mouse chromosome 4 and human chromosome 1p. Proc. Natl. Acad. Sci. U.S.A. (In the press)

23. Chapman, V.M.: 6-phosphogluconate dehydrogenase (PGD) genetics in the mouse: linkage with metabolically related enzyme loci. Biochem. Genet. 13, 849-856 (1975)

24. Lalley, P.A., Minna, Francke, U.: Conservation of autosomal gene synteny groups in mouse and man. Nature (In the press)

25. Nichols, E.A., Ruddle, F.H., Petras, M.L.: Linkage of the locus for serum albumin in the house mouse, Mus musculus. Biochem. Genet. 13, 551-555 (1975)

26. Womack, J.E., Hawes, N.L., Soares, E.R. and Roderick, T.H.: Mitochondrial balate dehydrogenase (MOR-1) in the mouse: linkage to chromosome 5 markers. Biochem. Genet. 13, 519-525 (1975)

27. Minna, J.D., Bruns, G.A.P., Krinsky, A.H., Lalley, P.A., Francke, U., and Gerald, P.S.: Assignment of a Mus musculus gene for triosephosphate isomerase to chromosome 6 and for glyoxalase-1 to chromosome 17 using somatic cell hybrids. Somat. Cell Genet. (In the press)

28. Hutton, J.J.: Linkage analyses using biochemical variants in mice. I. Linkage of the hemoglobin beta-chain and glucose phosphate isomerase loci. Biochem. Genet. 3, 507-515 (1969)

29. Shows, T.B., Chapman, V.M., Ruddle, F.H.: Mitochondrial malate dehydrogenase and malic enzyme. mendelian inherited electrophoretic variants in the mouse. Biochem. Genet. 4, 707-718 (1970)

30. Nichols, E.A., Ruddle, F.H.: Polymorphism and linkage of glutathione reductase in Mus musculus. Biochem. Genet. 13, 323-329 (1975)

31. Kozak, C., Nichols, E., Ruddle, F.H.: Gene linkage analysis in the mouse by somatic cell hybridization: assignment of adenine phosphoribosyltransferase to chromosome 8 and alpha-galactosidase to the X chromosome. Somat. Cell Genet. 1, 371-382 (1975)

32. Nichols, E.A., Chapman, V.M., Ruddle, F.H.: Polymorphism and linkage for mannose phosphate isomerase in Mus musculus. Biochem. Genet. 8, 47-53 (1973)

33. Delorenzo, R.J., Ruddle, F.H.: Glutamate-oxalate transaminase (GOT) genetics in Mus musculus: linkage, polymorphism, and phenotypes of the GOT-2 and GOT-1 loci. Biochem. Genet. 4, 259-273 (1970)

34. Russell, E.S. and Bernstein, S.E.: Mouse globin alpha chain. Mouse News Lett. 50, 44 (1974)

35. Breen, G.A.M., Lusis, A.J., and Paigen, K.: Linkage of genetic determinants for mouse B-galactosidase electrophoresis and activity. Genet. 85, 73-84 (1977)

36. Mishkin, J.J., Taylor, B.A., Mellman, W.J.: GLK a locus controlling galactokinase activity in the mouse. Biochem. Genet. 14, 635-640 (1976)

37. Kozak, C.A., Ruddle, F.H.: Assignment of the genes for thymidine kinase and galactokinase to Mus musculus chromosome 11 and the preferential segregation of this chromosome in chinese hamster/mouse somatic cell hybrids. Somat. Cell Genet. 3, 121-134 (1977)

33. Eicher, E.M., Womack, J.E.: Chromosomal location of soluble glutamic-pyruvic transaminase-1 (GPT-1) in the mouse. Biochem. Genet. 15, 1-8 (1977)

39. Bach, F.H.: Genetic control of major complex histocompatibility antigens. Genetics 79, 263-275 (1975)

40. Meo, T., Douglas, T., Rijnbeek, A.M.: Glyoxalase I polymorphism in the mouse: a new genetic marker linked to H-2. Science 198,311-313 (1977)
41. Chapman, V.M., Shows, T.B.: Somatic cell genetic evidence for X-chromosome linkage of three enzymes in the mouse. Nature 259,665-667 (1976)
42. Lusis, A.J., West, J.D.: X-linked inheritance of a structural gene for alpha-galactosidase in Mus musculus. Biochem. Genet. 14,849-855 (1976)
43. Bennett, D., Boyse, E.A., Lyon, M.F., Mathieson, B.J., Scheid, M., Yanagisawa, K.: Expression of H-Y (male) antigen in phenotypically female TFM/Y mice. Nature 257, 236-238 (1975)

Table 2. Comparison of the Mouse and Human Gene Maps Where Homologous Markers have been identified

Chromosome Assignment Mouse	Human	Enzyme or Marker	E.C. Number	McKusick Number	Mouse Mode	Gene Assignment References
1	2q	ID-1 (IDH-1)	1.1.1.42	14770	F,S	19,20
	1q	DIP-1 (PEP-C)	3.4.11.*	17000	F,S	19,21
2	9q	AK-1	2.7.4.3	10300	S	19
4	1p	PGM-2 (PGM-1)	2.7.5.1	17190	F,S	19,21
	1p	AK-2	2.7.4.3	10302	S	22
	1p	ENO-1	4.2.1.11	17245	S	22
	1p	PGD (6PGD)	1.1.1.44	17220	F,S	19,23
	1	PGD-1 (GDH)	1.1.1.47		F	20
5	4	PGM-1 (PGM-2)	2.7.5.1	17200	F,S	19,20
	4	PEP-S	3.4.11.*	17020	S	24
	4	ALB-1		10360	F	25
	7	GUS	3.2.1.31	25322	F	26
	7	MOR-1 (MDH-M)	1.1.1.37	15410	F	26
6	12p	TPI-1	5.3.1.1	19045	S	27
?	12p	LDH B	1.1.1.27	15010	F	20
7	19	GPI-1	5.3.1.9	17240	F,S	19,20
	19	PEP-D	3.4.13.9	17010	S	24
	11p	LDH A	1.1.1.27	15000	S	24
	11	Hbb (beta globin)		14190	F	28
	?	MOD-2 (ME-M)	1.1.1.40	15427	F	29
	15q	ID-2 (IDH-M)	1.1.1.42	14765	S	24
8	8p	GR-1 (GSR)	1.6.4.2	13830	F,S	19,30
	16q	APRT	2.4.2.7	10260	S	19,31
	6	GOT-2 (GOT-M)	2.6.1.1	13815	F	33
9	15q	MPI-1	5.3.1.8	15455	F,S	19,31,32
	15q	PK-3	2.7.1.40	17905	S	24
	6q	MOD-1 (ME-1)	1.1.1.40	15425	F,S	19,20,31
	3	BGE (B-GAL)	3.2.1.22	23050	F	35
10	10	HK-1	2.7.1.1	14260	S	24
	10	PP		17903	S	24
	12q	TRIP-1 (PEP-B)	3.4.11.*	16990	S	19
11	16	Hba (alpha globin)		14180	F	34
	17q	GLK (GALK)	2.7.1.6	23020	F,S	36,37
	17q	TK-1	2.7.1.21	18830	S	37
12	1	AMY-1,AMY-2	3.2.1.1	10465	F	18
	2p	ACP-1	3.1.3.2	17150	S	19
14	14q	NP-1 (NP)	2.4.2.1	16405	F,S	11,19
	13q	ES-10 (ES-D)	3.1.1.1	13328	F,S	11,19
15	?	GPT-1 (GPT-S)	2.6.1.2	13820	F	38
17	6p	H-2 (MHC)		14280	F	39
	6p	GLO-1	4.4.1.5	13875	F,S	27,40
18	18q	DIP-2 (PEP-A)	3.4.11.*	16980	S	19
19	10q	GOT-1	2.6.1.1	13818	F	33

Table 2. (continued) Comparison of the Mouse and Human Gene Maps Where Homologous Markers Have Been Identified

Chromosome Assignment Mouse	Human	Enzyme or Marker	E.C. Number	McKusick Number	Mouse mode	Gene Assignment References
X	Xq	PGK	2.7.2.3	31180	S	41
	Xq	A-GAL	3.2.1.2	30150	F,S	19,31,42
	Xq	HPRT	2.4.2.8	30800	S	19,41
	Xq	G6PD	1.1.1.49	305905	S	5,41
Y	Y	H-Y antigen			F	43

240

List of Contributors

D.P. Aden, The Wistar Institute, 36th Street at Spruce, Philadelphia, PA 19104/USA

J. Andersson, Biomedicum, University of Uppsala, Uppsala, Schweden

R. Asofsky, National Institute of Allergy and Infectious Diseases, Laboratory of Microbial Immunity, Bethesda, MD 20014/USA

F.H. Bach, Immunobiology Research Center, Wisconsin University Medical School, 1150 University Avenue, Madison, WI 53706/USA

P. Beverly, University College, London WC1E 6BS, U.K.

P.H. Black, Laboratory of Cellular and Molecular Research Cardiac Unit, Massachusetts General Hospital, Boston, MA 02114/USA

E. Bohrer, Medical Research Council, Clinical Research Centre, Watford Road, Harrow, Middlesex HA1 3UJ, U.K.

F.J. Bollum, Biochemistry Division, Uniformed Services University of the Health Sciences, 4301 Jones Bridge Road, Bethesda, MA 20014/USA

G.W. Butcher, Agricultural Research Council, Institute of Animal Physiology, Babraham, Cambridge CB2 4AT, U.K.

G. Buttin, Institut de Recherches en Biologie Moleculaire du C.N.R.S., Unite de Genetique Cellulaire, Tour 43-2, Place Jussieu, 75221 Paris, Cedex 05, France

P.A. Cazenave, Institut Pasteur, 28, Rue du Dr. Roux, 75015 Paris, France

R. Ceppellini, Basel Institute for Immunology, 487 Grenzacherstraße, CH-4005 Basel, Switzerland

J. Claflin, Dept. of Microbiology, University of Michigan Medical School, 6643 Medical Sciences II, Ann Arbor, MI 48109/USA

B. Clevinger, Washington University School of Medicine, 660 South Euclid Avenue, St. Louis, MO 63110/USA

M. Cohn, The Salk Institute for Biological Studies, San Diego, CA 92112/USA

D. Collavo, Genetic Unit and Laboratory of Experimental Oncology, Institute of Pathological Anatomy, University of Padua, Padua, Italy

C.M. Croce, The Wistar Institute, 36th Street at Spruce, Philadelphia, PA 19104/USA

J.M. Davie, Washington University School of Medicine, Dept. of Microbiology and Immunology, St. Louis, MO 63110/USA

R.I. DeMars, Immunobiology Research Center, Wisconsin University Medical School, 1150 University Avenue, Madison, WI 53706/USA

K. Denis, Dept. of Pathology, G-3, University of Pennsylvania, The School of Medicine, Philadelphia, PA 19174/USA

B. Diamond, Dept. of Cell Biology, Albert Einstein College of Medicine, 1300 Morris Park Avenue, New York, N.Y. 10461/USA

J. Dilley, Division of Medical Oncology, Stanford University School of Medicine, Stanford, CA 94305/USA

G. DiPauli, The Salk Institute for Biological Studies, San Diego, CA 92112/USA

R. DiPauli, The Salk Institute for Biological Studies, San Diego, CA 92112/USA

H.D. Engers, Dept. of Immunology, Swiss Institute of Experimental Cancer Research, CH-1066 Epalinges, Switzerland

M. Feldmann, Tumor Immunology Unit, Dept. of Zoology, University College, London WC1E 6B5, U.K.

F.W. Fitch, Dept. of Pathology, University of Chicago, 950 East 59th Street, Chicago, Ill. 60637/USA

W. Fitzpatrick, Medical Research Council, Clinical Research Centre, Watford Road, Harrow, Middlesex HA1 3UJ, U.K.

G. Galfre, Harvard Medical School, Dept. of Pathology, 25 Shattuck Street, Boston, MA 02115/USA

W. Geckeler, The Salk Institute for Biological Studies, San Diego, CA 92112/USA

W. Gerhard, The Wistar Institute, 36th Street at Spruce, Philadelphia, PA 19104/USA

J.W. Goding, Dept. of Genetics, Stanford University School of Medicine, Stanford, CA 94305/USA

R.A. Goldsby, Dept. of Genetics, Stanford University School of Medicine, Stanford, CA 94305/USA

D.I. Gottlieb, Dept. of Anatomy, Washington University School of Medicine, 660 South Euclid Avenue, St. Louis, MO 63110/USA

J. Greve, Dept. of Anatomy, Washington University Medical School, 660 South Euclid Avenue, St. Louis, MO 63110/USA ·

E. Gronowicz, Dept. of Medicine, Division of Immunology, Stanford University School of Medicine, Stanford, CA 94305/USA

R. Grützmann, Institute for Genetics, University of Cologne, D-5000 Cologne, FRG

G. Guarnotta, National Institute of Medical Research, The Ridgeway, Mill Hill, London NW7 1AA, U.K.

E. Haber, Laboratory of Cellular and Molecular Research Cardiac Unit, Massachusetts General Hospital, Boston, MA 02114/USA

G.J. Hämmerling, Institute for Genetics, University of Cologne, D-5000 Cologne, FRG

U. Hämmerling, Sloan-Kettering Institute, 410 East 68th Street, New York, N.Y. 10021/USA

D. Hansburg, Washington University School of Medicine, 660 South Euclid Avenue, St. Louis, MO 63110/USA

H. Hengartner, Basel Institute for Immunology, 487 Grenzacherstraße, CH-4005 Basel, Switzerland

Leonore A. Herzenberg, Dept. of Genetics, Stanford University School of Medicine, Stanford, CA 94305/USA

Leonard A. Herzenberg, Dept. of Genetics, Stanford University School of Medicine, Stanford, CA 94305/USA

C. Höhmann, Institute for Genetics, University of Cologne, D-5000 Cologne, FRG

J.C. Howard, Agricultural Research Council, Institute of Animal Physiology, Babraham, Cambridge CB2 4AT, U.K.

T. Imanishi-Kari, Institute for Genetics, University of Cologne, D-5000 Cologne, FRG

G.M. Iverson, Dept. of Genetics, Stanford University School of Medicine, Stanford, CA 94305/USA

P.P. Jones, Dept. of Genetics, Stanford University School of Medicine, Stanford, CA 94305/USA

R. Kennet, Dept. of Pathology, University of Pennsylvania Medical School, 36th Street and Hamilton Walk, Philadelphia, PA 19174/USA

G. Kim, National Institute of Allergy and Infectious Diseases, Laboratory of Microbial Immunity, Bethesda, MD 20014/USA

K.J. Kimm, Laboratory of Microbiology and Immunology, Rockville, MD 20852/USA

G. Klein, Institute for Tumor Biology, Karolinska Institute, S-104 01 Stockholm 60, Sweden

N. Klinman, University of Pennsylvania Medical School, 36th Street and Hamilton Walk, Philadelphia, PA 19174/USA

B. Knowles, The Wistar Institute, 36th Street at Spruce, Philadelphia, PA 19104/USA

G. Köhler, Basel Institute of Immunology, 487 Grenzacherstraße, CH-4005 Basel, Switzerland

S. Kontiainen, University College, London WC1E 6BS, U.K.

S. Koskimies, Institute for Tumor Biology, Karolinska Institute, S-104 01 Stockholm 60, Sweden

H. Koprowski, The Wistar Institute, 36th Street at Spruce, Philadelphia, PA 19104/USA

243

R. Kucherlapati, Division of Medical Oncology, Stanford University School of Medicine, Stanford, CA 94305/USA

S.-P. Kwan, Dept. of Cell Biology, Albert Einstein College of Medicine, 1300 Morris Park Avenue, Bronx, N.Y. 10461/USA

L. Lampson, Division of Medical Oncology, Stanford University School of Medicine, Stanford, CA 94305/USA

R. Laskov, Laboratory of Microbiology and Immunology, Rockville, MD 20852/USA

C. LeGuern, Institute Pasteur, 28, rue du Dr. Roux, 75015 Paris, France

H. Lemke, Institute for Genetics, University of Cologne, D-5000 Cologne, FRG

R. Levy, Division of Medical Oncology, Stanford University School of Medicine, Stanford, CA 94305/USA

E.C.C. Lin, Dept. of Microbiology and Molecular Genetics, Harvard Medical School, 25 Shattuck Street, Boston, MA 02115/USA

A.L. Luzzati, Basel Institute for Immunology, 487 Grenzacherstraße, CH-4005 Basel, Switzerland

O. Mäkelä, University of Helsinki, Dept. of Serology and Bacteriology, Haartmaninkatu 3, SF-00290 Helsinki 29, Finland

A. Mandel, Dept. of Genetics, Stanford University School of Medicine, Stanford, CA 94305/USA

L.A. Manson, The Wistar Institute, 36th Street at Spruce, Philadelphia, PA 19104/USA

J. Martinis, The Wistar Institute, 36th Street at Spruce, Philadelphia, PA 19104/USA

T.J. McKearn, Dept. of Pathology, University of Chicago, 950 East 59th Street, Chicago, Ill. 60637/USA

L. Medrano, Institut de Recherches en Biologie Moléculaire du C.N.S.R., 2, Place Jussieu, 75221 Paris Cedex 5, France

J.F. Miller, The Walter and Eliza Hall Institute for Medical Research, Post Office Royal Melbourne Hospital, Victoria 3050, Australia

C. Milstein, Harvard Medical School, Dept. of Pathology, 25 Shattuck Street, Boston, MA 02115/USA

J.D. Minna, Veterans Administration Medical Oncology Branch, National Cancer Institute, 50 Irving Street, NW, Washington, DC 20042/USA

M. Nabholz, Swiss Institute for Experimental Cancer Research, CH-1066 Epalinges, Switzerland

M. North, Genetic Unit, Swiss Institute of Experimental Cancer Research, CH-1066 Epalinges, Switzerland

V.T. Oi, Dept. of Genetics, Stanford University Medical Center, Stanford University, Stanford, CA 94305/USA

244

B.A. Osborne, Dept. of Genetics, Stanford University School of Medicine, Stanford, CA 94305/USA

R.M.E. Parkhouse, National Institute of Medical Research, The Ridgeway, Mill Hill, London, NW7 1AA, U.K.

L. Phalente, Institut de Recherches en Biologie Moléculaire du C.N.R.S., 2, Place Jussieu, 75221 Paris Cedex 5, France

M. Potter, Division of Cancer Biology, and Diagnosis, Laboratory of Cell Biology, National Cancer Institute, Bethesda, MD 20014/USA

K. Rajewsky, Institute for Genetics, University of Cologne, D-5000 Cologne, FRG

W.C. Raschke, The Salk Institute for Biological Studies, San Diego, CA 92112/USA

M. Reth, Institute for Genetics, University of Cologne, D-5000 Cologne, FRG

N.H. Ruddle, Dept. of Epidemiology, Yale University Medical School, 415 L.E.P.H., 60 College Street, New Haven, CT 06510/USA

N. Saravia, Immunobiology Research Center, Wisconsin University Medical School, 1150 University Avenue, Madison, WI 53706/USA

M. Sarmiento, Dept. of Pathology, University of Chicago, 950 East 59th Street, Chicago, Ill. 60637/USA

M. Scharff, Dept. of Cell Biology, Albert Einstein College of Medicine, Yeshiva University, 1300 Morris Park Avenue, Bronx, N.Y. 10461/USA

M.H. Schreier, Basel Institute for Immunology, 487 Grenzacherstraße, CH-4005 Basel, Switzerland

D. Secher, Harvard Medical School, Dept. of Pathology, 25 Shattuck Street, Boston, MA 02115/USA

M. Shander, The Wistar Institute, 36th Street at Spruce, Philadelphia, PA 19104/USA

M.J. Shulman, Basel Institute for Immunology, 487 Grenzacherstraße, CH-4005 Basel, Switzerland

K. Sikora, Division of Medical Oncology, Stanford University School of Medicine, Stanford, CA 94305/USA

E. Simpson, Medical Research Council, Clinical Research Centre, Watford Road, Harrow, Middlesex, U.K.

D. Solter, The Wistar Institute, 36th Street at Spruce, Philadelphia, PA 19104/USA

S.E. Spedden, Laboratory of Cellular and Molecular Research, Cardiac Unit, Massachusetts General Hospital, Boston, MA 02114/USA

R. Sporer, The Wistar Institute, 36th Street at Spruce, Philadelphia, PA 19104/USA

T.A. Springer, Dept. of Pathology, Harvard Medical School, 25 Shattuck Street, Boston, MA 02115/USA

M. Steinitz, Institute of Tumor Biology, Karolinska Institute, S-104 01 Stockholm 60, Sweden

J.W. Stocker, Basel Institute for Immunology, 487 Grenzacherstraße, CH-4005 Basel, Switzerland

F.P. Stuart, Dept. of Pathology, University of Chicago, 950 East 59th Street, Chicago, Ill. 60637/USA

D. Suri, Dept. of Genetics, Stanford University School of Medicine, Stanford, CA 94305/USA

M. Taniguchi, Laboratories for Immunology, School of Medicine, Chiba University, 1-8-1 Inohana Chibacity, Chiba, Japan 280

A. Torano, University College, London WC1E 6BS, U.K.

M.M. Trucco, Basel Institute for Immunology, 487 Grenzacherstraße, CH-4005 Basel, Switzerland

A.S. Tung, Dept. of Pathology, G-3, University of Pennsylvania Medical School, 36th Street and Hamilton Walk, Philadelphia, PA 19104/USA

E. Verastegui-Cerdan, The Wistar Institute, 36th Street at Spruce, Philadelphia, PA 19104/USA

P. Vogt, Medical Research Council, Clinical Research Centre, Watford Road, Harrow, Middlesex HA1 3UJ, U.K.

R. Wallich, Institute for Genetics, University of Cologne, D-5000 Cologne, FRG

N.L. Warner, Dept. of Pathology, University of New Mexico, School of Medicine, Albuquerque, N.M. 87131/USA

A. Weiss, Dept. of Pathology, University of Chicago, 950 East 59th Street, Chicago, Ill. 60637/USA

T. Wiktor, The Wistar Institute, 36th Street at Spruce, Philadelphia, PA 19104/USA

J. Williams, Dept. of Genetics, Stanford University School of Medicine, Stanford, CA 94305/USA

K. Williams, Dept. of Microbiology, University of Michigan Medical School, Ann Arbor, MI 48109/USA

D.E. Yelton, Dept. of Cell Biology, Albert Einstein College of Medicine, 1300 Morris Park Avenue, Bronx, N.Y. 10469/USA

V.R. Zurawski, Jr., Laboratory of Cellular and Molecular Research Cardiac Unit, Massachusetts General Hospital, Boston, MA 02114/USA

Current Topics in Microbiology and Immunology

Volume 80

31 figures, 16 tables. IV, 170 pages. 1978
ISBN 3-540-08587-4

Contents: J.J. Bullen, H.J. Rogers, E. Griffiths: Role of Iron in Bacterial Infection. — J.H.L. Playfair: Effective and Ineffective Immune Responses to Parasites: Evidence from Experimental Models. — G.R. Pearson: In Vitro and in Vivo Investigations on Antibody-Dependent Cellular Cytotoxicity. — S. Cohen, G.H. Mitchell: Prospects for Immunisation Against Malaria. — C. Scholtissek: The Genome of the Influenza Virus.

Volume 79

35 figures, 29 tables. III, 309 pages. 1978
ISBN 3-540-08587-4

Contents: H. Fan: Expression of RNA Tumor Viruses at Translation and Transcription Levels. — M. Mussgay, O.-R. Kaaden: Progress in Studies on the Etiology and Serologic Diagnosis of Enzootic Bovine Leukosis. — K.L. Beemon: Oligonucleotide Fingerprinting with RNA Tumor Virus RNA. — M.B. Gardner: Type C Viruses of Wild Mice: Characterization and Natural History of Amphotropic, Ecotropic, and Xenotropic MuLV. — J.A. Levy: Xenotropic Type C-Viruses. — R.R. Friis: Temperature-Sensitive Mutants of Avian RNA Tumor Viruses: A Review. — E. Hunter: The Mechanism for Genetic Recombination in the Avian Retroviruses.

Volume 78

45 figures, 11 tables. III, 248 pages. 1977
ISBN 3-540-08499-1

Contents: H. zur Hausen: Human Papilloma Viruses and Their Possible Role in Squamous Cell Carcinomas. — G.G.B. Klaus, A.K. Abbas: Antigen-Receptor Interaction in the Induction of B-Lymphocyte Unresponsiveness. — T. Hohn, I. Katsura: Structure and Assembly of Bacteriophage Lambda. — S.A. Plotkin: Perinatally Acquired Viral Infections. — J. Collins: Gene Cloning with Small Plasmids. — H.A. Nash: Integration and Excision of Bacteriophage 2. — G. Wengler: Structure and Function of the Genome of Viruses Containing Single-Stranded RNA as Genetic Material: The Concept of Transcription and Translation Helices and the Classifications of these Viruses into Six Groups. — A.M. Skalka: DNA Replication — Bacteriophage Lambda.

Volume 77

19 figures. III, 168 pages. 1977
ISBN 3-540-08401-0

Contents: B.E. Butterworth: Proteolytic Processing of Animal Virus Proteins. — K. Kano, F. Milgrom: Heterophile Antigens and Antibodies in Medicine. — W.E. Rawls, S. Bacchetti, F.L. Graham: Relation of Herpes Simplex Viruses to Human Malignancies. — W. Hengstenberg: Enzymology of Carbohydrate Transport in Bacteria. — A.E. Butterworth: The Eosinophil and Its Role in Immunity to Helminth Infection.

Volume 76

26 figures. III, 214 pages. 1977
ISBN 3-540-08238-7

Contents: W. M. Kuehl: Synthesis of Immunoglobulin in Myeloma Cells. — C.R. Pringle: Enucleation as a Technique in the Study of Virus-Host Interaction. — D. Richter, K. Isono: The Mechanism of Protein Synthesis: Initiation, Elongation and Termination in Translation of Genetic Messages. — J. Jelinkova: Group B Streptococci in the Human Population. — J. Storz, P. Spears: Chlamydiales: Properties, Cycle of Development and Effect on Eukaryotic Host Cells.

Volume 75

22 figures. 202 pages. 1976
ISBN 3-540-08013-9

Contents: A. Globerson: In Vitro Approach to Development of Immune Reactivity. — S.C. Bansal, B.R. Bansal, J.-P. Boland: Blocking and Unblocking Serum Factors in Neoplasia. — R. Hausmann: Bacteriophage T7 Genetics. — P. Starlinger, H. Saedler: IS-Elements in Microorganisms. — L.G. Schneider, H. Diringer: Structure and Molecular Biology of Rabies Virus. — Cumulative Author and Subject Index. — Volumes 40—75.

Volume 74

22 figures. IV, 168 pages. 1976
ISBN 3-540-07657-3

Contents: A. Rimon: The Chemical and Immunochemical Identity of Amyloid. — V. Pirrotta: The λ Repressor and Its Action. — K. Geider: Molecular Aspects of DNA Replication in Escherichia coli Systems. — R. Dziarski: Teichoic Acids. — P.A. Sharp, S.J. Flint: Adenovirus Transcription.

Volume 73

22 figures. III, 219 pages. 1976
ISBN 3-540-07593-3

Contents: S.U. Emerson: Vesicular Stomatitis Virus: Structure and Function of Virion Components. — M.A. Martin, G. Khoury: Integration of DNA Tumor Virus Genomes. — A.J. Levine, P.C. van der Vliet, J.S. Sussenbach: The Replication of Papovavirus and Adenovirus DNA. — H. Westphal: In Vitro Translation of Adenovirus Messenger RNA. — R. Hehlmann: RNA Tumor Viruses and Human Cancer.

Volume 72

27 figures, 8 tables. III, 194 pages. 1975
ISBN 3-540-07564-X

Contents: H.A. Andersen, L. Rasmussen, E. Zeuthen: Cell Division and DNA Replication in Synchronous Tetrahymena Cultures. — B.W. Baldwin, R.A. Robins: Humoral Factors Abrogating Cell-Mediated Immunity in the Tumor-Bearing Host. — G.L. Asherson, M. Zembala: Inhibitory T-Cells. — A.T. Haase: The Slow Infection Caused by Visna Virus. — Y.H. Pilch, D. Fritze, S.R. Waldman, D.H. Kern: Transfer of Antitumor Immunity by "Immune" RNA.

Volume 71

17 figures, 12 tables. III, 173 pages. 1975
ISBN 3-540-07369-8

Contents: W. Doerfler: Integration of Viral DNA into the Host Genome. — C. Moscovici: Leukemic Transformation with Avian Myeloblastosis Virus: Present Status. — H.G. Purchase, R.L. Witter: The Reticuloendotheliosis Viruses. — A.J.L. Macario, E. Conway de Macario: Antigen-Binding Properties of Antibody Molecules: Time-Course Dynamics and Biological Significance.

Volume 70

14 figures. III, 124 pages. 1975
ISBN 3-540-07223-3

Contents: H.A. Blough, J.M. Tiffany: Theoretical Aspects of Structure and Assembly of Viral Envelopes. — J.G. Stevens: Latent Herpes Simplex Virus and the Nervous System. — P.A. Schaffer: Temperature-Sensitive Mutants of Herpes-viruses. — C. Scholtissek: Inhibition of the Multiplication of Enveloped Viruses by Glucose Derivatives.

 Springer-Verlag Berlin Heidelberg New York